COMPARATIVE PHYSIOLOGY OF THE
VERTEBRATE DIGESTIVE SYSTEM

COMPARATIVE PHYSIOLOGY OF THE VERTEBRATE DIGESTIVE SYSTEM

C. E. STEVENS

North Carolina State University
College of Veterinary Medicine

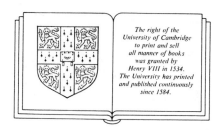

The right of the
University of Cambridge
to print and sell
all manner of books
was granted by
Henry VIII in 1534.
The University has printed
and published continuously
since 1584.

CAMBRIDGE UNIVERSITY PRESS

Cambridge
New York   New Rochelle   Melbourne   Sydney

Published by the Press Syndicate of the University of Cambridge
The Pitt Building, Trumpington Street, Cambridge CB2 1RP
40 West 20th Street, New York, NY 10011, USA
10 Stamford Road, Oakleigh, Melbourne 3166, Australia

First published 1988
First paperback edition 1990

Printed in Canada

Library of Congress Cataloging-in-Publication Data

Stevens, C. E. (Charles E.)
Comparative physiology of the vertebrate digestive system/ C. E. Stevens.
p. cm.
Bibliography: p.
Includes index.
ISBN 0-521-33044-0   ISBN 0-521-40803-2 (pbk.)
1. Vertebrates – Digestive organs.   2. Vertebrates – Physiology.
I. Title.
QP145.S78   1988
596'.0132–dc19

British Library Cataloging in Publication Data

Stevens, C. E.
Comparative physiology of the vertebrate digestive system.
1. Vertebrates. Digestive system.
Physiology
I. Title
596'.043

ISBN 0-521-33044-0 hardback
ISBN 0-521-40803-2 paperback

For ALVIN SELLERS,
who introduced me to the wonders of
the digestive system and its adaptations

Contents

Preface   *page* xi

1 General characteristics of the digestive system   1
  Invertebrates   2
  Vertebrates   6
    Headgut   9
    Foregut   10
    Midgut   14
    Pancreas and biliary system   16
    Hindgut   17
    Cloaca   20
    Neurohumoral control   21
  Summary   21

**2 The digestive tract of fish, amphibians, reptiles, and
    birds**   22
  Fish   22
  Amphibians   30
  Reptiles   31
  Birds   35
  Summary   39

3 The mammalian digestive tract   40
  Monotremata   44
  Pholidota   44
  Tubulidentata   46
  Cetacea   46
  Macroscelidea   48
  Insectivora   48
  Scandentia   50
  Chiroptera   50
  Carnivora   50
  Marsupialia   52

Edentata   57
Rodentia   60
Primates   65
Artiodactyla   70
Dermoptera   80
Lagomorpha   80
Perissodactyla   80
Proboscidea   81
Sirenia   81
Hyracoidea   84
Summary   84

4   **Motor activity and digesta transit**   86
    Motor activity   86
        Esophagus   87
        Stomach   92
        Midgut   101
        Hindgut   104
    Digesta transit   107
    Selective retention of digesta fluid or particles   116
    Summary   123

5   **Carbohydrate, fat, and protein digestion by
    endogenous enzymes, and absorption of end products**   125
    Carbohydrates   128
        Components of food   128
        Digestion   131
        Absorption   140
    Lipids   140
        Components of food   140
        Digestion   141
        Absorption   143
    Protein   144
        Dietary and endogenous protein   144
        Digestion   144
        Absorption   152
    Nucleic acids   154
    Summary   155

6   **Microbial fermentation and synthesis of nutrients, and
    absorption of end products**   159
    Carbohydrate fermentation   160
        Stomach   160

Hindgut   166
Absorption of volatile fatty acids   171
Nitrogen recycling and protein synthesis   175
   Stomach   175
   Hindgut   177
Synthesis of vitamins   181
Coprophagy   181
Evolution of herbivores   185
Summary   188

7 Secretion and absorption of electrolytes and water   191
Electrolyte and water balance   191
Electrolyte transport mechanisms   194
Secretion and absorption   196
   Salivary glands   196
   Stomach   199
   Pancreatic and biliary secretion   202
   Intestine   204
The pH of gastrointestinal contents   210
Enterocirculation   215
Summary   218

8 Neurohumoral control   220
Nervous control   220
Endocrine control   228
   Hormones   229
   Evolution of gut peptides   234
Summary   237

9 Conclusions   238

References   243
Index   291

Preface

This book is designed principally for use as a reference text for zoologists, physiologists, veterinarians, and others who are interested in the vertebrate digestive system, and as an auxiliary text for courses on comparative physiology and nutrition. A basic understanding of comparative anatomy and physiology is necessary for the transfer of information from one species to another. Studies of comparative physiology also offer a very useful approach to the understanding of the basic mechanisms that are involved. However, information on the digestive system of vertebrates is extensive and widely distributed in the literature of many disciplines, and much of this information is derived from the study of a limited number of laboratory and domesticated species. The objectives of this book are to describe the general anatomical and physiological characteristics of the digestive system in relation to the diet, environment, and other characteristics of animals in each of the major groups of vertebrates. The first three chapters deal with general characteristics and major variations in structure. Subsequent chapters focus on variations in the principal functions of this system. These discussions do not attempt to provide either an encyclopedic coverage or an exhaustive review of available literature; references to more detailed or extensive treatment of these subjects have been included wherever possible.

I am grateful to many people for their help in collecting, preparing, and reviewing this material. Much of the initial information was derived from the term papers of students enrolled in a course, taught at Cornell University, on the Comparative Physiology of the Digestive System. Many colleagues provided the specimens or photographs used for the series of line drawings comparing the digestive tract of different species, and Erica Melack deserves special thanks for the preparation of most of these illustrations. Cherril Wallen and Carolyn Smoak contributed a great deal to the preparation of the manuscript, and I am very grateful to Margaret Hemingway for her help and patience in the preparation of its numerous drafts. The comments on various chapters by Robert A. Argenzio, Peter J. Bentley, William D. Heizer, and Donald W. Powell resulted in many additions and improvements. The general review by Malcolm C. Roberts was especially helpful for the integration and presentation of information. Completion of the process would have been difficult without the encouragement and sufferance of my wife Barbara and my family.

1 General characteristics of the digestive system

The major purpose of the digestive system is to provide for the assimilation of nutrients required for growth, maintenance, energy, and reproduction. Digestion consists of a number of physical and chemical processes. Food is procured, broken down into smaller particles, mascerated, mixed with digestive enzymes, and propelled through the digestive tract by the motor or muscular activities of the digestive tract. Secretions of the digestive tract and its associated organs collectively provide mucus for lubrication, enzymes that aid in digestion, and fluids that establish the optimum pH for this process. Digestive enzymes hydrolyze carbohydrate, protein, fat, and other components of the diet into a limited number of compounds suitable for absorption. Microorganisms indigenous to the digestive tract can provide additional nutrients by the fermentation of carbohydrates that are not subject to attack by the above enzymes and by synthesizing amino acids and vitamins essential to the host animal.

The digestive tract is one of the most readily accessible routes for substances to enter the body. Therefore, it requires reasonably fail-safe mechanisms for the careful selection of the substances that will be allowed entry. This is accomplished through a variety of mechanisms, including food selection (palatability), rapid rejection of irritants by emesis (vomiting) or increased rate of passage, and degradation of substances before they have access to the more permeable intestinal tract. However, one of the most important protective mechanisms is the selective permeability of the digestive tract's epithelial barrier. The epithelial lining of the upper digestive tract is impermeable to most substances. Even those parts of the intestine that serve as primary sites for absorption are relatively impermeable to the passive diffusion of most water-soluble substances, including some of the minerals, vitamins, monosaccharides, and amino acids required as nutrients. The latter are largely absorbed via mechanisms that provide for highly selective active transport across the intestinal mucosa.

Many of the processes responsible for digestion and absorption are common to most species. Others have resulted from adaptations to the diet, environment, or other physiological characteristics of the animal, through divergence from a common or more primitive form, or by convergence—the appearance of similar structures or functions in completely unrelated

1

species. An understanding of these adaptations is essential for the proper care and maintenance of domesticated and captive wild animals, and for the preservation of wild species. Studies of these adaptations also can provide a better understanding of the basic mechanisms that are involved.

Invertebrates

Although the subject of invertebrates is beyond the scope of this book, brief mention of their digestive systems provides some necessary perspective. Barnes (1974) has pointed out that 95% of all described species in the Animal Kingdom are invertebrates and some of these are more closely related to vertebrates than to other invertebrate groups. He concludes that a taxonomist less biased than man might provide a better division of animals into arthropods (75% of all species) and nonarthropods (25%). Therefore, it is not surprising that many of the basic characteristics of the vertebrate digestive system have evolved in various invertebrate species. The reader is referred to Barnes (1974) and Andrew (1959) for broad discussions of invertebrate digestive systems, Wigglesworth (1972, 1984) for similar discussion of insects, and Barnard and Prosser (1973) for comparison of invertebrates and vertebrates. Vonk and Western (1984) also provide an excellent review of digestive enzymes in both groups of animals.

Protozoa and parazoa (sponges) do not have a digestive tract. Some protozoa absorb nutrients across their body cell membranes, but many ingest food by phagocytosis at a specific site or various points on their cell membrane. Ingested material is taken up into food vacuoles, which undergo prolonged passage through the cell while their contents are subjected to digestion. Nutrients are then transported across the vacuolar membrane, and waste products are evacuated from the cell. Many of the digestive enzymes found in vertebrates have been isolated from various species of protozoa. In some such as *Paramecium*, intravacuolar digestion is associated with a sequential acidification and alkalinization, in a manner similar to that seen in the digestive system of most vertebrates.

Most other invertebrates have a mouth and either a blind digestive cavity or a digestive tract that terminates in an anus. The digestive tract of many advanced invertebrates can be divided into a headgut (oral cavity and throat), foregut (esophagus and stomach), and intestine. In a number of these, the intestine can be further subdivided into a midgut, which serves primarily for digestion and absorption, and a hindgut, which aids in the absorption of inorganic ions and water secreted by the digestive system and excreted into the gut by the Malpighian tubules of most insects and some other arthropods.

Intracellular digestion is the principal form of digestion in sponges and the coelenterates such as the *Hydra*. However, the coelenterates and higher forms of invertebrate demonstrate an increasing dependence on extracellular digestion. Cells lining the digestive cavity may secrete mucus or digestive enzymes, absorb nutrients, or serve for food storage. These functions, which are carried out by multipurpose cells in the lower forms of invertebrates, become the property of specialized cells in more advanced species. Phagocytosis and intracellular digestion within food vacuoles are absent in many invertebrates. Kermak (cited by Barrington, 1962) demonstrated that gut epithelial cells of *Arenicola marina*, a marine-burrowing worm, which ingests much sand to obtain organic matter, phagocytize food particles and then transfer them to amoebocytes for digestion in a manner similar to that of many primitive metazoa, echinoderms, and other invertebrates with similar feeding habits. However, *Turbellaria lapidaria*, an annelid more selective in its feeding habits, has intestinal cells that lack food vacuoles and demonstrate a brush border of microvilli on their lumen surface. Absorptive cells of the insect midgut also have microvilli. These increase the surface area available for absorption, and contain enzymes that complete the digestion of carbohydrates and protein.

Cells that are specialized for secretion of mucus, enzymes, and other substances, are located in ceca or glands along the digestive tract of many of the more advanced species of invertebrate. For example, the intestinal ceca of trematodes contain both absorptive and secretory cells. Salivary glands are highly developed in many mollusks and arthropods. Some species of annelids such as the earthworms have a well-developed system of extracellular digestion with digestive glands and a general arrangement of alimentary organs and tissue similar to that of vertebrates.

The accessory digestive glands of mollusks are especially interesting, as the various species in this phylum appear to demonstrate a broad spectrum of the functions normally associated with the salivary glands, pancreas, and liver of vertebrates. Phylum Mollusca, which includes the clams, oysters, mussels, snails, slugs, squid, and octopus, contains the largest number of species of any phylum other than the arthropods. It contains carnivores, omnivores, herbivores, scavengers, parasites, and animals that live in a marine, freshwater, or terrestrial environment. Digestion is at least partly extracellular in all species, and enzymes may be secreted by salivary glands, esophageal pouches, portions of the stomach, intestinal digestive glands, or a combination of these.

Digestive diverticula in the stomach of bivalves (clams, oysters, mussels) contain phagocytic, vacuolated cells, which serve largely for absorption and intracellular digestion. These cells eventually undergo fragmentation to form spheres that contain the vacuoles with their content of undigestible material, excretory products, and residual enzymes. These spheres are released into the gut lumen and could serve as an important source of the extracellular enzymes. Although the latter may apply to the gastropod, it is less likely in bivalves, as the efflux from the diverticula is not remixed with stomach contents (Barrington 1962). Therefore, dissolution of wandering phagocytes has been suggested as a more probable source of their weak gastric protease and lipase activity. However, some species that demonstrate these extracellular enzymes have few phagocytes, and at least one bivalve genus, a small clam (*Nucula*), lacks the mechanism for a two-way flow of digesta in the diverticula ducts. Therefore, its cells secrete and excrete, but do not absorb. This specialization of the digestive glands proceeds further in the gastropods to the point where in the snail, *Helix*, digestive glands secrete a large variety of extracellular enzymes, but appear to have relatively little absorptive capacity. These glands have been referred to as the "hepatopancreas" in a number of species and called the "pancreas" and "liver" in cephalopods (squid and octopus).

The development of hepatic functions similar to those of vertebrates appears as another prominent feature in many species of this phylum. The "liver" appears as a gland of many cell types, which can function for absorption, intracellular digestion, secretion, excretion, and storage. In many species, food particles are seen in the vacuoles of its cells. In the squid, however, food must be absorbed and reach these cells via the blood (Campbell and Burnstock 1968). The squid "pancreatic" duct empties into the "liver" duct, and the combined secretions flow from a common duct, which can be directed into the stomach or its cecum. A sphincter between the hepatic and common duct prevents reflux of lumen contents into the liver. The "liver" produces carboxypeptidase, aminopeptidase, and dipeptidase, and the latter two, plus lipase, also are synthesized by the pancreas.

Phylum Arthropoda includes the horseshoe crab, crustaceans, arachnids, and insects. Digestion is almost exclusively extracellular, except for the final stages in the brush border of intestinal cells, and confined primarily to the midgut. Salivary glands, which are highly developed in many insects, usually empty into the buccal cavity. They may secrete mucus, enzymes, anticoagulants, agglutinins, venomous spreading agents, or silk in various species. The "hepatic ceca" of the horseshoe crab consist of two large glands, which function for both digestion and absorption. The midgut of crustaceans has a pair of ceca modified to form a group of ducts with blind secretory tubules. These provide a site for absorption, digestive enzyme secretion, and storage of glycogen, fat, and Ca. They also have been labeled as the hepatopancreas.

Van Weel (1974) concluded that the terms "hepatopancreas," "pancreas," and "liver" are inappropriate when applied to either mollusks or crustaceans, and that these structures should be referred to simply as midgut glands. Bidder (1976) agreed with respect to the cephalopods and proposed that the terms "digestive glands" and "digestive gland appendages" be used to designate the liver and pancreas, respectively. Gibson and Barker (1979), however, concluded that the digestive glands of decapod crustaceans are " . . . rightly and properly named" the hepatopancreas.

Digestion and the production of nutrients are further aided by indigenous, symbiotic microorganisms in many invertebrate species. Buchner (1965) reviewed the widespread distribution of endosymbiosis and the historical development of this subject. He pointed to the presence of algae in the cells lining the digestive tract of a wide range of invertebrates and the evidence that these organisms serve to provide their host with O_2 and carbohydrates, a site for food storage, and/or a mechanism for the excretion of CO_2, PO_4, and nitrogenous waste. In some species such as *Paramecium bursa*, they even allow the animal to survive in the absence of its normal food supply, if sufficient light is provided for maintenance of the algae. There is a similar inclusion of bacteria within protozoa and cells lining the digestive system of other invertebrates, especially certain insects. Although the algae were noted by early workers, who only questioned their importance, the bacteria were at first believed to be specialized cellular organelles. Proof that these were bacteria serving to synthesize vitamins or fix N_2 came only when means were developed for separation of the symbiont from its host. The relationship between the amoeba, *Pelomyxa*, and its bacterial symbionts has been examined fairly extensively. Each species appears to have a specific type or types of bacteria throughout the endoplasm and, occasionally, the ectoplasm. The ability of these protozoa to digest filter paper is believed due to bacterial cellulase. In more advanced species, large numbers of bacteria or protozoa may be located at specific sites along the digestive tract. Microbial fermentation and synthesis of nutrients have been well documented in many species, including annelids, mollusks, echinoderms, and insects.

Efficient digestion of food requires that it be presented in particles small enough for ingestion and for attack by enzymes. Various parts of the headgut filter out food particles in the microphagous filter-feeders. Other invertebrates accomplish this by regurgitation of enzymes into their prey (scorpions) or its tissues (mites). Some echinoderms completely evert their stomach to engulf their prey. However, in most invertebrates, food is broken down into smaller particles by physical action in the headgut or foregut. Once particles have been reduced to appropriate size, they need to be propelled through the digestive tract to allow for the series of episodic events associated with digestion, absorption, and the excretion of waste products.

Movement of food and fluids is accomplished by ciliated cells in the digestive system of many invertebrates. Transport of digestive secretions also may be aided by the movement of cilia in ceca or glandular ducts. However, the mechanical breakdown of food requires muscular activity. Furthermore, as the size of food particles and the diameter of the digestive tract increase, cilia lose their efficiency. Therefore, in many of the more advanced species, the motor activity required for ingestion and the transport of digesta is accomplished by muscular contraction. Both cilia and muscles are used for this purpose in many invertebrates, but cilia are absent from the digestive tract of others, such as the nematodes and insects.

Increased complexity of the digestive system also requires a means for integrating its separate processes. This is accomplished by the nervous system, which is first seen in sponges and is well developed in annelids, mollusks, and arthropods. Although there also is evidence for humoral control of the digestive system, the contribution of hormones is poorly understood.

Phylum Chordata contains the marine, freshwater, and terrestrial species, that possess pharyngeal clefts, a notochord, and a dorsal hollow nerve cord during at least one stage of life. The earliest members of this phylum are the prochordates such as amphioxis (Subphylum Cephalochordata) and the sea squirts (Subphylum Tunicata), from which the vertebrates are believed to have evolved approximately 500 million years ago. However, our discussion will be confined to Subphylum Vertebrata, which contains approximately 41,700 of these species.

Vertebrates

The various functions of the digestive system can be discussed in the order that events occur along the digestive tract or according to specific types of activity such as secretion, digestion, absorption, motor activity, and neurohumoral control. The first approach allows for better understanding of sequential events and the relationship between the digestive system of an animal and its diet, other physiological characteristics, and environment. The second approach provides for a more direct comparison of similarities and differences among species. Therefore, both of these approaches are used in the following chapters.

Figure 1.1 shows the phylogenetic origins and relationships of the vertebrates. The major groups that have survived to the present time are listed in Table 1.1. These can be divided into the Agnatha (cyclostomes) and the Gnathostomata, which includes the remaining two classes of fish plus the amphibians, reptiles, birds, and mammals. The bony fishes, Osteichthyes, are referred to as teleosts in some classifications. Although all vertebrates have a digestive tract and accessory organs, various parts of this system are not necessarily homologous, analogous, or even present in all species. Therefore, broad comparisons can best be made under the listings of headgut, foregut, midgut, hindgut, pancreas, and biliary system. References for much of the following information can be found in subsequent chapters, where individual subjects are discussed in more detail. However, for general references that include subjects not covered later, the reader is referred to An-

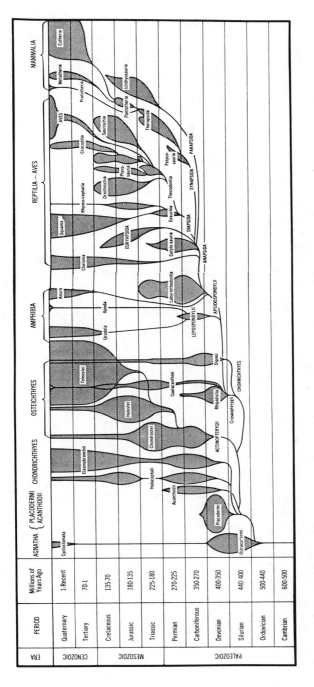

FIGURE 1.1 Phylogenetic origins of the various groups of vertebrates. (From Torrey, 1971.)

TABLE 1.1 Classification of vertebrates

SUPERCLASS: Agnatha
    CLASS: Cephalaspidomorphi. Lampreys
    CLASS: Myxini. Hagfish

SUPERCLASS: Gnathostomata
  GRADE: Pisces
    CLASS: Chondrichthyes. Cartilaginous fishes
      SUBCLASS: Holocephali. Chimaeras (1 order)
      SUBCLASS: Elasmobranchii. Sharks, skates, and rays (5 orders)
    CLASS: Osteichthyes. Bony fishes
      SUBCLASS: Dipneusti. Lungfish (2 orders)
      SUBCLASS: Crossopterygii. Lobe-finned fishes (1 order)
      SUBCLASS: Brachiopterygii. Gar, sturgeon, paddlefish, and bowfin (1 order)
      SUBCLASS: Actinopterygii. Ray-finned fishes (38 orders)
  GRADE: Tetrapoda
    CLASS: Amphibia
      ORDER: Gymnophionia (Apoda). Limbless, short-tailed amphibians
      ORDER: Caudata (Urodeles). Salamanders, newts, congo eels, and related species
      ORDER: Salientia (Anura). Frogs and toads
    CLASS: Reptilia
      ORDER: Crocodilia. Crocodiles, alligators, caimans, and gavials
      ORDER: Testudinata (Chelonia). Turtles, tortoises, and terrapins
      ORDER: Squamata
        SUBORDER: Ophidia (Serpentes). Snakes
        SUBORDER: Sphenodontia. Tuatara
        SUBORDER: Lacertilia (Sauria). Lizards
        SUBORDER: Amphisbaenia. Worm lizards
    CLASS: Aves (28 orders)
    CLASS: Mammalia
      SUBCLASS: Prototheria. Egg-laying mammals
        ORDER: Monotremata. Echidna and duck-billed platypus
      SUBCLASS: Theria. Mammals bearing their young alive
        INFRACLASS: Metatheria. Pouched mammals
        ORDER: Marsupialia. Kangaroos, opossums, etc.
        INFRACLASS: Eutheria. Placental mammals
      ORDER: Insectivora. Shrews, moles, etc.
      ORDER: Scandentia. Tree-shrews
      ORDER: Macroscelidea. Elephant shrews
      ORDER: Dermoptera. Flying lemurs
      ORDER: Chiroptera. Bats
      ORDER: Primates. Lemurs, monkeys, apes, humans
      ORDER: Carnivora. Cats, dogs, weasels, bears, etc.
      ORDER: Hyracoidea. Conies
      ORDER: Proboscidea. Elephants
      ORDER: Sirenia. Manatees, dugongs
      ORDER: Cetacea. Whales, porpoises, dolphins
      ORDER: Perissodactyla. Horses, tapirs, rhinoceros
      ORDER: Artiodactyla. Cattle, sheep, pigs, etc.
      ORDER: Edentata. Sloths, armadillos, anteaters
      ORDER: Tubulidentata. Aardvarks

TABLE 1.1 (*Continued*)

ORDER: Lagomorpha. Hares, rabbits, pika
ORDER: Rodentia. Rats, mice, squirrels, etc.
ORDER: Pholidota. Pangolins

Note: Extinct groups of animals are excluded. Fish are classified according to Nelson (1984); amphibians, according to Duellman and Trueb (1986); reptiles, according to Evans (1986); and mammals, according to Vaughan (1986). For classification of birds, see Storer (1971a).

drew (1959), Reeder (1964), Cloudsley-Thompson (1972), Ziswiler and Farner (1972), Harder (1975a,b), Luppa (1977), and Skoczylas (1978).

Headgut

The headgut is the cranial portion of the digestive tract that includes the oral or buccal cavity and the throat or pharynx. In fish and larval amphibians, it also includes the gill cavity and is referred to as the orobrancheal cavity. Aside from its respiratory functions in fish and amphibian larvae, the headgut serves primarily as a means for the capture and preparation of food for deglutition. Articulated jaws are present in all vertebrates other than the cyclostomes (lampreys and hagfish). Lips or teeth may be used for prehension of food. With the exceptions of the chelonians (turtles, tortoises, and terrapins) and birds, most vertebrates have teeth, which are used to grasp, position, puncture, tear, and/or triturate food.

The tongue of most vertebrates is attached to the floor of the mouth throughout most of its length and is used primarily to aid in deglutition and, in mammals, the positioning of food for mastication. However, it may be mobile and used for other purposes, including the capture of prey by some species of fish, adult amphibians, birds, and mammals, or as a sensing organ in reptiles. The headgut of some species of fish, amphibians, birds, and mammals contains elaborate filtering mechanisms that allow for separation of small plants and animals from water or larger particles. These filters may be located in the gills or mouth, or at the lateral margin of the jaw. The gill flaps, mouth, or tongue provides the pump for these filters.

Lubrication of the oral cavity is necessary for the deglutition (swallowing) of food, especially by terrestrial species. It is provided by mucus-secreting cells in the oral epithelium of fish and multicellular glands in more advanced vertebrates. The salivary glands of reptiles, birds, and mammals can be very complex in structure and perform additional functions as well. The glands of adult frogs and toads, swifts, woodpeckers, and mammalian anteaters secrete an adhesive material that aids swifts in building their nest and, when applied to the tongue, aids the other species in capturing prey. Digestive enzymes are found in the secretions of some adult amphibians, reptiles, birds, and mammals. The serous (watery) component of mammalian saliva is used as an aid to evaporative heat loss in the dog and cat, and provides the large quantities of bicarbonate and phosphate required for the buffering of fermentation end products in the stomach of ruminants and a number of other mammalian herbivores. The secretions of oral glands also can contain toxins and other substances. Kochva (1978) reviewed the various types of gland and glandular secretions found in reptiles and commented on their possible evolution. Protein and glycoprotein components of mammalian saliva were discussed by Ellison (1967). Junqueira and de Moraes (1965) compared the major salivary glands of vertebrates and concluded that these, unlike pancreas, evolved gradually during vertebrate phylogenesis and reached their maximal complexity in mammals.

Foregut

Esophagus

The major function of the vertebrate esophagus is to transfer ingesta from the mouth to the stomach, or to the intestine of those species that lack a stomach. However, it performs additional important functions in some animals. For example, the esophagus of some egg-eating snakes and turtles is organized to crush the shell. In some fish and reptiles, it serves as a temporary site for storage of partially swallowed prey that are awaiting gastric digestion. In most birds, the crop is the major site of food storage. The avian crop serves further as a source of nutrients for the young in some species. The esophagus is organized for controlled regurgitation in some vertebrates. Crop contents are regurgitated for the feeding of young by some birds. Others, such as the hawks and owls, regurgitate undigested parts of their prey from the gizzard. Regurgitation and remastication of food also are part of the complex and highly integrated cycle of rumination in ruminant mammals.

In most vertebrates, the esophagus is lined with a stratified squamous epithelium containing goblet (mucus-secreting) cells. Esophageal epithelium also contains ciliated cells in some fish, adult amphibians, and reptiles. Glands are found in the esophagus of many species. Glands that secrete pepsinogen have been described in the terminal esophagus of some fish (Kapoor, Smit, and Verighina 1975), adult frogs and toads (Reeder 1964), and chelonians (Luppa 1977). Esophageal glands of some amphibians such as the red-legged pan frog (*Kassina maculata*) are said to secrete more pepsinogen than the stomach (Hirji 1982). The bat, *Plecotus auritus*, has a relatively long segment of abdominal esophagus with oxyntic (acid-secreting) cells in its mucosa (Botha 1958). It is interesting to note that Burns, Flores, Moshyedi, and Albacete (1970) describe four types of aberrant esophageal epithelium seen in humans: 1) cardiac glandular mucosa, 2) proper gastric glandular mucosa containing oxyntic cells, 3) ciliated epithelium, and 4) epithelium resembling that of the small intestine.

With the possible exception of the ciliated esophagus of embryonic fish and larval amphibians, ingesta is transferred through the vertebrate esophagus, either primarily or entirely by muscular activity. The esophagus generally is invested with an inner layer of circular muscle and an outer layer of longitudinal muscle. However, this arrangement can be reversed at the terminal esophagus of fish and the longitudinal layer may be complete only near the gastroesophageal junction in adult amphibians and reptiles. Furthermore, the circular muscle is actually oblique in segments of the esophagus of some mammals. It forms a well-defined sphincter at the termination of the esophagus of some fish and all adult amphibians and reptiles. A similar sphincter can be found in many mammalian species such as moles, bats, rodents, rabbits, domestic pigs, and horses (Botha 1962). However, this area is delineated only by a constriction, with no anatomical sphincter, in other species such as dogs, cats, ferrets, cattle, and primates.

Esophageal muscle is striated in fish but smooth in amphibians, reptiles, and birds. Mammals show considerable species variation in the presence and distribution of these two types of muscle (Ingelfinger 1958; Doty 1968). Both layers of muscle are striated throughout the length of the esophagus in dogs, ferrets, mice, rats, sheep, cattle, and elephants. In rabbits and domestic pigs, the longitudinal muscle is striated, but the circular muscle becomes smooth immediately above the gastroesophageal junction. The esophagus of domestic cats contains smooth muscle over approximately the terminal 8% of its longitudinal and 16% of its circular muscle. Both layers of muscle in the human esophagus consist of striated muscle over the initial fourth of its length. There is then a transition to entirely smooth muscle in the terminal third. This more extensive section of smooth muscle in the terminal esophagus is seen also in pinniped carnivores (seals, sea lions, walruses), whales, opossums, equine species, and many nonhuman primates.

The remainder of the digestive tract consists of smooth muscle. The only apparent exceptions are stomachless fish, which may have a short segment of striated muscle in the upper intestine.

Stomach

With the exceptions of some fish and the larval toads, all vertebrates have a stomach or analogous organs. In most species, it serves as a site for storage, masceration, and some physical breakdown of food. It also contributes to the trituration of food in most birds, as well as in some fish, reptiles, and mammals. One other major function, in most vertebrates, is the initiation of protein digestion by the action of pepsin and HCl. This particular process of protein digestion appears to be an innovation of vertebrates that is shared by all that have a stomach, with the exception of some amphibian larvae and a few mammalian species.

The stomach of fish, amphibians, reptiles, and many mammals is a relatively simple, tubular, or asymmetric expansion of the digestive tract. However, in birds these same functions are carried out by the crop (storage), proventriculus (pepsinogen and HCl secretion), and gizzard (trituration), and the stomach can be both complex in shape and voluminous in some bats, the whales, and a variety of mammalian herbivores.

Different regions of the human stomach have been characterized as the cardia, body, fundus, and pylorus. The cardiac region is that nearest the gastroesophageal junction, and thus the heart. The body is the distended portion of the stomach, and the term "fundus" (bottom) refers to the outpocketed area. The pyloric region is the terminal segment, which often has a thicker layer of circular muscle and includes the pyloric valve. Although these terms are often applied to the stomach of other vertebrates, they are useless in describing the external characteristics of the more complex stomachs of birds and many mammals. Furthermore, they show no necessary correlation with the structural or functional characteristics of the gastric epithelium lining the lumen. Therefore, for comparative purposes, the regions of the stomach can best be described by the characteristics of their epithelial lining.

The stomach of most vertebrates contains a region of proper gastric glandular mucosa, which secretes pepsinogen and HCl. Both are secreted by the same glandular cell in fish, adult amphibians, reptiles, and birds, but by separate cells in mammals. A cardiac glandular region is absent in fish, but present near the gastroesophageal junction of reptiles, some adult amphibians, and most mammals. A pyloric glandular region is described in fish, adult amphibians, reptiles, and mammals. Mucus is secreted by both glandular and surface epithelium in each of these regions. The cardiac and pyloric glands have been shown to secrete HCO_3, in at least some species. Although neither cardiac nor pyloric glandular regions are described in birds, Ziswiler and Farner (1972) state that the avian proventriculus contains centrally located glands that secrete only mucus and fluid. The avian gizzard also contains tubular glands, but these secrete a substance that provides the horny material lining this organ.

Figure 1.2 illustrates some variations in the distribution of gastric epithelium in mammals. Starting at the gastroesophageal junction, the stomachs of humans, dogs, and many other mammals show successive areas of cardiac, proper gastric, and pyloric glandular mucosa. However, the stomachs of many species show an additional region of nonglandular stratified squamous epithelium extending from the gastroesophageal junction. Furthermore, the cardiac glandular mucosa, which forms only a narrow band in the human and dog stomach, occupies much of the body in the stomach of the pig. In Camelidae such as the camel and llama, it forms islands within the areas of stratified squamous epithelium of the first two compartments and occupies a major segment of the third compartment. The entire forestomach of cattle is

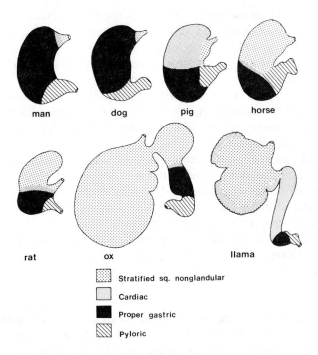

man   dog   pig   horse

rat   ox   llama

▢ Stratified sq. nonglandular

□ Cardiac

■ Proper gastric

▨ Pyloric

FIGURE 1.2 Some variations in the distribution of gastric mucosa. The stomach of each species demonstrates regions of cardiac, proper gastric, and pyloric glandular mucosa. The pig and llama stomachs contain a relatively large region of cardiac glandular mucosa. The pig, horse, rat, llama, and ox stomachs also show increasingly large areas of stratified squamous nonglandular epithelium. Stomachs are not represented on the same scale: for example, the volume capacity of the adult bovine stomach is approximately 70 times that of the human stomach, or 14 times the capacity per kilogram body weight. (Adapted from Stevens 1973.)

lined with stratified squamous epithelium, and glandular mucosa is limited to the final compartment of the stomach, the abomasum. Similar variations in the distribution of these epithelial regions in species belonging to many of the mammalian orders are illustrated in Chapter 3.

It has been suggested that the nonglandular epithelium protects the stomach of insectivores from insects and that of herbivores from plant roughage. This epithelium also has been shown to have important secretory and absorptive functions in ruminants. Observation of its presence and distribution in a wide range of mammalian orders led Oppel (1897) and Bensley (1902-3) to the conclusion that it represented regression of the highly specialized proper gastric glandular mucosa to the less complex cardiac glandular mucosa and, finally, to nonglandular stratified squamous epithelia.

The musculature of the vertebrate stomach generally consists of an inner layer of circular muscle, which forms a valve or sphincter at the junction with the intestine, and an outer layer of longitudinal muscle. A third, oblique layer of muscle is present near the gastroesophageal junction of some mammals. The outer longitudinal layer is almost lacking from the gizzard of birds, and is concentrated in a few bands on the stomach of kangaroos and some herbivorous primates, drawing the stomach into a series of saccules. The arrangement of these muscle layers is extremely complex on the forestomach of ruminants.

The foregut of some lower vertebrates also has developed adaptations that aid in performing functions other than those of digestion. Ducts to the swim bladder, a hydrostatic organ for the adjustment of specific weight, open into the esophagus of many fish. The stomach of many amphibians must shut down its secretory activity during hibernation to help prevent autodigestion. That of the gastric brooding frog ceases both its secretory and motor activity to serve as a "uterus" for the hatching and development of larvae until they are juvenile frogs (Fanning, Tyler, and Shearman 1982).

Midgut

The vertebrate midgut is the major site for the digestion of carbohydrate, fat, and protein. The midgut is also the major site for the absorption of the end products of this digestion, as well as the vitamins and minerals required as nutrients. It is difficult to distinguish the midgut from the hindgut in many fish, where there is often no clear demarcation with respect to epithelial morphology, a change in diameter, or the presence of a sphincter or valve. The same applies to larval forms of amphibians. In the adult amphibians, it can usually be distinguished by its smaller diameter, but separation of the midgut and hindgut by a valve seems to be limited to frogs. The terminal limit of the midgut is generally quite evident in reptiles, birds, and most mammals, but it is as difficult to distinguish in some mammalian species as it is in many fish.

The luminal surface of the midgut may be expanded by ceca, folds, or ridges of various kinds, including pyloric ceca or a spiral valve in some species of fish (Harder 1975a), intricate folds and ridges in some reptiles (Parsons and Cameron 1977), and orad-facing pouches in beaked whales (Flower 1872). In salamanders, birds, and mammals, the intestinal absorptive surface is increased further by villi—macroscopic projections of epithelial and subepithelial tissue. These contain capillaries and, in salamanders and mammals, lacteals for the removal of absorbed fat. The surface is further increased by the presence of microvilli, the brush border, on the apical border of intestinal absorptive cells. Kanou (1984) made a morphological comparison of microvilli in a few species of fish, amphibians, reptiles, birds, and mammals.

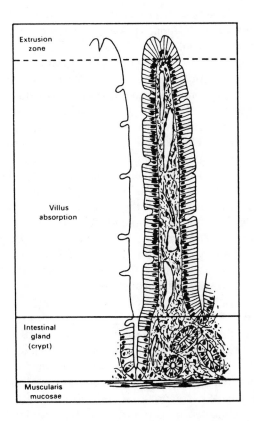

Extrusion
zone

Villus
absorption

Intestinal
gland
(crypt)

Muscularis
mucosae

FIGURE 1.3 Intestinal villus and crypt of Lieberkühn. (From Bell, Emslie-Smith, and Paterson 1980.)

The villi of mammals are in close association with tubular glands called the crypts of Lieberkühn. Figure 1.3 represents a cross section of a mammalian villus and crypt. The crypts contain a variety of cells. The argentaffin cells are believed to secrete serotonin, or its precursor, and possibly other gut hormones. Paneth cells are found in the crypts of rodents, primates, and ruminants, but are rare in the intestine of Carnivora and the domestic pig. These cells contain secretory granules, but their function is unknown. The crypts also contain mucus-secreting goblet cells and undifferentiated cells. The latter undergo rapid mitosis and are the precursors of the absorptive cells, migrating up the crypt and onto the villi as they develop the brush border and other characteristics. Both the goblet and absorptive cells migrate to the tip of the villus where they are sloughed into the lumen.

Although most fish do not have intestinal glands that extend into the submucosa, the Gadidae (Jacobshagen 1937) and Macrouridae (Geistdoerfer

1973) have glands at the base of surface folds throughout the length of the intestine. These have been designated crypts of Lieberkühn, but they contain no cell types different from those of surface epithelium (Harder 1975a). Crypts of Lieberkühn also have been described in the midgut of salamanders (Reeder 1964), some reptiles (Luppa 1977), and some birds (Ziswiler and Farner 1972). The crypts found at the base of intestinal folds in reptiles are less developed than those of birds and mammals, and their epithelium is similar to that of the surface. Both the surface and crypts contain goblet, argentaffin, and Paneth cells in some species. The crypts found in birds also vary among species, ranging from those that contain only absorptive and goblet cells to those that contain only cells with basophilic granules.

Clearly defined zones of cell proliferation such as those at the base of the mammalian crypts are not seen in the midgut of larval and adult lampreys (Youson and Langville 1981; Youson and Horbert 1982), larval amphibians (Marshall and Dixon 1978), or the freshwater painted terrapin, *Chrysemys scripta* (Wurth and Musacchia 1964). Zones of cell proliferation at the base of intestinal folds have been described in more advanced species of fish (Hyodo-Taguchi 1970; Gas and Noaillac-Depeyre 1974; Stroband and Debets 1978) and adult amphibians (Andrew 1963; Martin 1971; McAvoy and Dixon 1977).

The mammalian midgut or small intestine can be further subdivided into the duodenum, jejunum, and ileum. The proximal small intestine of many mammals contains subepithelial glands called Brunner's glands, which are believed to secrete an alkaline fluid and mucus. They extend a short distance along the duodenum of carnivores, a longer distance in omnivores, and the greatest distance in herbivores. Brunner's glands are said to be absent in non-mammalian vertebrates (Andrew 1959), but Ziswiler and Farner (1972) noted glands of similar structure at the gastroduodenal junction of some birds.

The cyclostome intestine has a very thin musculature, consisting of an oblique muscle in lampreys, and digesta are mixed and transported with the aid of cilia. However, in other vertebrates, these functions are carried out by inner circular and outer longitudinal layers of muscle.

Pancreas and biliary system

Although the pancreas and liver are not part of the midgut, their embryonic derivation and their contributions to digestion center on this segment of the tract. The exocrine pancreas secretes enzymes important to the digestion of carbohydrate, fat, and protein. These enzymes will be discussed in Chapter 4.

The pancreas of cyclostomes appears to be represented in a primitive stage as ceca. Barrington (1957) stated that the latter are closely associated with follicles, which are believed to represent the islets of Langerhans of the endocrine pancreas. In the more advanced classes of fish, the pancreas can vary widely in its form and distribution. It is often diffusely distributed along the intestinal wall and may even extend into the liver. The pancreas is a compact organ in the elasmobranches, lungfish, some Siluridea (freshwater sheatfish), and all higher classes of vertebrates. The pancreas of birds is said to be relatively large in insectivores, piscivores (fish eaters), and omnivores, but smaller in carnivores. The avian pancreas usually consists of three lobes with individual ducts—an arrangement that has proved useful in experiments designed to study partial pancreatic fibrosis (Bensadoun and Rothfeld 1972). All three ducts enter the intestine, one of them after joining the bile duct.

The vertebrate liver is a distinct, compact organ in all species. It serves a number of functions, but the one primarily associated with digestion is bile secretion. In most vertebrates, bile is stored in a gallbladder and released when high concentrations of fat are present in the midgut. The gallbladder is absent in some fish, and both the gallbladder and hepatic ducts disappear in lampreys after completion of metamorphosis or the adult feeding stage. The gallbladder is absent also in horses, deer, seals, rats, and a few other rodent species.

Hindgut

The hindgut of most vertebrates serves principally for the final storage of digesta and for retrieval of inorganic ions and water secreted into the upper digestive tract. In animals that excrete urine into a cloaca, it also can aid in the resorption of urinary ions and water. In addition, the hindgut of herbivorous reptiles and birds, and most herbivorous mammals, serves as the major site for microbial digestion.

As stated earlier, the hindgut of many fish cannot be readily distinguished from the midgut. Larval amphibians lack a distinct hindgut, but the hindgut of adults is lined with columnar epithelium and is usually larger in diameter than the midgut. With the exception of frogs, there appears to be no valve or sphincter separating the two. The reptilian hindgut is separated from the midgut by a valve or sphincter and is enlarged in many species. The initial segment contains a cecum in a few species; this plus a segment of the proximal hindgut is partially compartmentalized by mucosal folds in some herbivores. The avian hindgut generally consists of a relatively short, straight, enlarged extension of the intestine. The origin is distinguished by a sphincter or valve and, usually, a pair of ceca. This segment of gut is often called the rectum (because it is usually straight), but its function, is more closely akin to that of the mammalian proximal colon.

The hindgut of mammals tends to recapitulate vertebrate phylogeny, even among the adults of various species. Species belonging to a number of mammalian orders show no clear distinction between a midgut and hindgut with

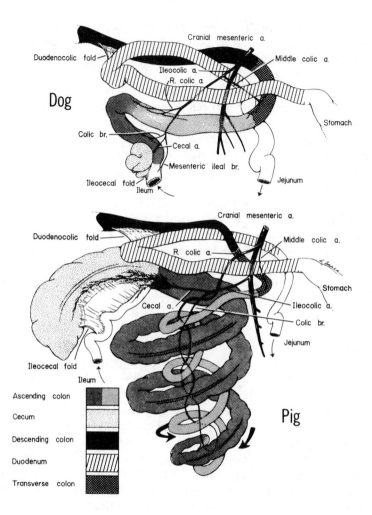

FIGURE 1.4 The large intestine of the dog and pig. (From de Lahunta and Habel 1986.)

respect to gross structural characteristics. Other species have an enlarged hindgut, but no indication of a separating valve or sphincter. However, most mammals have a distinct hindgut, consisting of a large intestine and often a cecum, which is paired in a few species, at its valvular junction with the small intestine. The cecum and large intestine are voluminous and usually sacculated in many mammalian herbivores. The mammalian hindgut is lined with a columnar epithelium, which lacks microvilli, and numerous goblet cells.

The musculature of the vertebrate hindgut is similar to that of the midgut

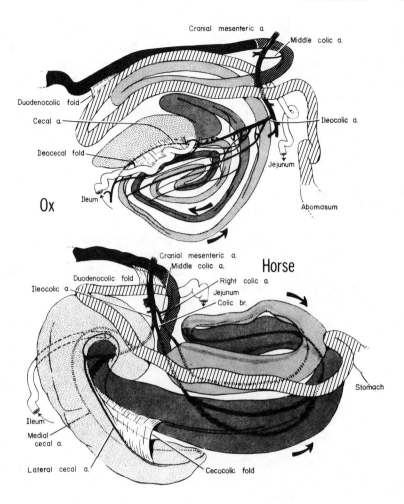

FIGURE 1.5 The large intestine of the ox and horse. (From de Lahunta and Habel 1986.)

in most species—that is, a thin layer in cyclostomes and a double layer of circular and longitudinal muscle in other vertebrates. The circular layer is heavier than that of the midgut in some species; and, in many mammals, the longitudinal muscle is formed into bands similar to those described for the stomach of some mammalian species. These bands serve the similar purpose of drawing the included segment into saccules or haustra, which can help delay the rate of digesta passage.

As with the stomach, comparative descriptions of the gross structure of the mammalian large intestine have suffered from attempts to apply terminology used in human anatomy. The human large intestine can be divided into a cecum, colon, and rectum. As the colon leaves its junction with the small intestine and cecum, it can be further subdivided into ascending, transverse, and descending segments, according to the direction it takes in the abdominal cavity. Although the transverse colon can be compared to similar segments in other species on the basis of its mesenteric attachment and the loop that it and the duodenum form around the cranial mesenteric artery, the "ascending" colon can vary considerably in length, volume, and the course it takes in the abdominal cavity (Figs. 1.4 and 1.5). Furthermore, as we will see in Chapters 4 and 7, the motor, secretory, and absorptive characteristics of different segments of the colon do not necessarily correlate with with these anatomical subdivisions.

Cloaca

The hindgut of many vertebrates terminates in a cloaca. Ducts from the urinary and genital tracts of fish usually exit from the body at a point separate from that of the digestive tract. However, a cloaca, which accepts these ducts, is found at the termination of the intestine of embryos of some species and in adults of the lungfish, the crossopterygian *Latimeria*, the Elasmobranchii, and the hagfish. A cloaca also is present in adult amphibians, reptiles, birds, and a few mammals.

The reptilian cloaca consists of a coprodeum, urodeum, and proctodeum. The first two compartments are lined with an epithelium similar to that of the rest of the hindgut, and the proctodeum is lined with stratified squamous epithelium. The urinary and reproductive tracts enter the urodeum. The cloaca of some terrapins is capable of acting as a pump, and bilateral cloacal bursae aid in the adjustments for buoyancy (Jackson 1971). These animals can regulate their resting lung volume. The change in the lung volume is accompanied by an equal, but opposite, exchange of water between the cloacal bursae and the environment. Water is aspirated into or released from the bursae by the cloacal pump.

The avian cloaca consists of three chambers similar to those of reptiles. The coprodeum is often represented only as an enlargement of the intestinal lumen. This is usually partially separated by a septa from the urodeum, which receives the ducts from the ovary or testes and the kidney. The urodeum is less distinctly separated from the proctodeum, which terminates at the anus. Stratified squamous epithelium is limited to the caudal compartments in young birds. The dorsal wall of the cloaca also contains the cloacal bursa, a relatively solid lymphoid organ that has provided significant contributions to our present understanding of immunology.

Among mammals, the monotremes, marsupials, some Insectivora species, and at least one species of rodent (the mountain beaver, *Aplodontia rufa*) retain the cloaca. In the absence of a cloaca, the hindgut of most mammals does not aid in the retrieval of urinary salts and water. It does, however, continue to serve as a major site for the recovery of the electrolytes and water secreted into the upper digestive tract, as well as those of dietary origin.

Neurohumoral control

The various episodic events that are associated with ingestion and digestion of food, and with the transit and absorption or excretion of digesta, are under nervous and/or endocrine control. The functions served by neurotransmitting agents and the peptide hormones are discussed in Chapter 8.

Summary

Many of the tissues, organs, and functions of the vertebrate digestive system are homologous or analogous to those seen in various invertebrate species. These include the general organization of the digestive tract and its associated glands or organs, and many of the basic mechanisms responsible for digestion of food and absorption of nutrients. However, some major characteristics seem to be limited to the digestive system of vertebrates, including the digestion of protein by pepsin at an acid pH and the development of the pancreas as an organ with specific functions. The major classes of vertebrate show a number of basic similarities involving the midgut, pancreas, and biliary system. However, the headgut, foregut, and hindgut can demonstrate some considerable differences, and members within each class have developed highly specific and often elaborate adaptations to their diet or environment. The following two chapters describe the structural characteristics of the digestive tract of these animals with respect to their taxonomic group, diet, and environment.

2

The digestive tract of fish, amphibians, reptiles, and birds

This chapter will concentrate on variations in the digestive tract of fish, amphibians, reptiles, and birds with reference to diet, feeding behavior, and ecology. As animals have been classified by taxonomists according to their presumed genetic relationships, a general familiarity with taxonomy is necessary for the understanding of adaptations. Pitts (1968) has pointed out that although engineers might devise a human kidney that did not require filtration of 180 liters of plasma, followed by resorption of most of this, in order to excrete a small quantity of waste, their design need not consider the original prototype nor that adaptations would need to be made in a working model and with consideration of other body functions. The same is true for the digestive system. Passage of large volumes of water through the oral cavity and gills of fish, for respiratory purposes, limits the alternatives by which food can be ingested in small particles. Termination of the urinary, reproductive, and digestive tracts in the cloaca of amphibians, reptiles, and birds provides both limitations and opportunities for adaptations that are not shared by most fish and mammals. Control of body temperature in birds and mammals decreases the influence of ambient temperature on feeding, but it also requires a more regular and efficient system for the assimilation of nutrients.

Fish are classified according to Nelson (1984), and amphibians are classified according to Duellman and Trueb (1986). The taxonomy of reptiles follows that used in *Biology of the Reptilia* (Gans 1969; Gans and Gans 1978), and the taxonomy of birds follows that used in *Avian Biology* (Farner, King, and Parkes 1971).

However, many adaptations of the digestive system that are seen in different species show a closer correlation with their diet or environment than with their taxonomic classification. Therefore, knowledge of the diet, feeding characteristics, and environment of an animal is often equally necessary for an understanding of this system.

Fish

The four classes of fish are the Cephalaspidomorphi, Myxini, Chondrichthyes, and Osteichthyes (see Table 1.1). The first two classes contain

22

the lowest craniate vertebrates, the cyclostomes (lamprey and hagfish). The cartilaginous fish can be classified into two subclasses: the Elasmobranchii, which includes sharks, skates, and rays; and the Holocephali (*Chimaera* sp.), which includes fish with a blunt snout and threadlike tail. The Osteichthyes (bony fish) can be subdivided into four subclasses: Crossopterygii (lobe-finned fish), Dipneusti (lungfish), Brachiopterygii (gar, sturgeon, paddlefish, and bowfin), and Actinopterygii. The Dipneusti have both gills and lungs with a pulmonary circulation. The Actinopterygii comprise a large number of the present-day fishes. The family Cyprinidae is of special interest because its members are distributed throughout the world (1000 species), and it contains the majority of the freshwater fish of North America including the carp, minnows, chubs, and barbs. The anatomy of the fish digestive system has been reviewed by Andrew (1959), Harder (1975a,b), and Kapoor, Smith, and Verighina (1975). The physiology of the digestive system is reviewed in its various aspects by Barrington (1957), Campbell and Burnstock (1968), Barnard and Prosser (1973), and Hoar (1983).

Fish, like many other groups of animals, include species that are carnivores, omnivores, or herbivores. They may feed on dead or living material. They may be micro- or macrophagous. However, each of these characteristics is seen in lower phyla and even within single classes such as Insecta. One unusual development is that some species of fish can adapt their diet to almost any of the above forms, or, as Steven (1930) has stated, they " . . . eat what they can get." For example, Barrington (1957) has pointed out that the roach (*Rutilus rutilus*) can be a carnivore or herbivore, according to necessity, and some species of fish can vary their diet from plankton in the summer to fish in the winter, sometimes even utilizing the bacteria and algae of the water as a source of food.

The increasing importance of aquaculture has stimulated a great deal of interest in the digestive system of fish during their larval stage of development and subsequent transformation into adult forms. The digestive system of fish larvae differs from that of the adult, and its ontogeny also differs in different taxa (Govoni, Boehlert, and Watanabe 1986). The alimentary canal of larvae is less complex and usually remains unchanged during the larval period, which can last from months to a year. Most fish larvae are raptorial plantivores, regardless of the feeding habits of the adult form.

The cyclostomes have no jaws. Species belonging to more advanced classes have movable jaws, but muscles are absent in their lips and cheeks. The tongue usually lacks muscle and shows little mobility. Exceptions include the archerfish (*Toxotes jaculator*), which uses a muscular tongue to squirt streams of water at insect prey, and several other species that have a mobile, tooth-bearing tongue. The orobranchial cavity of fish shows a wide variety of arrangements for the capture, sorting, and trituration of food. Teeth, used for capture, tearing, or crushing of food, are present in most species. These may consist of sharp teeth for capture of prey (sharks and piranhas) or grazing on algae attached to rocks or coral (Scaridae). Teeth may be located on the jaws, tongue, pharynx, or almost any surface of the orobranchial cavity. Cyprinids have only pharyngeal teeth, which are used to break up hard food or hold and swallow prey. A number of microphagous species such as the parrot fish *Scarus radicans* have pharyngeal pockets, which collect algae during grazing for trituration prior to swallowing. Because continuous flow of water through the mouth and gills is required for the respiratory and osmoregulatory functions, escape of food past the gills may be prevented by gill rakers, projections that serve as a sieve. These may be spaced quite far apart, for fish that feed on large prey; alternatively, they may form a fine meshwork, for microphagous feeders such as the menhaden, shad, or basking shark (*Cetorhinus maximus*).

The esophagus of fish is generally short, wide, and straight. It is lined with a multilayer of squamous epithelium containing a large number of mucous cells in freshwater fish. Al-Hussaini (1946, 1947) found that the esophagus of some marine teleosts was lined with complex, highly vascularized mucosal folds with columnar epithelium and few mucosal cells. Yamamoto and Hirano (1978) suggested that this may be a common feature in marine teleosts, associated with the osmotic regulatory function of their esophagus (see Chapter 7). The posterior segment of the esophagus contains glands in some species (Kapoor et al. 1975). Ciliated epithelium is found in the esophagus of embryonic forms, as well as adult cyclostomes, perch, and some Elasmobranchii. The esophagus of teleosts terminates in a sphincter, which may serve to prevent excessive swallowing of freshwater.

The stomach is absent in cyclostomes (Fig. 2.1) and a number of more advanced species belonging to various orders and families. Most of these fish are microphagous with well-developed pharyngeal teeth for trituration of food (Fig. 2.2). When present, the general form of the stomach can be classified as straight, siphon (U)-shaped, or Y-shaped with a gastric cecum. The straight stomach, as seen in pike (Fig. 2.3), is rare. The siphon-shaped stomach, depicted in the sturgeon (Fig. 2.4) and trout (Fig. 2.5), is common in Elasmobranchii and Osteichthyes. The Y-shaped stomach, such as that shown in the eel (Fig. 2.6), has a blind sac or cecum that arises from the curvature. The circular muscle of the stomach can be well developed, as in mullet (*Mugil*), menhaden (*Brevoortia*), and gizzard shad (*Dorosoma*), where this, plus a tough epithelial lining, provides a gizzardlike action. A valve or sphincter of circular muscle, a mucous membrane fold, or both are present at the juncture of the stomach and intestine.

The intestinal tract of fish can vary from one that is relatively short and straight to one that is long and arranged in spirals and loops. The length does not necessarily correlate with feeding habits, but it tends to be longest in herbivorous species and especially in fish that ingest large amounts of un-

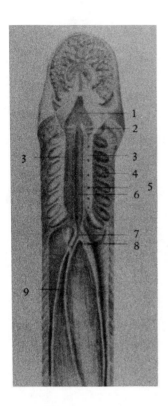

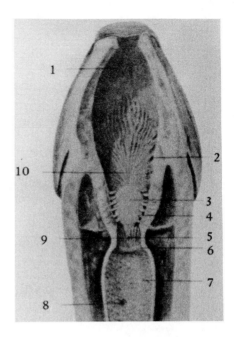

FIGURE 2.1 [Left] Sea lamprey (*Petromyzon marinus*). 1) Prebranchial foregut, buccal (pharyngeal) cavity; 2) velum, valve between buccal cavity and branchial foregut; 3) branchial foregut; 4) gill-sac; 5) ductus branchialis internus; 6) epibranchial foregut; 7) metabranchial foregut; 8) anterior end of midgut; 9) spiral valve = typhlosolis. (After Pernkopf, from Pernkopf and Lehner 1937.)

FIGURE 2.2 [Right] Chub (*Leuciscus cephalus*). Note the absence of a stomach in this cyprinid fish. 1) Buccal cavity; 2) gill slits and arches; 3) masticating plate; 4) pharyngeal teeth [3 and 4 form the chewing apparatus of the branchial foregut]; 5) ductus pneumaticus; 6) pylorus; 7) midgut; 8) opening of ductus choledochus; 9) metabranchial foregut; 10) branchial foregut. (After Pernkopf, from Pernkopf and Lehner 1937.)

digestible mud or plant material. The mucosal surface of the intestine is increased by folds of various design, and by a spiral valve in species belonging to all groups except the hagfish and the teleosts. The spiral valve is present in fish with a short, straight intestine and usually those in which the stomach is small or absent. It consists of folds of mucosa and submucosa, which project into the intestinal lumen and vary in their number of turns (Fig. 2.7). In some

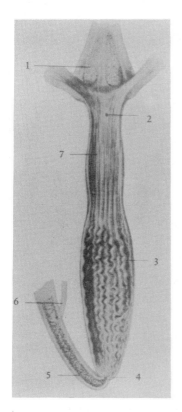
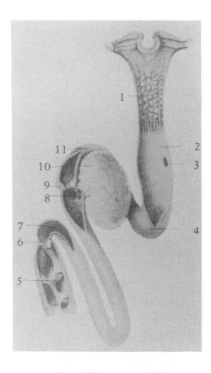

FIGURE 2.3 [Left] Pike (*Esox lucius*). 1) Branchial foregut; 2) ductus pneumaticus; 3) stomach; 4) pylorus; 5) midgut; 6) ductus choledochus; 7) esophagus. (After Pernkopf, from Pernkopf and Lehner 1937.)

FIGURE 2.4 [Right] Sturgeon (*Acipenser sturio*). Foregut, intermediate gut, and anterior portion of spiral valve. 1) Papillae of esophagus; 2) stomach; 3) opening of ductus pneumaticus; 4) stomach bend; 5) spiral valve; 6) valve between intermediate gut and spiral valve; 7) intermediate gut, without spiral valve; 8) opening of pyloric appendages; 9) pylorus; 10) appendices pyloricae; 11) pars pylorica of stomach. (After Pernkopf, from Pernkopf and Lehner 1937.)

species, these leave a space free for direct passage of digesta. In others, they come in contact at the center, forming a drill-bit arrangement, or fold over in the center to form a conelike structure. This remarkable structure, which is unique among vertebrates, serves both to delay the passage of digesta and to increase the absorptive surface.

Another arrangement of the upper intestinal tract, unique among vertebrates but found in a wide range of fish that have a stomach, is the pyloric

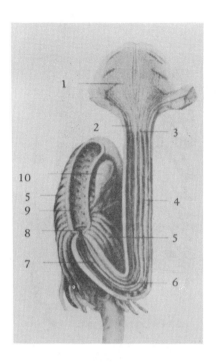

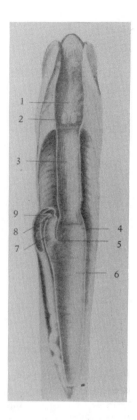

FIGURE 2.5 [Left] Trout (*Salmo fario*). 1) Branchial foregut; 2) esophagus; 3) ductus pneumaticus; 4) stomach; 5) appendices pyloricae; 6) stomach bend; 7) pars pylorica of stomach; 8) pylorus; 9) anterior end of midgut; 10) gallbladder. (After Pernkopf, from Pernkopf and Lehner 1937.)

FIGURE 2.6 [Right] Eel (*Anguilla anguilla*). 1) Branchial foregut; 2) horny pads; 3) esophagus; 4) stomach, showing the circular muscles; 5) pars angularis; 6) gastric cecum with independent circular muscles; 7) pars pylorica of stomach; 8) first pyloric valve; 9) second pyloric valve. (After Pernkopf, from Pernkopf and Lehner 1937.)

ceca. These can vary in size and shape from small evaginations of the intestinal wall to tubular, branched, or tuftlike structures (Figs. 2.8 and 2.9). Their number can vary among species from 1 to 1000. They are lined with cells similar to those of the remaining intestinal wall, and their presence does not seem to correlate with the length of the intestinal tract or the diet of the species. They appear late in ontogenic development and could act both to delay passage and to increase the absorptive surface. Therefore, they have

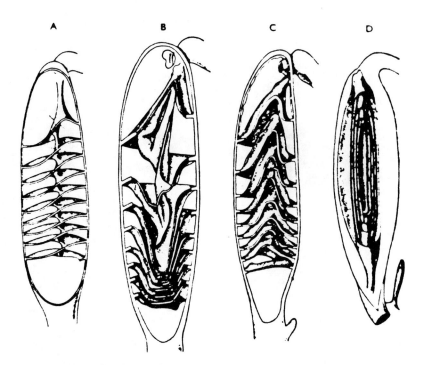

FIGURE 2.7 Different types of spiral valves in Elasmobranchii. Spiral valve: A) with medial column, B) with caudad directed funnels, C) with craniad directed funnels, D) with cylindrically wound-up mucosal fold = scroll-type valve (*Sphyrna*). Dorsal wall of gut removed; intermediate intestine at upper right. (After Parker, from Bertin 1958.)

been suggested as a site for microbial digestion. However, a recent study of rainbow trout (*Salmo gairdneri*), cod (*Gadus morhua*), largemouth bass (*Micropterus salmoides*), and striped bass (*Morone saxatilis*) showed that the ceca absorbed glucose and amino acid at rates similar to that of the remainder of the proximal intestine (Buddington and Diamond 1987). Studies of digesta transit in the trout also showed that small particles passed through the ceca at rates similar to that of the major digesta flow. Therefore, it was concluded that these ceca serve principally as an additional absorptive surface.

The hindgut of fish is often difficult to distinguish from the midgut. However, the terminal segment of the Elasmobranchii intestine has a thicker layer of circular muscle, and a small cecum has been described in some individuals of catfish (*Bagarius bagarius*), knifefish (*Notopterus notopterus*), and cod (*Raniceps raniceps*). Lungfish, and at least some Crossopterygii lack any valvular separation between mid- and hindgut, but an ileorectal valve is

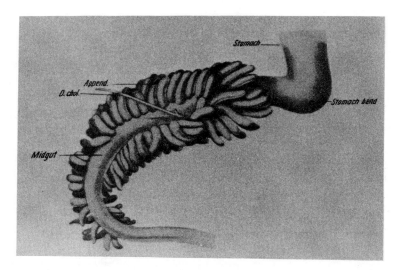

FIGURE 2.8 Stomach of a salmonid (*Hucho*) with about 200 appendices pyloricae. Append., appendices pyloricae; D. chol., ductus choledochus. (After Pernkopf, from Pernkopf and Lehner 1937.)

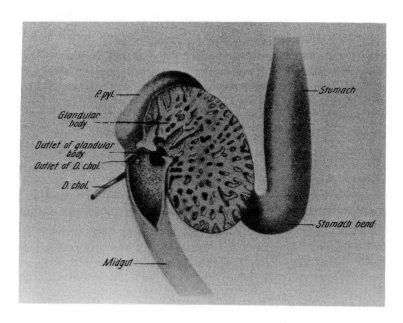

FIGURE 2.9 Stomach of the sturgeon (*Acipenser sturio*) with homogeneous glandular body of appendices pyloricae, cut open. P. pyl., pars pylorica; D. chol., ductus choledochus. (After Pernkopf, from Pernkopf and Lehner 1937.)

present in many teleosts, and an annulospiral septum consisting of circular muscle and glandular epithelium is found in the hindgut of the trout *Salmo facia*.

In many fish, the epithelium of the hindgut is similar to that of the adjacent midgut, except for an increase in goblet cells and sometimes cilia. The terminal gut of the cyprinids, a family that includes carp and minnows, has no distinguishing characteristics, other than an increase in zymogen-secreting cells (Al-Hussaini 1949a,b). However, studies of the scup (*Stenostomus chrysops*) indicated that cells in the posterior midgut have special phagocytic capabilities (Strauss and Ito 1969). Barrington (1957) also cited the work of Young and Fox, showing that the cells of the hindgut of surfperch (Embiotocidae) can concentrate and possibly excrete pigments contained in their diet of shrimp. Therefore, it would appear that the hindgut has special excretory, absorptive, and motor functions in some species of fish.

Amphibians

As implied by its name, class Amphibia represents the transition of vertebrates from water to land. Most amphibians begin life as free-living, aquatic larvae, which transform into terrestrial adults. The degree of transformation may differ among the three orders: Gymnophionia (wormlike burrowing amphibians), Caudata (salamanders, newts, and congo eels), and Salientia (frogs and toads). The Salientia are the most highly specialized and undergo the greatest metamorphosis. General information on the digestive system of amphibians is provided by Reeder (1964) and Andrew (1959).

Amphibian larvae include carnivores, omnivores, and herbivores. Many feed on plankton, bacteria, or detritus. Others graze on plants. Salientian larvae have a horny beak and labial teeth. Water is pumped by muscular activity of the mouth through an elaborate pharyngeal filtering system, gill slits, and spiracle (breathing tube). Food particles are trapped in mucus, covering a set of filter plates, and transferred by cilia to and through the esophagus. A stomach serves for proteolytic digestion in salamander larvae, but provides only a storage function in the larvae of frogs and is absent from the foregut of toad larvae. The intestine of amphibian larvae is longer than the body and is often coiled with no distinct division into midgut and hindgut.

Metamorphosis of the amphibian larvae into adults is associated with major changes in structure and function of the digestive system over a relatively short time. There are marked changes in the buccal cavity and feeding apparatus. Glands develop in the stomach and in the esophagus of some species (Janes 1934). Intestinal epithelium degenerates and is then replaced from subepithelial nests of cells. This process is associated with an increase in proteolytic enzyme activity, which is believed to aid in the breakdown of cells, followed by a disappearance and then a reappearance of this activity. The terminal segment of the intestine develops into the enlarged hindgut in most species.

All adult amphibians are carnivores that swallow their prey whole (Erdman and Cundall 1984; Larsen 1984). Dentition tends to be weak and used only for grasping and positioning of prey, although Caecilians (tropical burrowing amphibians) have sharp and relatively long teeth that are directed toward the rear of the mouth. The tongue is generally well developed and acts in frogs and toads as an organ for seizure of prey. Multicellular glands in the mouth produce mucus, which aids in deglutition and imparts an adhesive quality to the tongue. These glands appear to produce no enzymes except for reports of a weak amylolytic activity. The mouth and esophagus are usually lined with ciliated and mucus-secreting cells. The esophagus is relatively short and large in diameter.

The stomach of adult amphibians tends to be tubular in shape (Fig. 2.10). The intestine is relatively short in comparison to that of amphibian larvae. However, the hindgut often can be differentiated from the midgut by its larger diameter. They are separated by a valve in the frog, but there appears to be no valve or sphincter at this point in most amphibians, and there appears to be no distinct hindgut in the salamander, *Necturus* (Claypole 1897). The hindgut is generally lined with a columnar epithelium containing goblet cells. Although the amphibian skin, kidney, and bladder have been carefully examined for their contributions to water absorption, the contribution of the large intestine appears to have received relatively little attention. This may be due to the fact that amphibians do not usually drink water, and it is assumed that much of the electrolyte-water balance is effected by the skin, kidney, and urinary bladder.

Reptiles

The reptiles are air-breathing, scaled, cold-blooded animals, which first appeared in the Carboniferous Period, and reached their greatest height (in both senses) in the Mesozoic Era. If the insect invasion of land was the most spectacular in numbers, the reptiles certainly established the record for size of individual terrestrial animals, but relatively few species presently remain.

Class Reptilia contains three orders: Crocodilia, Testudinata, and Squamata. Order Crocodilia includes the crocodiles, alligators, caimans, and gavials. The Testudinata, often referred to as chelonians, can be subdivided according to habitat into turtles (marine), terrapins (freshwater), and tortoises (land). The Squamata consists of the snakes, lizards, worm lizards, and tuatara, a New Zealand reptile. The anatomy of the reptilian digestive system has been reviewed by Parsons and Cameron (1977), Luppa (1977), and Ot-

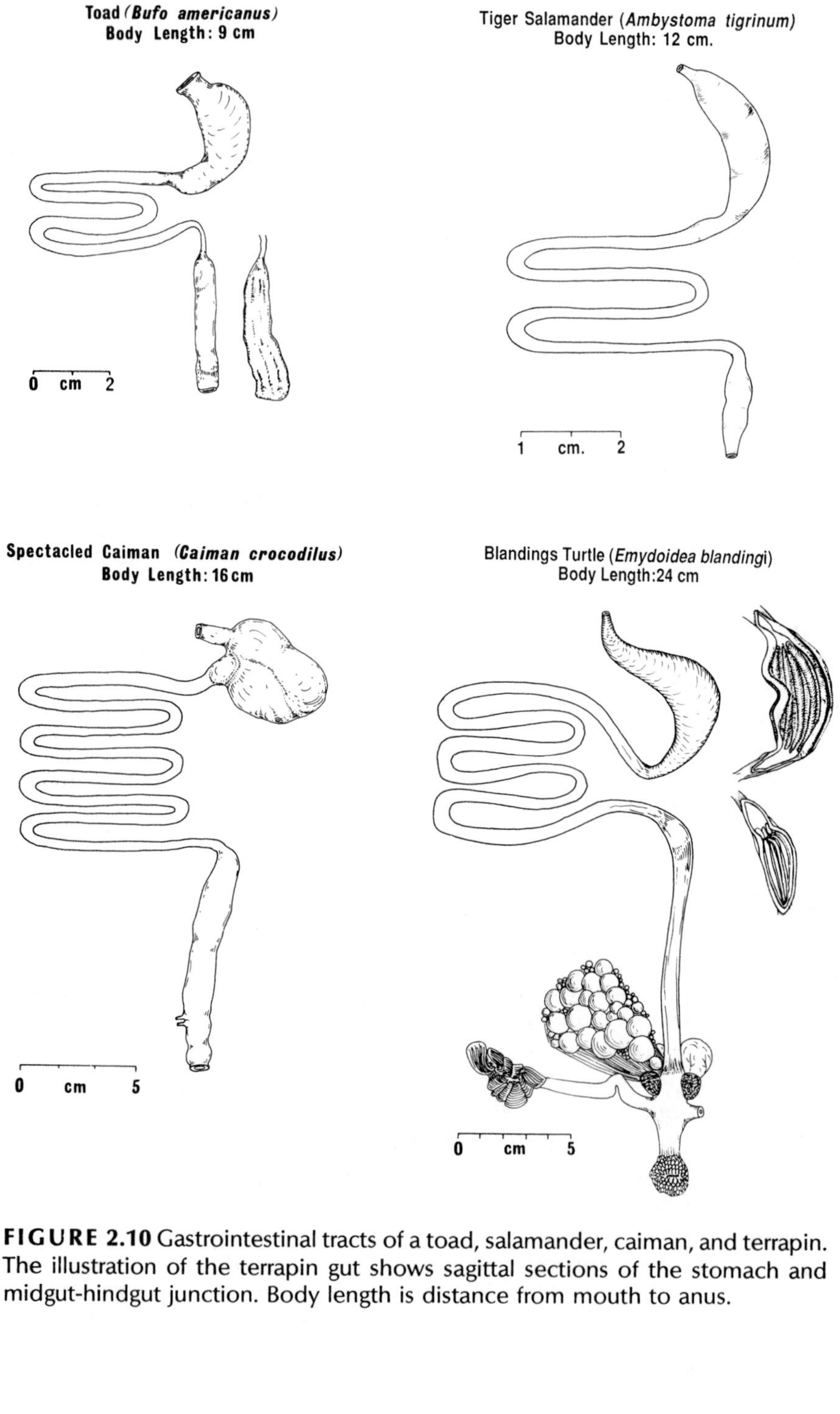

FIGURE 2.10 Gastrointestinal tracts of a toad, salamander, caiman, and terrapin. The illustration of the terrapin gut shows sagittal sections of the stomach and midgut-hindgut junction. Body length is distance from mouth to anus.

taviani and Tazzi (1977). Skoczylas (1978) has reviewed information on the physiology of this system.

Most reptiles are either carnivores or omnivores, and many species subsist on insects during their early or entire life. Although the largest dinosaurs were herbivores, only the tortoises, some turtles, and about 40 of the present 2500 species of lizards are herbivorous, and the latter are usually omnivores until they reach a body size of 50-300 g (Pough 1973). With the exception of the chelonians, which obtain their food by means of a horny beak, all reptiles have teeth. These are used in grasping and tearing, and they are continuously replaced and often added to as the jaw lengthens. The fang teeth of snakes may also be used to inject venom or digestive enzymes. The teeth of mollusk-eating lizards are modified for crushing (Edmund 1969), and herbivorous lizards such as some iguanids have cusplike teeth (Hotton 1955). The jaws of some snakes are arranged for distention and even disarticulation to allow in-gestion of prey.

The oral cavity of reptiles contains ciliated and mucus-secreting cells. A number of species have a distensible tongue, similar to that of amphibians, but serving as a sensory organ. As the first group of vertebrates to generally adopt terrestrial living and subsistence on dry food, many species demon-strate complex oral glands. Salivary glands, as such, are usually absent and, when present, secrete only mucus. However, complex glands that secrete venoms and digestive enzymes are found in snakes and lizards (Kochva 1978).

The esophageal epithelium is often ciliated like that of amphibians, and contains goblet cells, which increase in number along its length, to the point, in alligators and snakes, of almost covering the terminal mucosa. Esophageal glands also may be present. These are most frequently seen in chelonians, and in some species they appear to secrete pepsinogen. The reptilian esophagus can be very distensible, serving as a storage area during gastric digestion of large prey. In some egg-eating snakes, the ventral surface of the vertebral column is attached to the dorsal wall of the esophagus and provides a surface for the crushing of egg shells.

The stomach of reptiles tends to be tubular in shape (Figs. 2.10 and 2.11). The stomach of Crocodilia (Fig. 2.10) is more outpocketed, with a very mus-cular pylorus. The pylorus of the alligator stomach is separated from the remainder of the stomach by a constriction, and from the intestine by semilunar valves. Pyloric muscle also is extremely well developed in the boa constrictor and Florida indigo snake (Blain and Campbell 1942).

Gastroliths (stones, gravel, or sand) have been reported in the stomach of crocodilians, chelonians, and both insectivorous and herbivorous lizards. These appear to be most prevalent and most studied in the crocodilians. In a study of *Crocodylus nyloticus*, Corbet (1960) reported a 73% incidence of gastroliths in individuals up to 1 m in body length, 88% in those between 1

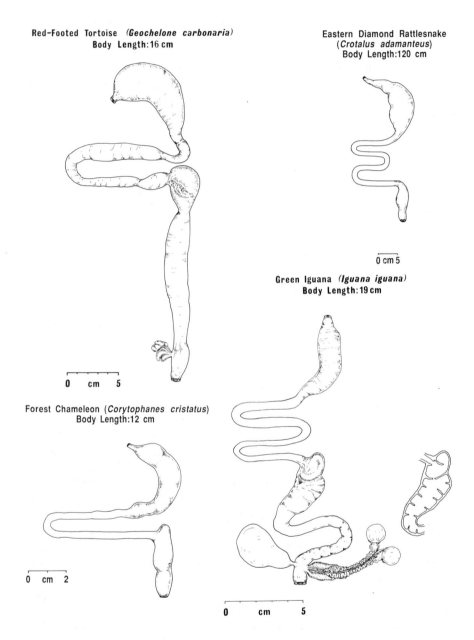

Red-Footed Tortoise *(Geochelone carbonaria)*
Body Length: 16 cm

Eastern Diamond Rattlesnake
(*Crotalus adamanteus*)
Body Length: 120 cm

0 cm 5

Green Iguana *(Iguana iguana)*
Body Length: 19 cm

0   cm   5

Forest Chameleon (*Corytophanes cristatus*)
Body Length: 12 cm

0   cm   2

0   cm   5

FIGURE 2.11 Gastrointestinal tracts of an herbivorous tortoise (from Guard 1980), a snake, an insectivorous lizard, and an herbivorous lizard (iguana, from Guard 1980). The latter illustration shows a sagittal section of the proximal hindgut. Note mucosal folds extending into the cecum and proximal colon.

and 2 m in length, and 100% in larger specimens. It has been suggested that they may serve as ballast, but their presence in the fossils of herbivorous terrestrial dinosaurs and present-day lizards suggests a triturating function similar to that seen in the gizzard of birds.

The intestinal tract of lizards tends to be longest in herbivores and shortest in carnivores, whereas the opposite is true for both the length and volume of the hindgut. Skoczylas cites the study of 40 species by Lönnberg (1902), which found the midgut and hindgut of herbivores to be approximately two times and one time the body length, respectively, as compared to 0.87 and 0.34 times the body length in carnivorous species. Parsons and Cameron (1977) described the internal relief of the midgut of a wide range of reptiles. An ileocolic valve or sphincter separates the midgut and hindgut, which are of equal diameter in some species. However, the hindgut is enlarged in most reptiles, and in some herbivores, a cecum is present at its junction with the midgut (Fig. 2.11). The cecum and proximal colon of herbivorous lizards in the families Agamidae and Iguanidae are compartmentalized by mucosal folds.

Birds

Birds differ from other nonmammalian lower vertebrates in their ability to control body temperature, and from all nonextinct vertebrates in their cover of feathers and (with the exception of bats) the modification of their forelimbs for flight. Modifications for flight are accompanied by a redistribution of weight, which is associated with an absence of teeth and decreased weight of the jaw skeleton and its muscles, as well as acquisition of the gizzard as the organ for trituration. Storer (1971a) reviewed the problems associated with the taxonomic classification of birds and listed 28 orders of existing species. Many of the common species are often referred to as either "galliform" (cocklike) or "passeriform" (sparrowlike). The galliforms include the pheasant, partridge, grouse, quail, and common domestic fowl. The passeriforms include more than one-half (7000) of all avian species. The general characteristics of the avian digestive system have been described by Ziswiler and Farner (1972). Andrew (1959) and Barnard and Prosser (1973) provide additional comparative information.

The horny beak or bill and associated mouth parts show a wide range of modifications in carnivores, piscivores, insectivores, and carrion eaters, as well as birds that feed primarily on fruit, seeds, pollen, nectar, leaves, or roots (Storer 1971b; Kear 1972; Cloudsley-Thompson 1972). The bill of flamingos contains marginal projections, which provide a filter for feeding on small vertebrates or, in the lesser flamingo, blue-green algae and diatoms. In the pelican and some passerine species, the floor of the mouth cavity is distensible and serves for food storage. The tongue shows as many modifications as the bill, but it contains muscle only in the parrots. Although salivary glands were reported to be absent in the snake bird (*Anhinga anhinga*) (Antony 1920), they usually are present and highly developed. They function primarily for mucigenous lubrication. However, as noted in Chapter 1, they secrete an adhesive material in swifts and woodpeckers, and are said to secrete amylase in some species.

The foregut, midgut, and hindgut of a number of species are illustrated in Figures 2.12 and 2.13. The esophagus tends to be long and large in diameter and is usually dilated into a unilateral, bilateral, or spindle-shaped crop, which serves for storage of food. The crop of some species also serves for storage of food, which is regurgitated to feed the young. In the hoatzin (*Opisthocomus hoazin*), it is said to have taken over the major functions of the proventriculus and gizzard. These variations and the special secretion of "milk" by the pigeon crop are discussed in a number of the above references. In the chicken, the crop is isolated by a sphincter that opens only after the gizzard is filled. The crop then fills and undergoes periodic rhythmic contractions that empty it over the next few days.

Other functions of the conventional vertebrate stomach are carried out by the proventriculus (gastric secretion) and the ventriculus (trituration, maceration, and pumping). The proventriculus is a spindle or cone-shaped structure. Its size can vary. For example, predacious carnivores such as the hawk, petrel, heron, or gull tend to have a highly distensible proventriculus, but in galliform, passeriform, and certain other species the proventriculus appears to function only for the secretion of gastric juice during the passage of food into the gizzard.

The ventriculus, or gizzard, is a muscular organ lined with koilen, a horny material consisting of protein and carbohydrates, which is periodically molted by many species. The gizzard can vary a great deal in its relative size and degree of musculature. The single layer of circular muscle is thin walled in most carnivores and frugivores. In the fruit-eating tanager, it has been reduced to an insignificant band. Conversely, it shows the greatest development in the granivorous and most herbivorous species, capable of producing intraluminal pressures of 200 mm Hg in the domestic fowl and 280 mm Hg in the goose. The gizzard often serves the primary function of trituration, and swallowed grit increases its efficiency, although it is not essential.

Although the gizzard is generally considered to be the compensatory organ for lack of teeth, it also serves as a site for food storage, acid-pepsin digestion of protein, and propulsion of food into the intestine (Ziswiler and Farner 1972). There is even evidence of pepsinogen secretion by the gizzard of the common kestrel (*Falco tinnunculus*, a European falcon) and the buzzard (*Buteo buteo*). In birds such as the owls, hawks, kingfishers, and shrikes, the gizzard also serves as a filter for less digestible parts of the prey, which are then formed into pellets and regurgitated.

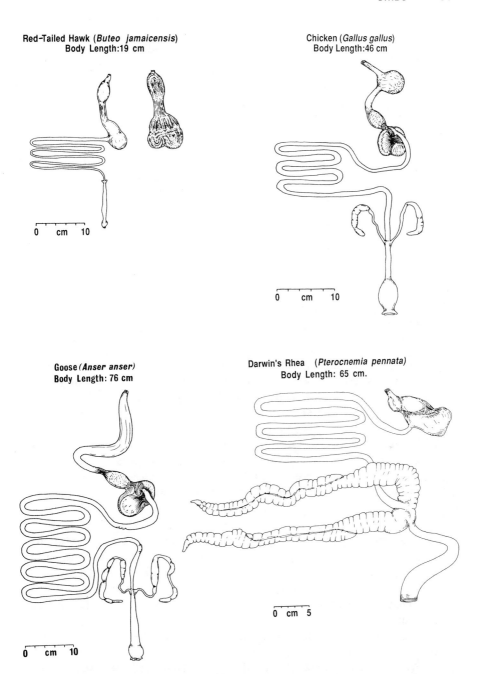

Red-Tailed Hawk (*Buteo jamaicensis*)
Body Length:19 cm

Chicken (*Gallus gallus*)
Body Length:46 cm

Goose *(Anser anser)*
Body Length: 76 cm

Darwin's Rhea   (*Pterocnemia pennata*)
Body Length: 65 cm.

FIGURE 2.12 Gastrointestinal tracts of a hawk (with sagittal section of foregut), chicken, goose, and rhea. (The gastrointestinal tract of the goose is from Clemens, Stevens, and Southworth 1975b.)

Grass Parakeet (*Melopsittacus undulatus*)
Body Length: 9 cm.

Ruffed Grouse (*Bonasa umbellus*)
Body Length: 29 cm.

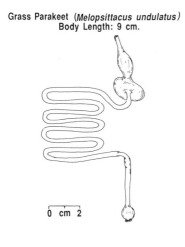

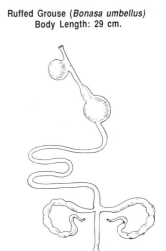

FIGURE 2.13 Gastrointestinal tracts of a budgerigar and a grouse.

An additional chamber, sometimes called the "pyloric stomach," is present in penguins, grebes, pelicans, and many storks, as well as some ducks, geese, and rails. Its function is uncertain, but its presence appears to correlate with diets high in water content and, at least in the grebe, it may serve as a filter.

The avian midgut or small intestine is usually considered to be comprised of a duodenum and ileum. The wide variation in number and arrangement of intestinal loops was illustrated by Mitchell (1901). Leopold (1953) discusses the relationship between intestinal morphology and diet in gallinaceous birds.

The hindgut of birds generally consists of the ceca and a relatively short, straight extension of the intestine. Its origin is distinguished by a sphincter or valve, and the ceca are separated from the remainder of the hindgut by valves. They are usually paired, although some birds such as the heron have only one, and others (hummingbirds, swifts, and some pigeons and woodpeckers) have none. In the parrot and budgerigar, they are absent in both the embryo and adult bird.

Ceca are most highly developed in herbivores and omnivores (but not necessarily granivores), and especially well developed in the African ostrich (*Struthio* sp.), rheas (Fig. 2.12), and the grouse (Fig. 2.13). Mitchell (1901) classified avian ceca into four different types: 1) *primitive*, which are well developed and thin walled (e.g., duck, goose, grebe, loon); 2) *enlarged*, which are characterized by a greater amount of lymphoid tissue and apparently a greater role in cellulose digestion (e.g., galliform species); 3) *lymphoepithelial*, which are characterized by apparently little digestive function, (e.g., passeriform species); and 4) *functionless vestigial* (or absent), which are seen in the penguin, hawk (Fig. 2.12), and budgerigar. The ceca are usually histologically similar to the intestine, except for lymphoid tissue. The African ostrich (*Struthio camelus*) has an extremely long colon, over ten times the length of the ceca (Skadhauge, Warüi, Kamau, and Maloiy, 1984).

Summary

Many characteristics of the digestive system of fish, amphibians, reptiles, and birds represent adaptations to diet. This is especially true for the buccal cavity and pharynx, but it can involve the foregut as well. These include adaptations of the teeth or tongue for procurement of food, and adaptations of the teeth, pharynx, esophagus, or stomach for reducing its particle size. Microphagous animals often have elaborate mechanisms that allow for the separation and ingestion of small food particles. Macrophagous species require adaptations that allow for intermittent feeding and the storage of food, as well as its reduction to a size suitable for digestion.

The midgut shows numerous structural adaptations that serve to increase its mucosal surface area, and the hindgut of terrestrial vertebrates demonstrates various adaptations that allow for retention of digesta. These may be related to the diet, as is the case with respect to the hindgut of herbivorous reptiles and birds. However, another advantage of increasing the surface area or digesta retention time is that it allows for the conservation of dietary, excretory, and secretory fluids.

The following chapter discusses some similar adaptations of the mammalian digestive tract. It also describes additional adaptations of the teeth, jaw, foregut, and hindgut, which extend the general efficiency of the digestive tract with respect to the physical breakdown of food, microbial digestion, and conservation of electrolytes and water.

3

The mammalian digestive tract

Mammals are believed to have first appeared in the Jurassic Period of the Mesozoic Era, but the Tertiary Period in the Cenozoic Era provides the earliest evidence of their success. The evolution of mammals has been traced back to the therapsids, Triassic Period reptiles, which are believed to have evolved into the Prototheria or monotremes of today. Marsupials and eutherian (placental) mammals appear to have evolved from the same source, but via the Pantotheria, present in the Jurassic and Cretaceous periods (Waring, Moir, and Tyndale-Biscoe 1966). Although marsupials and eutherians have many common features, which include converging patterns of digestive system, the marsupials are said to have retained more of the reptilian characteristics. The earliest mammals were animals small in body size. They are believed to have fed on insects, other small vertebrates, and the young of amphibians, reptiles, and birds (Crompton 1980). Their success, in the presence of numerous reptilian predators and competitors for food, may have been due to their ability to control their body temperature, which allowed for nocturnal feeding (Crompton, Taylor, and Jagger 1978).

The length and capacity of the mammalian digestive tract increase during prenatal development. Musculature develops in a craniocaudal sequence, circular muscle first. In species with complex stomachs such as ruminants, compartmentalization may be evident quite early in gestation. The large intestine tends to develop later. It initially terminates with the renal system in a cloaca like that of lower terrestrial vertebrates, but outlets for the renal and digestive systems eventually separate, prior to the birth of most mammals.

Mammals differ from other vertebrates in that they suckle their young, which are born at an early stage of development. Although the composition of milk varies with species (Oftedal 1980), it tends to be high in fat and low in carbohydrates. The diet following weaning is usually the reverse of this. The practice of nursing the young is associated with some major changes in the digestive system, during the period between birth and weaning (Koldovsky 1970). These include the eruption of teeth generally after birth, absorption of intact milk protein for a period shortly after the birth of some species, and changes in both secretory activity and the composition of digestive enzymes during this stage of development.

Table 3.1 lists the mammalian orders according to the diets of inclusive

40

TABLE 3.1 Mammalian orders listed according to diets of inclusive species

Order	Animal	Animal & plant or plant concentrates	Plant fiber
Monotremata	+		
Pholidota	+		
Tubulidentata	+		
Cetacea	+		
Macroscelidea	+		
Insectivora	+	+	
Scandentia	+	+	
Chiroptera	+	+	
Carnivora	+	+	+
Marsupialia	+	+	+
Edentata	+	+	+
Rodentia	+	+	+
Primates	+	+	+
Dermoptera		+	
Artiodactyla		+	+
Lagomorpha			+
Perissodactyla			+
Proboscidea			+
Sirenia			+
Hyracoidea			+

Source: Modified from Stevens (1980).

species. Thirteen of the 20 orders contain species that feed principally on other animals, and each of these 12 includes species that feed on insects or other small invertebrates. Ten orders contain omnivores, which feed on plants and animals, or species that feed principally on plant concentrates, such as seeds, nuts, fruit, nectar, or pollen. The third category contains herbivorous species that can subsist largely on a diet containing the fibrous portion of plants. Eleven orders contain these species, most of which demonstrate some major adaptations of the stomach and/or large intestine in association with this capability.

One major characteristic of mammals, generally absent in other vertebrates, is their ability to masticate or chew their food. Davis (1961) and Crompton and Parker (1978) discussed the evolution of the mammalian masticatory apparatus. As previously noted, the teeth of most reptiles are used for prehension, puncturing, and tearing of food, which is then swallowed whole or in relatively large pieces. New teeth erupt between earlier ones, as replacements throughout the life of the animal. Most mammals differ from reptiles and other vertebrates in having large premolars and molars in the upper and lower jaw, which grow opposite each other with uneven surfaces that fit together during occlusion. The presence of muscles in the cheeks and a more muscular and mobile tongue aid in the placement of food between the shearing, grinding surfaces of these teeth, as well as in the development of the suction required for nursing. Furthermore, the muscles attached to the lower jaw (mandible) form a complex sling, which provides for horizontal as well as vertical movement.

Mammals show a wide variation in their feeding apparatus in relation to diet. Cloudsley-Thompson (1972) classified animals according to feeding habits and described the mouth parts associated with these characteristics. Anteaters tend to have weak jaws, simple teeth, and a tongue adapted for this purpose. Examples are seen in the orders Pholidota, Tubulidentata, Edentata, and Marsupialia. Baleen whales, which include the largest of all fossil or living animals, are filter feeders. The filter apparatus consists of two rows of horny baleen plate, each less than 0.5 cm thick, which hang from each side of the upper jaws. These may be long and narrow (right whales) or much shorter (rorquals) and can number from 250 to 400 in various species. However, even those, such as the sei whale, *Balaenoptera borealis*, that have a fine baleen strainer will eat cuttlefish in areas where krill is less profuse, and some baleen whales feed on fish in part (rorqual and humpback) or almost exclusively (Bryde's whale). Other carnivorous mammals tend to have well-developed incisors and canine teeth. However, in herbivores the canine teeth are often absent, and the incisors are usually adapted for cropping of vegetation. In most ruminants, the upper incisors are absent, and the area is covered by a horny pad. Incisors also are absent in the nectivorous, hog-nosed bat, *Choeronycteris mexicana*, which facilitates extention of the tongue during feeding (Tuttle 1987). The grinding surface of molars in herbivorous mammals tends to have a square and elaborate surface. Sanson (1978) discussed the evolution of mastication in the macropod marsupials.

Although many excellent illustrations of the gastrointestinal tract of vertebrates exist, it is difficult to obtain photographs or drawings that depict the entire tract in a manner that allows comparison of dimensions and gross structural characteristics. Therefore, many of the previous and following illustrations were prepared to allow comparison under similar conditions with special attention to details of the stomach and large intestinal structure. Removal of the gastrointestinal tract from the animal can change its dimensions, including the length of intestinal segments. Arrangement of the intestine in the pattern chosen also destroys a number of anatomical characteristics such as duodenal loops or spirals of the colon. Moreover, the shape and dimensions of the stomach and cecum can be affected to an even greater degree by the time between feeding and examination, as demonstrated in Figure 3.1.

Discussion of the gastrointestinal tract will proceed by mammalian orders

Rat (*Rattus norvegicus*)
Body Length:17 cm

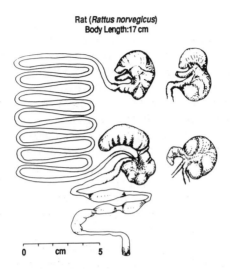

FIGURE 3.1 Gastrointestinal tract of a rat. Animals were fed at 12-hour intervals for two weeks prior to sacrifice. The specimen on the left was obtained 4 hours after feeding. The stomach and cecum to the right were obtained immediately after feeding. Note the effect of time after feeding, on the volume and shape of these two organs. After sacrifice, the digestive tract was immediately dissected free of its attachments, arranged in the pattern indicated, and drawn to scale, with special attention to the structural characteristics of the stomach and large intestine. Body length represents the distance from mouth to anus. (From Stevens 1977.) Specimens of the gastrointestinal tract obtained from other species and arranged in the same manner are depicted in subsequent figures.

as listed in Table 3.1 and, except where otherwise indicated, by the taxonomic classifications listed by Vaughan (1986) for orders and families and Nowak and Paradiso (1983) for genera and species. This aids in finding additional reference material. It also underlines the point that animals that appear closely related can show extreme variations in gastrointestinal anatomy and, conversely, that species with no immediate genetic relationship can demonstrate marked similarities. General characteristics of mammals belonging to each of these orders are described by Nowak and Paradiso (1983). The following discussion will deal primarily with gross structural characteristics such as the relative dimensions of the major gut segments and the degree of compartmentalization or sacculation of the stomach and large intestine. The early studies of Flower (1872) and Mitchell (1905) provide information on the gastrointestinal tract of a wide range of mammals. The comparative review of gastric mucosa by Bensley (1902-3) provides an additional source of information on this particular subject. References more limited in scope or recent in publication are given in the text.

Monotremata

The monotremes are listed as a single order under subclass Prototheria, the egg-laying mammals. This subclass and order are now represented only by the echidna (spiny anteaters) and the duck-billed platypus, which are restricted to Australia and its associated areas. The echidna (Fig. 3.2) are terrestrial anteaters resembling the anteaters in the order Edentata, described later in this chapter, with respect to food habits and associated adaptations. The stomach is lined with stratified squamous epithelium. Harrop and Hume (1980) and Krause (1970) stated that the pH of its contents was 6.2-6.5 after a meal of termites. The pylorus is muscular and somewhat elongated. The intestinal tract of *Echidna hystrix* was said by Flower (1872) to show no evidence of a sphincter or any other marked distinction between the small and large bowel other than the presence of a small vermiform cecum lined with lymph glands.

The duck-billed platypus (Fig. 3.2) is an aquatic animal that feeds on insects, mollusks, and worms. Its stomach also is lined with stratified squamous epithelium plus Brunner's glands near its junction with the intestine. As in the echidna, the small and large intestine are not separated by a sphincter or constriction, but a somewhat longer vermiform cecum is present.

The general pattern of the monotreme intestinal tract is similar to that of the reptile and demonstrates only two major differences from that of birds (Mitchell 1905). Most birds have a specialized duodenal loop, not seen in monotremes or most other mammals, and a straight short segment of hindgut. The hindgut of monotremes and most mammals extends a considerable distance in a series of arches, loops, or spirals.

Pholidota

Although the pangolins, or scaly anteaters, are placed in this order, they were originally classified as edentates and have many characteristics of the anteaters belonging to that order. This order includes eight species that inhabit parts of Africa and Asia, have horny scales, short legs, and a long body and tail similar to reptiles. The stomach of *Manis* species is small, round, and almost entirely lined with stratified squamous epithelium (Fig. 3.3). The glandular stomach appears to be entirely represented by two small regions of cardiac glands on the lesser and greater curvature and a glandular mass on the greater curvature. The latter consists of proper gastric gland follicles whose ducts empty into the stomach via a single orifice. The stomach has muscular walls and often contains stones, indicating that it may also serve a triturating function. The intestinal canal shows no division between the midgut and hindgut other than a gradual enlargement and includes no cecum (Mitchell 1905).

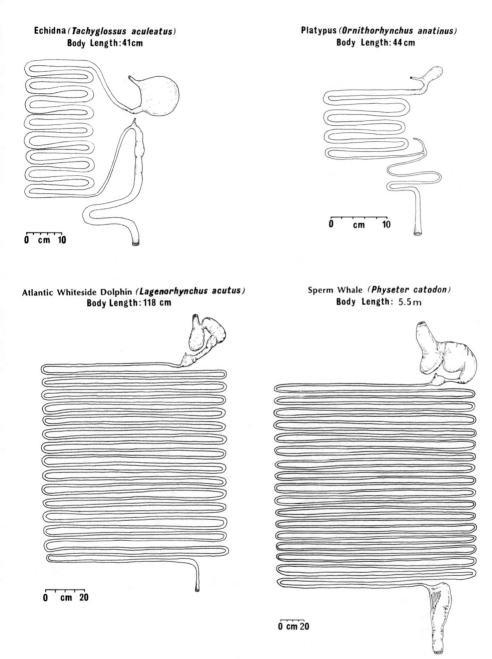

Echidna *(Tachyglossus aculeatus)*
Body Length: 41cm

Platypus *(Ornithorhynchus anatinus)*
Body Length: 44 cm

Atlantic Whiteside Dolphin *(Lagenorhynchus acutus)*
Body Length: 118 cm

Sperm Whale *(Physeter catodon)*
Body Length: 5.5m

FIGURE 3.2 Gastrointestinal tracts of an echidna (from Stevens, 1980), platypus, dolphin, and whale.

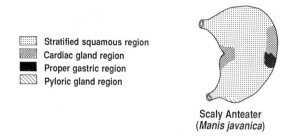

Stratified squamous region
Cardiac gland region
Proper gastric region
Pyloric gland region

Scaly Anteater
(*Manis javanica*)

FIGURE 3.3 Distribution of gastric epithelium in the scaly anteater. (From Bensley 1902-3.)

Tubulidentata

This order is represented by one living species, the aardvark, which feeds on ants and termites. Flower's examination of an aardvark specimen indicated that the large intestine had a prominent ileocecal valve and a large cecum (7 inches long and 3 inches in diameter at its apex). The colon was sacculated in its most proximal section, as was approximately one-fourth the length of the small intestine.

Cetacea

The cetaceans, which include 78 species of dolphins, porpoises, and whales, are referred to as either toothed (suborder Odontoceti) or baleen (suborder Mysticeti) whales. The various species can be subdivided according to their principle diet into four groups: 1) Sarcophagi, which include the killer whale or grampus, and feed on other mammals (e.g., seals); 2) Tenthophagi, which include the sperm whale, narwhal, and beluga, and feed primarily on cephalopods; 3) Ichthyophagi, which include the common porpoise, most dolphins, and finwhales, and feed on fish; and 4) Pteropodophagi, which include the right whales, and feed primarily on pteropods and small crustaceans. However, there also is evidence that microbial fermentation takes place in the stomach of some baleen whales (see Chapter 6).

The toothed whales have a mouth designed for rapidly ingesting and swallowing their prey, but the baleen whales lack true teeth and feed by opening their mouth and trapping a large volume of water, with its content of fish or other animals. The whale then partially closes its mouth and raises the tongue, like a piston, to force water out through the sieve formed by ridges of palatal mucous membrane—the whalebone or baleen. Finwhales have a very large mouth and feed on fish, but the Greenland whale, which only eats pteropods and crustaceans, has a gigantic mouth that is as large as its entire thoracic and abdominal cavity.

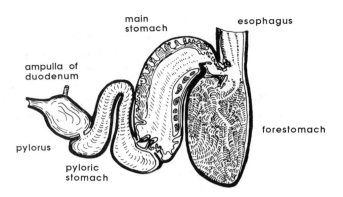

main
stomach

esophagus

ampulla of
duodenum

forestomach

pylorus

pyloric
stomach

FIGURE 3.4 Sagittal section of a bottle-nose dolphin stomach. (After Pernkopf, from Slijper 1962.)

The gastrointestinal tracts of two species of toothed whales are illustrated in Figure 3.2. The esophagus of whales is large in diameter. It opens into the first compartment of a large, multicompartmentalized stomach (Fig. 3.4). Slijper (1962) stated that the first compartment of most cetaceans was lined with a stratified squamous (noncornified) epithelium, whereas the second compartment was lined with proper gastric mucosa. The remaining compartments were lined with pyloric mucosa. However, he found that the compartment lined with stratified squamous epithelium was absent in beaked whales. From the presence of stones and the nature of the diets, he assumed that this compartment (when present) functioned primarily as a gizzard. According to Ridgway (1972), the first stomach compartment of toothed whales was lined with stratified squamous epithelium, whereas the second compartment was lined with simple tubular glands that contained mucus, chief, and oxyntic cells similar to those found in the proper gastric mucosa of other mammals. The subsequent pyloric compartments were lined only with mucous-type cells. Ridgway also noted that the proximal duodenum was dilated into an additional compartment that received the common bile duct.

Among the toothed whales, there was considerable variation in the ratio of intestinal length to body length: bottle-nose whales, 5.5; beaked whales, 6; beluga whales, 10; narwhal, 11; various species of dolphin, 7-14 (Slijper 1962). Baleen whales tended to show lower ratios (finwhale, 4; little piked and humpback whales, 5.5). These ratios did not necessarily correlate with the diet or size of the species.

The cecum was said to be absent from all toothed whales, with the exception of the Gangetic dolphin *Platanista gangetica* (Ridgway 1972). As villi were absent in the last 30 cm of intestine in the Pacific white-striped dolphin, this was designated as the hindgut. However, a young sperm whale

(Fig. 3.2) demonstrated a short but distinct hindgut. The intestine of baleen whales was said to differ in that it had a short, conical cecum. Flower concluded that this characteristic, plus others such as olfactory organs and rudimentary hindlimbs, suggested a closer relationship to terrestrial mammals. This would provide indirect support for the thesis that many of the adaptations seen in the hindgut of the terrestrial vertebrates came largely as an answer to the increased requirement for reabsorption of electrolytes and water present in the diet and gastrointestinal secretions.

Macroscelidea

The Macroscelidea (elephant shrews) are insectivores indigenous to Africa, and often included in the order Insectivora. They have a simple stomach, a long, cylindrical cecum, and a long colon. The latter is folded twice upon itself in the intact animal (Fig. 3.5).

Insectivora

This order contains six families and 406 species. Although the gastrointestinal tract varies among different species, it is generally simple in structure. The hedgehog (*Erinaceus europaeus*), which is actually an omnivorous animal, provides a good example (Fig. 3.5). Its stomach is simple and similar in shape to that of man, and the pylorus has a relatively thick layer of circular muscle. There is no external distinction between small and large intestine, although the intestine narrows in an area that would correspond to the ileum of other mammals. The gastrointestinal tract of the common English mole, family Talpidae, (Fig. 3.5), is similar to that of the hedgehog except that the stomach is very large with a well-developed fundus. The intestine of these animals appears to differ from that of all other mammals in that the rectum passes ventral to the pelvis, rather than through it.

The members of the shrew family (Soricidae) have a stomach rounder than that of other insectivores, and its cardiac inlet and pyloric outlet are close to each other. The intestine is extremely short—only three to four times the length of the shrew's body.

Tenrecs are small, prolific animals that are indigenous to Madagascar and live primarily on worms. The intestine of these animals shows no indication of any division into small and large bowel other than a slight enlargement of the terminal straight segment. Flower (1872) also noted the absence of a cecum from the digestive tracts of the Chrysochloridae (golden moles) and Solenodontidae.

Elephant Shrew (*Rhynchocyon chrysopygus*)
Body Length:25.5 cm

Hedgehog (*Erinaceus europaeus*)
Body Length:18 cm

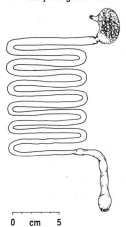

0   cm   5

0   cm   5

Mole *(Talpa europaea)*
Body Length:14 cm

Insectivorous Bat (*Myotis lucifugus*)
Body Length:7 cm

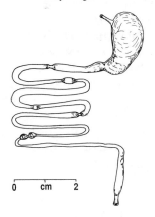

0   cm   5

0   cm   2

FIGURE 3.5 Gastrointestinal tracts of a hedgehog (from Clemens 1980), a mole (from Stevens 1980), an insectivorous bat (from Stevens 1980), and an elephant shrew.

Scandentia

This order is represented by a single family (Tupaiidae) of tree shrews, which resemble small, long-snouted squirrels. They are often included in order Insectivora or considered prosimian primates. They feed on insects, but also eat fruit and other animals. According to Flower (1872), they have a short cecum and short colon.

Chiroptera

The digestive system of bats (942 species) has attracted considerable interest due to their primitive nature and wide range of diets. The discussions of Flower (1872), McMillan and Churchill (1947), Brown (1962), Rouk and Glass (1970), Forman (1972), and Tuttle (1987) describe variations in the diet and digestive tract of insectivorous, fish-eating, frugivorous, nectivorous and blood-feeding bats. Variations in the compartmentalization and epithelial lining of the stomach are shown in Figure 3.6.

Insectivores tend to have a relatively simple globular stomach (Figs. 3.5 and 3.6). Some nectivorous and frugivorous species have a stomach that is both large and complex (Fig. 3.6). The stomach of *Desmodus rotundus* and *D. rufus*, South American blood-feeding species, is unique among mammals, consisting of a long, blind, sacculated tube (Figs. 3.6 and 3.7). This convoluted stomach is approximately twice the length of the animal. According to Rouk and Glass (1970), the stomach of *D. rotundus* contains acinar and tubuloacinar glands unlike those of other mammals. The studies of Ito and Winchester (1960, 1963) provide descriptions of mucosal structure.

The intestinal tracts of the insectivorous and vampire bats tend to be relatively short. The hindgut is short and straight and provided with a cecum in only a very few species—for example, *Rhinopoma harwicke* and *Megaderma spasma*. It is interesting to note that although the intestine of the fruit bats (e.g., *Pteropus*) is much longer than that of the insectivorous species, the opposite is true in birds, where frugivores tend to have the shortest, widest intestine (Mitchell, 1905). In most species of bat, there is no external indication to distinguish between the small and large intestine, other than a slight increase in diameter and mucosal changes from villiform to longitudinal folds.

Carnivora

The gastrointestinal tract of the 284 species in this order is characterized by a relatively simple stomach and a short intestinal tract. As indicated by their name, most of these animals are flesh eaters. The terrestrial Carnivora consist

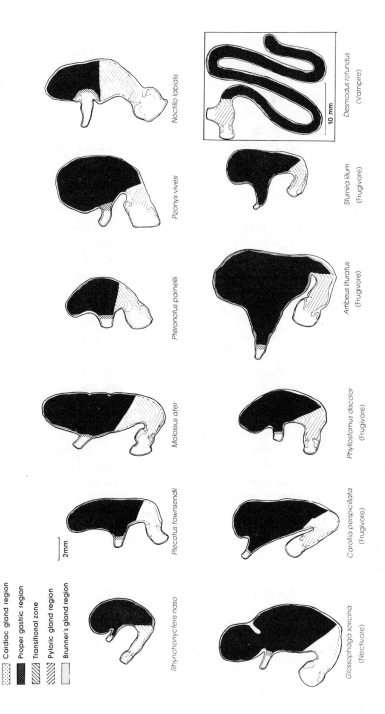

FIGURE 3.6 Distribution of gastric epithelium in various species of insectivorous, nectivorous, frugivorous, and vampire bats. Top row represents insectivores, except for *Pizonyx vivesi*, which is a fish-eating bat. (From Forman 1972.)

of the families Canidae (dogs, wolves, foxes, jackals), Felidae (cats), Viverridae (civits, genets, mongooses), Hyaenidae (hyenas, aardwolf), Ursidae (bears), and Procyonidae (raccoons, ringtail cats, etc.). Family Mustelidae (mink, skunks, ferrets, weasels, otters, wolverines) includes the semi-aquatic otters and the sea otter (*Enhydra lutris*), which is almost completely aquatic. The aquatic Carnivora include earless seals (F. Phocidae), the eared seals (F. Otaridae), and the walruses (F. Odobenidae). They include the largest species within this order, ranging from 90 to 3600 kg in body weight.

The intestinal tract of dogs is relatively short, and the large intestine is unsacculated (Fig. 3.7). The cecum consists of a coiled appendage located just distal to the ileocecal valve. Domestic cats have a gastrointestinal tract resembling that of dogs except for the cecum, which is not as coiled (Fig. 3.7). Some species of Felidae were said to have a sacculated section of proximal colon, and the gastrointestinal tracts of Viverridae and Hyaenidae were reported to be quite similar to that of the domestic cat, although the cecum was very rudimentary in one viverrine species and absent in another (Mitchell 1905).

The stomachs of Ursidae, Procyonidae, and Mustelidae are simple, and the distal segment of their intestine is marked only by a sudden change of the mucosa, with no cecum. Figures 3.7 and 3.8 illustrate the gastrointestinal tracts of a raccoon and mink. The simplicity and relatively small volume of the hindgut is somewhat surprising because raccoons and many bears are omnivorous, and some bears such as the panda are herbivores. Yet this appears to have resulted in no marked modifications of the digestive tract.

Aquatic carnivores (suborder Pinnipedia), feed entirely on fish or marine animals. The common seal (*Phoca vitulina*) has an elongated tubular stomach, sharply bent at the beginning of the pyloric antrum (Fig. 3.8). Its intestine is quite long. The cecum is small and the colon is approximately as long as the body. Other members of this group appear to demonstrate a similar digestive tract anatomy (Ridgway 1972).

Marsupialia

As previously noted, the evolution of marsupials and the eutherian (placental) mammals appears to have undergone an early separation. Because the 242 species of marsupials have many of the general features of reptiles and a limited ecological distribution, they are listed as a single order under Infraclass Metatheria and often considered as more primitive mammals. However, as Flower pointed out, the various families of Marsupialia could be considered as corresponding to the orders of eutherian mammals. Their digestive systems show as great a variation.

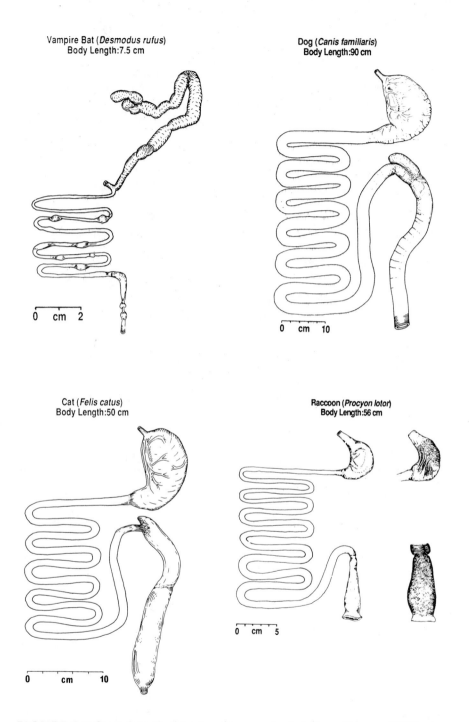

FIGURE 3.7 Gastrointestinal tracts of a vampire bat (from Stevens 1980), dog (from Stevens 1977), raccoon (from Argenzio and Stevens 1984), and cat.

Mink *(Mustela vison)*
Body Length: 42 cm

Harbor Seal *(Phoca vitulina)*
Body Length: 96 cm

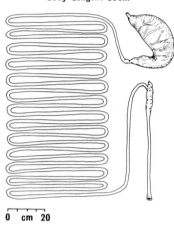

Bush-tailed Phascogale *(Phascogale tapoatafa)*
Body Length: 20 cm

Tiger Cat *(Dasyurus maculatus)*
Body Length: 50 cm

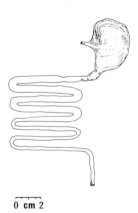

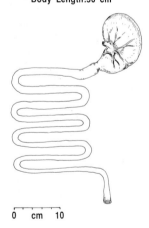

FIGURE 3.8 Gastrointestinal tracts of a mink (from Stevens 1977), seal, marsupial phascogale, and tiger cat.

The gastrointestinal tracts of marsupials and their relationship to diet, nutrition, and ecology are discussed by Hume (1982a). He divided the carnivores into two groups. The first group consists of two American families: the Caenolestidae (shrew-opossums) and the Didelphidae (opossums), although some of the latter are omnivores. The second group consists of four Australian families: the Dasyuridae, Myrmecobiidae, Notoryctidae, and Thylacinidae. Family Dasyuridae contains 49 species, ranging in size from the tiny shrew-sized planigales to the 10-kg Tasmanian devil. The others each contain only one species: the insectivorous numbat (*Myrmecobius fasciatus*), the marsupial mole (*Notoryctes typhlops*), and the Tasmanian tiger (*Thylacinus cynocephalus*), which is probably extinct.

The gastrointestinal tracts of two species of Dasyuridae are illustrated in Figure 3.8 (bottom). They consist of a simple stomach and relatively short intestine with no external distinction between midgut and hindgut. The gastrointestinal tracts of species belonging to the other marsupial families that have been discussed also have a relatively short intestine and, with the exception of the opossums, lack a cecum.

Marsupial omnivores were divided by Hume into three groups. The first includes the families Peramelidae (bandicoots) and Thylacomyidae (bilbies). The second consists of some members of Didelphidae. The third group includes Australian aboreal species such as the striped Leadbeater, the eastern pigmy and mountain pigmy possums, and the yellow-bellied and sugar gliders. Figure 3.9 illustrates the gastrointestinal tracts of a bandicoot and an American opossum. These animals also demonstrate a relatively short intestinal tract, but the hindgut is enlarged and includes a well-developed cecum.

The marsupial herbivores can be divided into nonmacropods and macropods. The nonmacropods include all members of families Vombatidae (wombats), Phalangeridae (cuscus, brushtail possum, and scaly-tailed possum), Phascolarctidae (koala), and some members of the family Petauridae (greater glider and ringtail possums). The stomach of the koala and wombat has, near the cardia, a glandular mass that contains proper gastric glandular mucosa. The gastrointestinal tracts of a wombat, koala, and greater glider are shown in Figures 3.9 and 3.10; those of the koala and greater glider have an extremely large cecum.

The macropod marsupials (family Macropodidae) have a complex voluminous stomach, which serves as the major site for microbial fermentation. These animals are contained in two subfamilies: Potoroinae (rat kangaroos) and Macropodinae (kangaroos and wallabies). The gastrointestinal tract of a kangaroo is illustrated in Figure 3.10. The stomachs of these animals consists of a long, tubular structure drawn into sacculation over much of its length by bands of longitudinal muscle. The distribution of stratified squamous and glandular mucosa is shown for three species in Figure 3.11. The lumen sur-

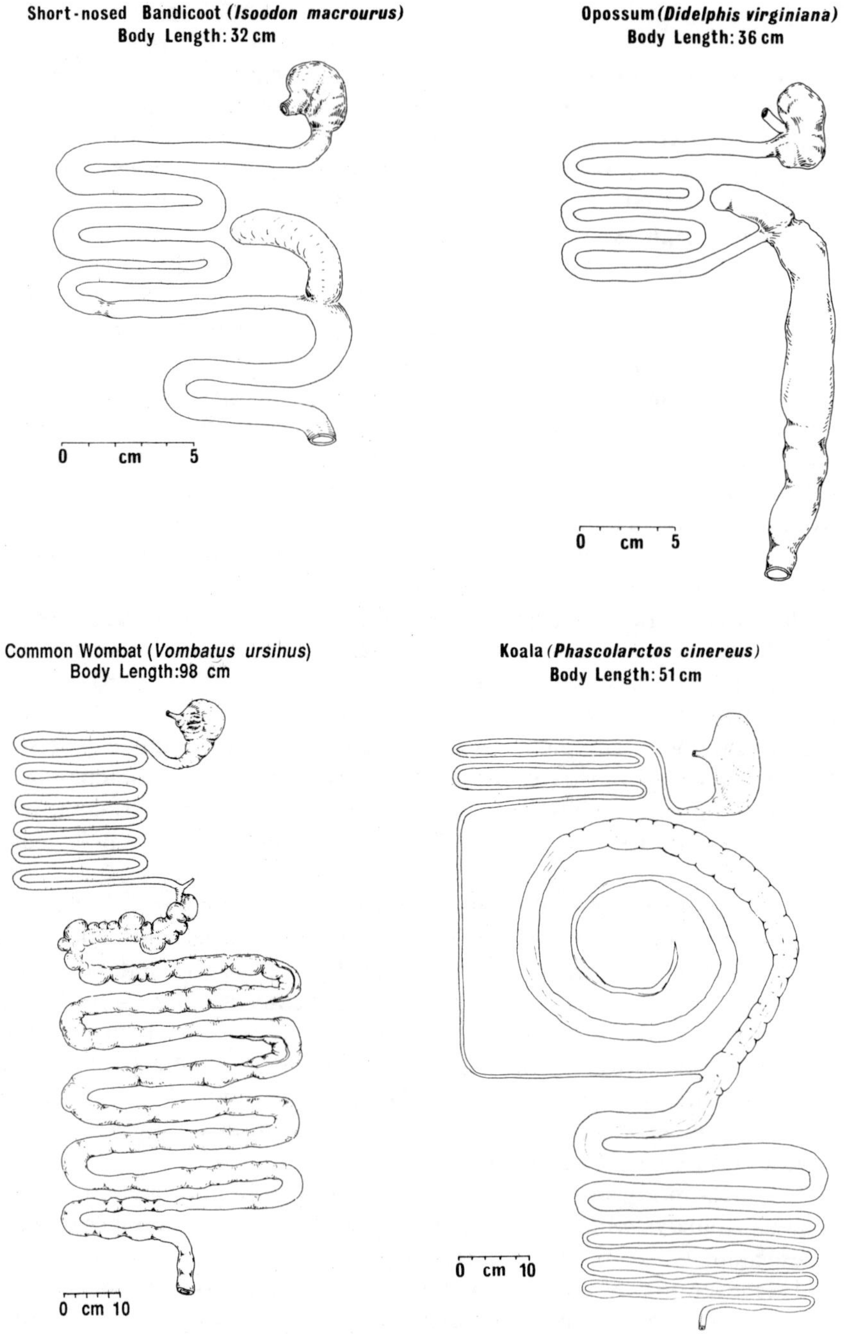

Short-nosed Bandicoot (*Isoodon macrourus*)
Body Length: 32 cm

Opossum (*Didelphis virginiana*)
Body Length: 36 cm

0 cm 5

0 cm 5

Common Wombat (*Vombatus ursinus*)
Body Length: 98 cm

Koala (*Phascolarctos cinereus*)
Body Length: 51 cm

0 cm 10

0 cm 10

FIGURE 3.9 Gastrointestinal tracts of a koala (from Harrop and Hume 1980), a bandicoot, a wombat, and an opossum.

face has a groove that runs along the lesser curvature of the stomach. This groove, which may serve for the rapid passage of milk through the stomach of neonates, will be noted in many mammals that have a voluminous compartmentalized stomach.

Edentata

The edentates (anteaters, armadillos, sloths) are believed to be a relatively primitive group of New World, placental mammals. Although their name indicates an absence of dentition, all but the anteaters have teeth in one form or another. This order, which includes three families and 31 species, represents two extreme cases of alimentary tract specialization according to diet. One (F. Myrmecophagidae) lives on ants and termites, whereas the sloths (F. Bradypodiae) feed only on leaves. The anteaters are generally characterized by their long head, long tongue, extremely large submaxillary gland, and a mouth that is toothless (although the inner surface of the cheek may contain ridges or spines, which could aid in mastication). The stomach is relatively simple, consisting of an expanded area containing proper gastric glands and a smaller pyloric area lined with a thicker epithelium and heavy muscular walls. According to Flower, the pylorus of some species such as the South American giant anteater (*Myrmecophaga tridactyla*) is lined with stratified squamous epithelium, giving it the characteristics of the gizzard of birds. The pylorus of the *Tamandua* anteater and little two-toed anteater (*Cyclopes*) also is thickened, but not to gizzardlike proportions.

In the giant anteater, the small intestine was reported by Flower to be about seven times the body length, and the large intestine about equivalent to the length of the body. Transition between the two was indicated by an abrupt enlargement of diameter, but no valve and only a poorly defined cecal pouch were noted. The small intestine of the *Tamandua* anteater was approximately 20 times the length of the colon. A valve and a short, round cecum opened into a colon, which was approximately twice the diameter of the ileum and enveloped in a thick muscular coat. The intestine of the two-toed anteater was distinguished by the presence of paired, foliated ceca that opened to the colon via very small apertures.

Although the digestive tract of armadillos (F. Dasypodidae; Fig. 3.10) shows many similarities to that of the anteaters, they feed on a wider variety of animal food and some plant material. The stomach appears to contain only stratified squamous epithelium and pyloric glandular mucosa (Fig. 3.11). The nine-banded armadillo (*Dasypus novemcinctus*) has an intestine similar to that of the giant anteater with a short canal and no cecum, but the intestine of the six-banded armadillo (*Dasypus septemcinctus*) was reported to have a valve and a pair of short, round ceca (Flower 1872).

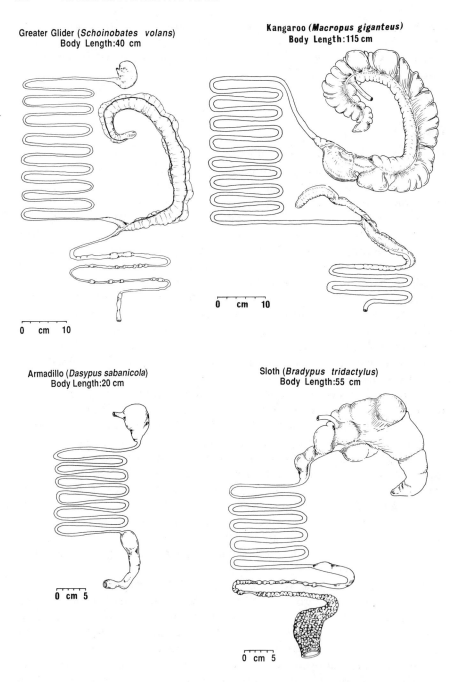

Greater Glider (*Schoinobates volans*)
Body Length:40 cm

Kangaroo (*Macropus giganteus*)
Body Length:115 cm

Armadillo (*Dasypus sabanicola*)
Body Length:20 cm

Sloth (*Bradypus tridactylus*)
Body Length:55 cm

FIGURE 3.10 Gastrointestinal tracts of a kangaroo (from Stevens 1977), an armadillo (from Stevens 1980), a sloth (from Stevens 1980), and a greater glider.

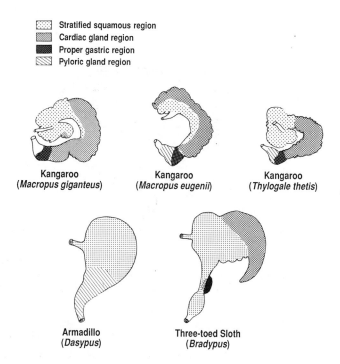

Stratified squamous region
Cardiac gland region
Proper gastric region
Pyloric gland region

Kangaroo
(*Macropus giganteus*)

Kangaroo
(*Macropus eugenii*)

Kangaroo
(*Thylogale thetis*)

Armadillo
(*Dasypus*)

Three-toed Sloth
(*Bradypus*)

FIGURE 3.11 Distribution of gastric epithelium in three macropod marsupials (from Langer, Dellow, and Hume 1980), and an armadillo and a sloth (from Bensley 1902-03).

The sloths are indigenous to South American jungles and are divided into two genera: *Bradypus*, the three-toed sloth; and *Choloepus*, which has only two toes on its front feet. They differ from the anteaters and armadillo in having a short, round face, a short tongue, normal-sized submaxillary glands (but slightly larger parotids), and a completely herbivorous diet (Montgomery and Sunquist 1978). The stomach of the sloth is as complex as that of any species, and the two genera differ from each other mainly in that the stomach of *Bradypus* is even more complicated than that of *Choloepus*. Figure 3.10 illustrates the gastrointestinal tract of *Bradypus tridactylus*. The stomach consists of three major divisions. The first is a large compartment, partially divided into anterior and posterior sacs (Fig. 3.12). The anterior sac has a prolonged cecal appendage. The esophagus enters the posterior sac, which is also subdivided into an upper and lower pouch. The second compartment is very small. A groove runs across its lesser curvature from the cardia to the third major compartment, in a manner similar to that seen in the kangaroo stomach. The third compartment is tubular. Stratified squamous epithelium is distributed over much of the first compartment, the entire second com-

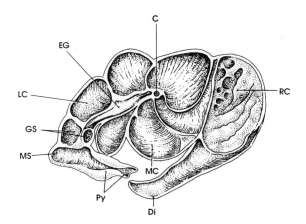

FIGURE 3.12 Stomach of *Bradypus tridactylus*. Cardia (C); esophageal (ventricular) groove (EG); left cecum (LC); middle cecum (MC); right cecum (RC); diverticulum (Di); muscular stomach (MS); pyloris (Py); gastric sulcus (GS). (After Pernkopf, from Grassé 1955.)

partment, and most of the third compartment (Fig. 3.11). The first compartment contains a large island of cardiac glandular mucosa, and the third compartment contains an oval patch of proper gastric glands on its greater curvature. The stomach of *Choloepus* was said to differ primarily in the fact that the anterior sac of the first compartment was shorter and less complicated in its arrangement of internal septa (Flower 1872).

The small intestine of the sloth joins the large by simply a funnel-shaped dilation. Although Mitchell (1905) described a pair of small ceca near the ileocecal junction of *Bradypus tridactylus*, Figure 3.12 shows no evidence of these ceca. Distention of the rectum in this specimen was due to fecal pellets. These animals defecate only at infrequent intervals.

Rodentia

The rodents represent a group of animals that have been very successful, as measured by their ecological distribution and the number of families (25), species (1750), and individuals. Rodents are generally herbivores, although some such as the rat are omnivores and some are carnivores (Landry 1970). The digestive system of rodents should be of interest to physiologists, if for no other reason than their widespread use of some species as experimental animals.

Rodents show a wide range of species variation with respect to the structure of their stomach and hindgut. The gastrointestinal tracts of a rat, guinea

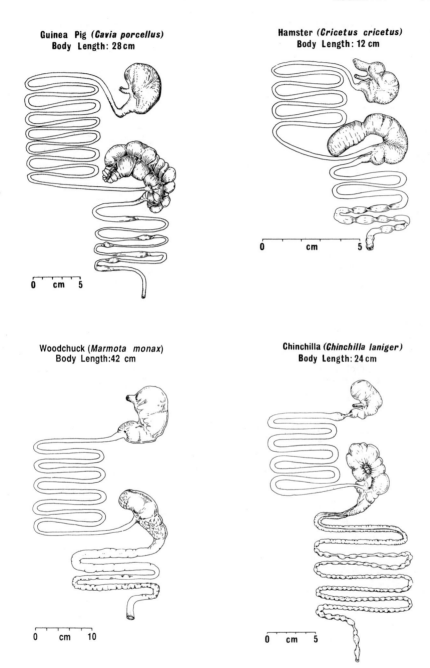

Guinea Pig *(Cavia porcellus)*
Body Length: 28 cm

Hamster *(Cricetus cricetus)*
Body Length: 12 cm

Woodchuck *(Marmota monax)*
Body Length: 42 cm

Chinchilla *(Chinchilla laniger)*
Body Length: 24 cm

FIGURE 3.13 Gastrointestinal tracts of a guinea pig, hamster, woodchuck, and chinchilla.

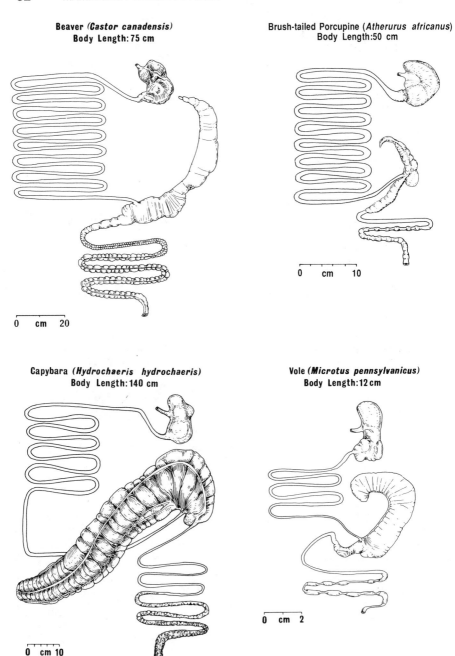

Beaver *(Castor canadensis)*
Body Length: 75 cm

0   cm   20

Brush-tailed Porcupine *(Atherurus africanus)*
Body Length: 50 cm

0   cm   10

Capybara *(Hydrochaeris hydrochaeris)*
Body Length: 140 cm

0   cm 10

Vole *(Microtus pennsylvanicus)*
Body Length: 12 cm

0   cm   2

FIGURE 3.14 Gastrointestinal tracts of a beaver, porcupine, capybara, and vole.

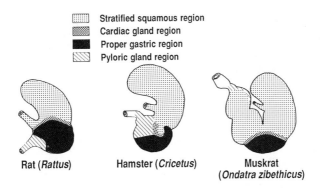

Stratified squamous region
Cardiac gland region
Proper gastric region
Pyloric gland region

Rat (*Rattus*)   Hamster (*Cricetus*)   Muskrat
(*Ondatra zibethicus*)

FIGURE 3.15 Distribution of gastric epithelium in three species of rodents. (From Bensley 1902-3.)

pig, woodchuck, hamster, chinchilla, beaver, porcupine, capybara (a large, South American rodent), and vole are illustrated in Figures 3.1, 3.13, and 3.14. The stomachs of guinea pigs, woodchucks, and capybara are simple in structure. The same is true of a number of other species including the European dormouse (*Glis glis*), squirrels, and beavers. However, the beaver stomach has a partition or fold attached to its greater curvature and a glandular mass near the cardia, and the stomachs of hamsters, chinchilla, the Norway rat (*Rattus norvegicus*), the pack rat (*Neotoma* sp.), the common dormouse (*Muscardinus avellanarius*), the muskrat (*Ondatra zibethicus*), lemmings, and voles are compartmentalized to varying degrees. Those of lemmings and voles are divided into three compartments with a groove passing along the lesser curvature, like that of the kangaroo and sloth stomach. The cranial portion of the stomach is lined with stratified squamous epithelium in many of these species (Fig. 3.15).

The hindgut of rodents also demonstrates a range of complexity, often to a degree inversely proportional to that seen in the stomach of a given species. The cecum is usually large in relation to the rest of the digestive tract. As noted in Figure 3.1, the cecum of the Norway rat is quite voluminous 4 hours after a meal. Jerboas and woodchucks have a cecal capacity less than that of rats, but guinea pigs, beavers, and capybara have a greater cecal capacity. The cecal contents of guinea pigs are said to equal 5-10%, and those of the capybara 12% of the animal's body weight.

The cecum of many rodents is sacculated with bands of longitudinal muscle. The initial segment of colon also is sacculated in a number of species. The cecum of the tree porcupine (*Coendou prehensilis*) terminates in what appears to be a long vermiform appendix like that of the rabbit (Fig. 3.27), and its relatively voluminous proximal colon draws comparison to the same segment of colon in some primates (Fig. 3.18). According to Mitchell

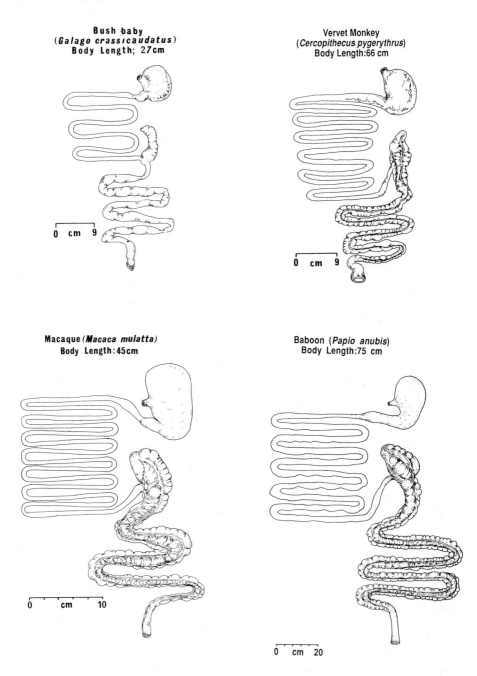

FIGURE 3.16 Gastrointestinal tracts of a galago and a vervet monkey (from Clemens 1980), a macaque, and a baboon.

(1905), the proximal colon of the common porcupine (*Histrix cristata*) seemed to demonstrate a pair of ceca (Mitchell 1905). The proximal colon of the vole and lemming forms a tight spiral (see Fig. 4.14). Behmann (1973) has provided a well-illustrated comparative review on the functional anatomy of the rodent cecum and colon.

Primates

This order of 166 species includes the prosimians, monkeys, apes, and man. Vaughan (1986) divides the primates into two suborders: Strepsirhini and Haplorhini. The strepsirhines include the five families Lemuridae (lemurs), Indriidae (woolly lemurs), Daubentoniidae (aye-aye), Lorisidae (lorises), and Galagidae (galagos). Lemurs are variously insectivorous, omnivorous, or herbivorous-frugivorous. Woolly lemurs are leaf-eaters. The lorises feed on insects and small vertebrates, and the aye-ayes and galagos are principally insectivorous.

The gastrointestinal tract of a galago, the bush baby, is illustrated in Figure 3.16. The following descriptions and scientific names of other strepsirhines are those of Flower and Mitchell. The cecum of the ruffed lemur was longer than that of the bush baby and unusually well developed. The small intestine of *Indri brevicaudatus* was considerably longer than that of the bush baby, and its cecum was twice the body length and slightly sacculated. The lorises are slow-moving animals found in Africa and Southeast Asia. The intestinal tracts of two species (*Nycticebus javanicus* and *N. tardigradus*) were reported to be shorter than that of the galago. However, the cecum was longer, slightly sacculated, and constricted at its apex into a vermiform appendage. The digestive tracts of two other African species (*Perodicticus potto* and *P. calabarensis*) and an inhabitant of Ceylon (*Loris gracilis*) were similar except for a much shorter cecum. The aye-ayes, *Daubentonia madagascarensis*, have teeth that resemble those of rodents to such a degree that they were once so classified. Their colon was much like that of most species of this order.

Suborder Haplorhini can be subdivided into five families: Tarsiidae (tarsiers), Cercopithecidae (Old World monkeys), Cebidae (New World monkeys), Pongidae (great apes and gibbons), and Hominidae (humans). The tarsiers inhabit the islands of the Indo-Malayan Archipelago, feeding on insects and small vertebrates. The colon of one species was found to be quite short, only about one-fifth as long as the small intestine. The cecum was approximately one-half the length of the colon and arranged in a spiral.

The Cercopithecidae includes many genera. Some of these are African monkeys, whereas others such as the *Macaca* (rhesus and pig-tailed monkeys) are almost entirely Asian species. They tend to be vegetarians or folivores. However, more recent evidence indicates that the baboon can develop the behavioral pattern of predatory carnivores in the wild state (Harding and Strum 1976). Most members of this family have cheek pouches (which aid in food storage), a simple stomach, a relatively short, small intestine, and a sacculated cecum. Figure 3.16 shows the gastrointestinal tract of a vervet monkey, a macaque, and a baboon. Three longitudinal muscle bands draw the colon of these animals into marked sacculations or haustra. The anatomy of many of the species in this family has been described by Hill (1966a,b).

The genera *Presbytis* (Asian monkey) and *Colobus* (African monkey) are of particular interest, since they have a remarkably complex stomach. The stomach wall contains two bands of longitudinal muscle, and the normal tonus of these muscles results in a series of pouches, similar to those of the kangaroo stomach. The early study of Owen (1835) concluded that the stomach of *Presbytis entellus* was subdivided into cardiac, middle, and pyloric compartments, normally separated by relatively small, constricted orifices (Fig. 3.17). There was no evidence of rumination in captive animals, and he concluded that the large, complex stomach allowed rapid ingestion and storage of food by these timid animals, but that they did not have the cheek pouches seen in related monkeys. The small intestine of *Presbytis* was about eight times the length of the body. The large intestine was twice the length of the body and the cecum one-fourth of the body length.

Pernkopf (1937) described stratified squamous epithelium covering the lumen surface of the first compartment and part of the second compartment of the *Presbytis* stomach. Hill (1952) concluded that the *Colobus* and *Procolobus* stomachs were quite similar, and he stated that the esophageal mucosa, evidenced by its "... firm, horny texture, opaque white appearance, and folded margin," continued some distance into the first compartment, especially along its dorsal and lateral walls, in a manner similar to that seen in *Presbytis* and in some species belonging to other orders. Therefore, it appears that the complex stomach of these monkeys also may include a relatively large area of stratified squamous epithelia. Figure 3.18 shows the gastrointestinal tract of *Colobus abyssinicus*.

Family Cebidae, which includes all New World monkeys, tends to be omnivores, although many species feed principally on fruit and leaves, and the night monkey, *Aotus trivirgatus*, is said to feed on insects and bats (Vaughan 1986). The gastrointestinal tracts of three species are illustrated in Figure 3.18, and those of a spider monkey (*Ateles* sp.), capuchin monkey (*Cebus capucinus*), howling monkey (*Alouatta* sp.), and two genera of the subfamily Callithricidae (*Midas* sp. and *Hapala penicillata*) were described by Flower (1872) and Mitchell (1905). These animals showed considerable variation in the relative length of the small intestine, cecum, and colon. The cecum was not sacculated in many of these species, but the colon was sac-

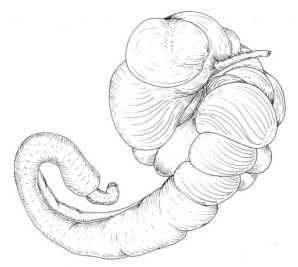

FIGURE 3.17 Stomach of *Presbytis entellus*. (From Owen 1835.)

culated as a result of one or more bands of longitudinal muscle in most of these animals. The proximal colon of the spider and capuchin monkeys showed an expansion similar to that illustrated in the night and woolly monkeys.

Family Pongidae is comprised of the anthropomorphous apes. The gorilla, chimpanzee, and orangutan are generally considered to be strictly vegetarians, although there is evidence that chimpanzees may hunt and eat termites and smaller monkeys (van Lawick-Goodall 1968). The gastrointestinal tract of a chimpanzee is illustrated in Figure 3.19. Its colon is sacculated by the presence of three longitudinal bands of muscle. These bands extend over most of its length and over the cecum, which terminates in vermiform appendage. The gastrointestinal tract of the gorilla is similar, except that the small intestine is relatively long and the large intestine is more voluminous (Mitchell 1905; Raven 1950). The cecum has a similar vermiform appendage, and three longitudinal bands of muscle pass over the cecum and first 10% of the colon.

Orangutans and gibbons are arboreal herbivores, although gibbons feed occasionally on insects and small vertebrates. Orangutans inhabit Borneo and Sumatra, and gibbons are found in Southeast Asia and the West Indies. The small intestine and colon of the orangutan are longer than that of chimpanzees, and its proximal colon is expanded (Figure 3.19). A specimen of the gastrointestinal tract of a gibbon (*Hylobates* sp.) suggested that it is similar to that of other apes, except for a shorter length of colon (Flower 1872).

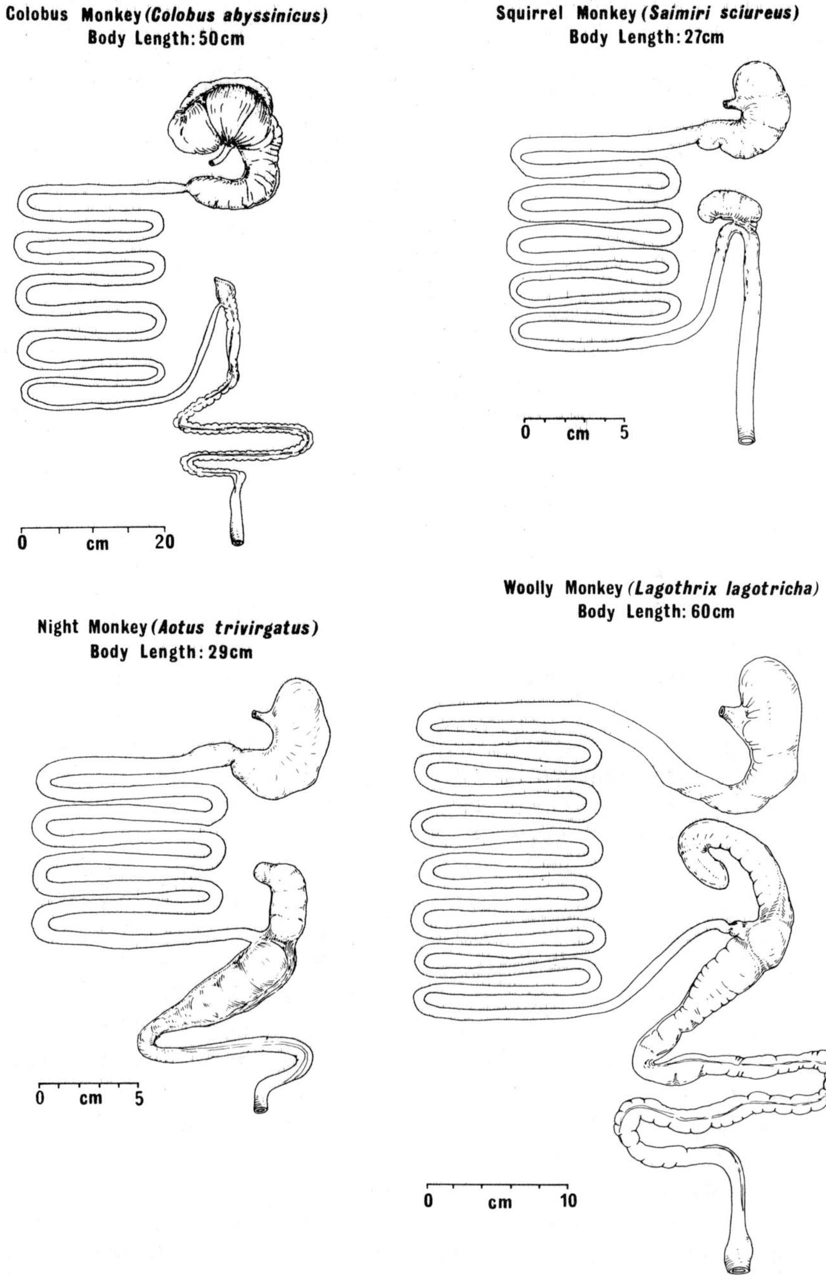

FIGURE 3.18 Gastrointestinal tracts of a colobus monkey (from Stevens 1983), squirrel monkey, night monkey, and woolly monkey.

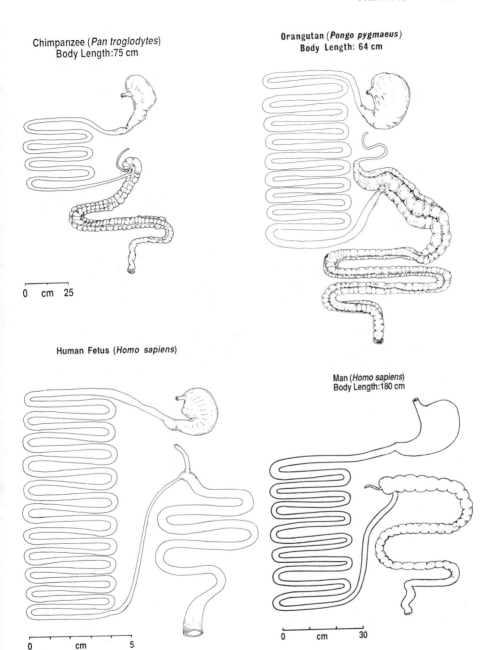

Chimpanzee (*Pan troglodytes*)
Body Length:75 cm

Orangutan (*Pongo pygmaeus*)
Body Length: 64 cm

0   cm   25

Human Fetus (*Homo sapiens*)

Man (*Homo sapiens*)
Body Length:180 cm

0   cm   5

0   cm   30

FIGURE 3.19 Gastrointestinal tracts of an adult human (from Wrong, Edmonds, and Chadwick 1981) and a chimpanzee, an orangutan, and a human fetus at 10 weeks.

Chivers and Hladik (1980) reviewed an extensive series of studies relating measurement of various parameters of the digestive tract to the diet of faunivorous, frugivorous, and folivorous primates. The faunivores had a simple stomach and large intestine and a long small intestine. Folivores had a larger stomach and/or an enlarged cecum and colon. Weight, volume, and surface area ratios between the stomach plus large intestine and the small intestine varied in a similar manner.

The human gastrointestinal tract is well described elsewhere, and only a few major features of the large intestine will be highlighted for comparison. The cecum appears in the fetal gut of man as a conical diverticulum (Fig. 3.19). The intestine then lengthens, but its terminal segment remains narrow in caliber. In the sixth month, faint longitudinal bands of muscle can be seen running from the apex of the cecum over the colon. At this stage, cecal growth is arrested, and this, along with the apparent shortening of its lateral longitudinal muscle bands and the continued growth of the colon, results in the development of the adult form (Fig. 3.19). Therefore, a bulge of colon forms what is really a false cecum, which does not correspond either to the apex of the fetal cecum or to the cecum of most other species. However, the vermiform appendage remains, and the colon is sacculated throughout almost its entire length, like that of the apes and many monkeys. The latter characteristic led Elliott and Barclay-Smith (1904) to suggest that the structure of the human large intestine is closer to that of a herbivore than an omnivore.

Artiodactyla

This order contains the ungulates (hoofed mammals) with an even number of toes. Although the domestic swine are omnivores, with a simple stomach, most artiodactyls are herbivores and have a stomach that is both voluminous and highly compartmentalized. Because members of this order contribute much of the food and fiber utilized by humans, their digestive systems have received extensive study, and much of our knowledge of digestion in herbivores is derived from these animals.

There has been much disagreement among taxonomists over whether the 171 species within this order should be classified according to their ability to ruminate (i.e., regurgitate, remasticate, and reswallow their food), the number of stomach compartments, or other characteristics. The present discussion will follow the classification of Simpson (1945), which subdivides the order into suborders Suiformes, Tylopoda, and Ruminantia. The Suiformes includes the families Suidae (swine), Tayassuidae (javelinas or peccaries), and Hippopotamidae (hippopotami). The Tylopoda consists of the family Camelidae (New and Old World camels). Ruminantia consists of the

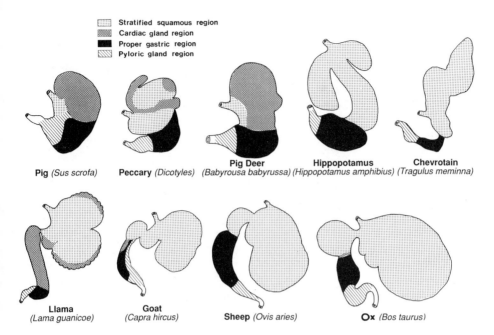

FIGURE 3.20 Distribution of gastric epithelium in various species of Artiodactyla. (From Bensley 1902-3; Langer 1974; Moir 1965; Sisson 1975; Cummings, Munnell, and Vallenas 1972.)

family Tragulidae (chevrotains or mouse deer) and the more advanced families of artiodactyls (i.e., Cervidae, Giraffidae, Antilocapridae, and Bovidae).

Figure 3.20 illustrates some variations in the distribution of the different types of gastric epithelium in the stomachs of artiodactyls. It also indicates the various degrees of gastric compartmentalization, ranging from the simple stomach of the domestic pig to the voluminous, multicompartmental stomachs of other species.

The Suidae includes the domestic pig, African warthog, African river hog, and Malayan pig deer. The gastrointestinal tract of a domestic pig is shown in Figure 3.21. Its stomach is relatively simple in form, with a small conical pouch projecting along the esophagus. On its mucosal surface the lesser curvature is marked by a longitudinal groove. The pyloric aperture is marked by a pad of tissue, which provides for a crescent-shaped outlet similar to that seen in some species of carnivore and edentate. The cranial one-half of the pig's stomach is lined with a very small area of nonglandular, stratified squamous epithelium, and a relatively large area of cardiac glandular mucosa

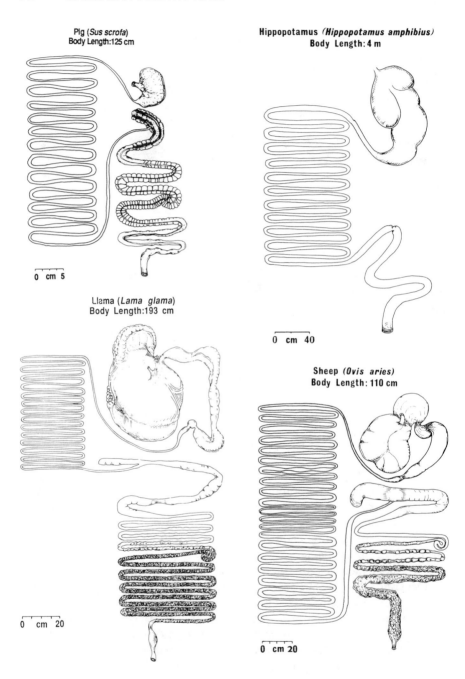

FIGURE 3.21 Gastrointestinal tracts of a pig (from Argenzio and Southworth 1974), sheep (from Stevens 1977), hippopotamus, and llama.

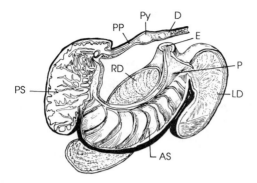

FIGURE 3.22 Stomach of *Hippopotamus amphibius*. Anterior stomach (AS); right diverticulum (RD); left diverticulum (LD); pillar (P); esophagus (E); pars pylorica (PP); pylorus (Py); posterior stomach (PS); duodenum (D). (From Frechkop 1955.)

(Fig. 3.20). The caudal portion of the stomach contains regions of proper gastric and pyloric glandular mucosa, which are approximately equal in size.

The intestine of the domestic pig is relatively long, and the colon forms about one-fourth of this. The cecum is conical and sacculated. The colon is sacculated throughout much of its length and arranged in a spiraling coil, which first diminishes and then, upon reversing its direction, increases the diameter of its spiral. The spiral colon is a characteristic of most artiodactyls.

The stomach of the African warthog (*Phacochoerus*) appears to have no cardiac pouch and less of a nonglandular area (Flower 1872). However, the African red river hog (*Potamochoerus*) has a stomach similar to that of the domestic pig, and Langer (1974) found that the babirusa (Malayan pig deer) showed an even greater development of the cranial pouch. The New World peccary (*Tayassu* sp.) has a very complex stomach. The cranial division is subdivided into a number of compartments. These are lined mostly with nonglandular stratified epithelium, but one contains islands of cardiac glands. The more caudal division contains the proper gastric and pyloric mucosa. A groove with elevated lips passes from the cardia into the pylorus. The intestinal tract of peccaries is similar to that of the pig.

The hippopotamus has one of the most complex stomachs of all suiform species (Fig. 3.22). The cranial portion is the most complex in its compartmentalization and is lined with nonglandular stratified epithelium. The caudal compartment contains the proper gastric glands and a terminal region of pyloric glandular mucosa. There appears to be no evidence of cardiac glandular mucosa. The large intestine differs from that of other artiodactyls in that it is short and includes no cecum (Fig. 3.21). The ileocolic junc-

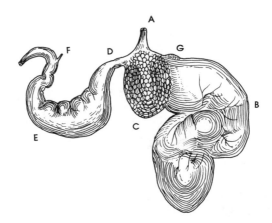

FIGURE 3.23 Stomach of the African water chevrotain, *Hyemoschus aquaticus*. Letters refer to A) esophagus, B) rumen, C) reticulum, D) rudimentary omasum, E) abomasum, F) entrance of the common bile duct, and G) spleen. (From Flower 1872.)

tion is marked primarily by the cessation of villi, the beginning of longitudinal folds of mucosa, and a large patch of lymph follicles covered by reticular mucous membrane. The colon comprises only one-tenth of the total intestinal length, and it is arranged in transverse folds rather than spirals. Therefore, the hippopotamus represents two extremes of the Artiodactyla gut—one of the most complex stomachs and the most simple and short large intestine.

All other members of order Artiodactyla ruminate. The stomach of chevrotains, small (2.3-4.6 kg), deerlike animals found in Southern Asia and West Africa, has a large nonglandular blind sac folded into a sigmoid pouch or rumen. The esophagus enters a partially separated, cranial diverticulum of this pouch called the reticulum because of the reticular pattern of its epithelial lining (Fig. 3.23). The reticulorumen is lined with nonglandular, stratified squamous epithelium. In the Asian genus, *Tragulus,* its outlet passes directly into a glandular compartment containing regions of proper gastric and pyloric glandular mucosa, but no evidence of cardiac glandular mucosa (Langer 1974). However, individuals of the African genus, *Hyemoschus*, have an intervening omasum (see below) containing one or two leaves (Moir 1968). The glandular stomach is referred to as the abomasum, because it appears homologous to the glandular abomasum of more advanced ruminants. The intestinal tract of a number of Indian species of Tragulina revealed a relatively long large intestine, which included a cecum and a spiral colon (Mitchell 1905).

The only living representatives of the Camelidae are the camels and their

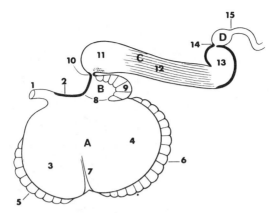

FIGURE 3.24 Schematic representation of the llama and guanaco stomach. A) First compartment; B) second compartment; C) third compartment; D) duodenal ampulla; 1) esophagus; 2) ventricular groove; 3) cranial sac; 4) caudal sac; 5) sacculated area in cranial sac; 6) sacculated area in caudal sac; 7) transverse sulcus; 8) entrance to second compartment; 9) glandular cells of second compartment; 10) tubular passage to third compartment; 11) initial one-fifth of third compartment (note reticulated area on lesser curvature); 12) middle three-fifths of third compartment (lines represent mucosal pleats); 13) terminal one-fifth of third compartment; 14) pylorus; 15) duodenum. (From Vallenas, Cummings, and Munnell 1971.)

South American relatives—the llama, alpaca, guanaco, and vicuna. New World camelids have a complex, three-compartment stomach. The first compartment is partially subdivided into a cranial and caudal sac by a transverse muscular pillar crossing its ventral surface (Figs. 3.24 and 3.25). The first and second compartments communicate via a relatively large opening. Both compartments are lined with stratified squamous epithelium except for their ventral portions, which contain a large number of recessed glandular pouches. A strong muscular sphincter separates the second compartment from the elongated third compartment, which is entirely lined with glandular epithelium. The glandular sacs of the first two compartments and the mucosal surface of the proximal four-fifths of the third compartment are lined with a cardiac glandular mucosa (Cummings, Munnell, and Vallenas 1972). Proper gastric and pyloric glands are confined to the terminal one-fifth of the third compartment. A groove with elevated lips (esophageal or ventricular groove) joins the esophagus to the third compartment. The camel and dromedary appear to have a stomach similar to that of the New World species in both its compartmentalization and distribution of epithelium.

Figure 3.21 illustrates the llama gastrointestinal tract. The intestinal tract

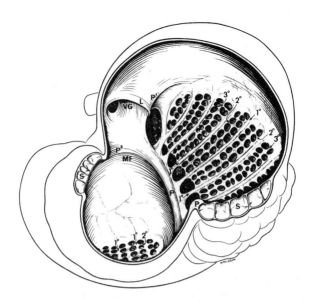

FIGURE 3.25 Left lateral longitudinal section of the llama stomach. A portion of the left wall of the first compartment has been removed to allow visualization of the mucosal surface and glandular pouches. E) Entrance to second compartment; L) lip of ventricular groove; MF) mucosal folds in cranial sac; P) transverse pillar; P¹) caudal limb of pillar; P²) cranial limb of pillar; S) glandular saccules; VG) ventricular groove; 1°) primary crests; 2°) secondary crests; 3°) tertiary crests. (From Vallenas et al. 1971.)

is relatively long in comparison with other Artiodactyla and almost equally divided between small and large intestine. The diameter of the cecum and the most proximal segment of colon are enlarged in comparison with the remainder of the intestine, but they do not appear to be sacculated. The intestinal tract of the dromedary (*Camelus dromedarius*) appears to differ only in having a much longer cecum and a shorter, more coiled spiral colon (Mitchell 1905).

The advanced ruminants include a large number of domesticated and wild species of ruminants—for example, cattle, sheep, goats, deer, antelope, elk, buffalo, yak, and giraffe. The structure and function of the highly compartmentalized stomach of cattle and sheep have been subjected to extensive study. The reticulorumen of domestic cattle is partially subdivided by a transverse fold and muscular pillars into a cranial reticulum and the dorsal and ventral sacs of the rumen (Fig. 3.26). The gastroesophageal opening is connected by a groove, with muscular lips, to the reticulo-omasal orifice. This orifice is located in relatively close proximity to the cardia and is provided with a sphincter. The omasum, which forms the second distinct compartment, is a globular-shaped organ. The groove continues along its lesser

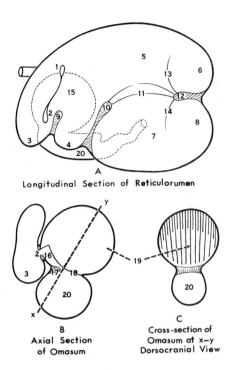

A

Longitudinal Section of Reticulorumen

B
Axial Section
of Omasum

C
Cross-section of
Omasum at x—y
Dorsocranial View

FIGURE 3.26 Diagrammatic sections of the bovine reticulorumen and omasum. Structures and compartments of major importance are numbered as follows: 1) cardia; 2) reticulo-omasal orifice; 3) reticulum; 4) cranial sac of rumen; 5) dorsal sac of rumen; 6) caudodorsal blind sac; 7) ventral sac of rumen; 8) caudoventral blind sac; 9) ruminoreticular fold; 10) cranial pillar; 11) right longitudinal pillar; 12) caudal pillar; 13) dorsal coronary pillar; 14) ventral coronary pillar; 15) omasum; 16) omasal canal; 17) omasal pillar; 18) omaso-abomasal orifice; 19) omasal lamina (leaf); 20) abomasum. (From Sellers and Stevens 1966.)

curvature, and the remainder of this organ contains long, thin plates of tissue, called omasal leaves, which project from the dorsal and dorsolateral curvature of the organ. The reticulorumen and omasum together are referred to as the forestomach. The omasum joins the third major compartment, the abomasum, via a relatively large orifice.

Both the reticulorumen and omasum are lined with nonglandular stratified squamous epithelium, whereas the abomasum contains proper gastric and pyloric glandular mucosa. Stomachs of sheep and goats are similar to that of cattle. Bensley (1902-3) found no cardiac mucosa in the abomasum of sheep, but Langer (1974) described a region of cardiac mucosa on the greater curvature and pyloric mucosa along the length of the lesser curvature of the goat abomasum. The relative size of the omasum varies in different species of ruminants. It appears to be largest in domestic cattle, where the surface area of its leaves can constitute one-third of the entire forestomach.

Hofmann (1968, 1973, 1983) examined the relationship between the structural characteristics of the forestomach and the diet in a wide range of East African ruminants. He found that these animals could be categorized as 1) selective feeders, which browsed on the more highly nutritious, succulent portions of the plant; 2) bulk feeders that were relatively nonselective in their browsing and grazing; and 3) an intermediate group of animals that varied in their feeding habits according to area or availability of food. The forestomach of selective feeders including the small ruminants such as the dik-dik, *Madoqua guentheri* (Fig. 3.27), which has a body weight of 3.5-5 kg, had a relatively small reticulorumen and omasum, as compared to the intermediate group and, particularly, the nonselective grazers.

The gastrointestinal tract of a sheep is shown in Figure 3.21. The small intestine is very long in relation to the large intestine, when compared to the llama or the pig. The cecum and a proximal segment of colon appear more voluminous than those of the llama, but show a similar lack of sacculation. According to Sack and Ballantyne (1965), the cecum of a musk-ox calf was approximately one-third the length of the animal's body. Therefore, ruminants may demonstrate wide variations in the length and capacity of their large intestine.

In summary, the stomach of individual artiodactyl species varies from a simple, nonvoluminous, noncompartmentalized organ to an extremely voluminous highly compartmentalized organ. There are major species variations in the distribution of the different types of gastric epithelium. However, the stomach of all species contains an area of nonglandular stratified squamous epithelium emanating from the gastroesophageal junction. This area varies in its relative size and, in some species, contains islands of cardiac mucosa. In others, it simply joins the zone of cardiac mucosa, and, in a few, cardiac mucosa may be entirely absent. A second common characteristic is the presence of a groove along the mucosal surface of the lesser curvature of the stomach. In young ruminants, the muscular lips of this groove close at the time of nursing, forming a tube to shunt milk past the forestomach and directly into the abomasum. It presumably serves a similar purpose in other mammalian species that have a multicompartmentalized stomach.

The intestinal tract of artiodactyls shows wide variation in the relative length of the small versus large intestine, as well as in large intestinal capacity and degree of sacculation. There appears to be an inverse relationship between the relative complexity and capacity of the stomach versus the large intestine. This suggests that herbivores with a voluminous, compartmentalized stomach have a lesser "need" for a voluminous, sacculated cecum and

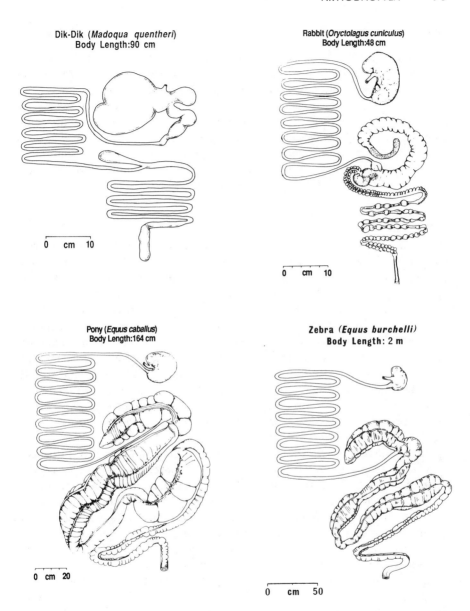

FIGURE 3.27 Gastrointestinal tracts of a rabbit and a pony (from Stevens 1977), a dik-dik, and a zebra.

colon. However, the length and complexity of the large intestine also correlates with the need for conservation of water, as evidenced by differences between the hippopotamus and camelid species.

Dermoptera

The colugos (F. Cynocephalidae) are often referred to as the flying lemurs because of their gliding ability and were once classified with prosimian primates. These include two species of animals about the size of a domestic cat that live in Southeast Asia. Their diet consists of leaves, buds, flowers, and fruit. The intestinal tract of *Cynocephalus volans* is nine times the body length, and the cecum, which is approximately four times the body length, is divided into two compartments (Wharton 1950). Flower (1872) stated that the colon is relatively large in diameter and sacculated in its more proximal segment, as a result of three longitudinal bands of muscle.

Lagomorpha

The rabbits, hares, and pika, which constitute this order, are herbivores presently represented by two families and 65 species that are similar to rodents and were once included in the same order. The gastrointestinal tract of a rabbit is shown in Figure 3.27. The stomach is simple, but relatively long with a well-developed oral pouch. The cardiac glandular region projects only a short distance into the stomach. Proper gastric glands line a large portion of the stomach, including the pouch, and the terminal one-fourth is lined with pyloric glandular mucosa. Therefore, these animals, which are strict herbivores, have a stomach lined entirely with glandular epithelium and largely with the proper gastric glands. The terminal ileum is dilated immediately above the ileocolic valve in a structure called the sacculus rotundus, which has lymph follicles in its wall. Another group of these follicles partially encircle the colon distal to the valve.

The rabbit cecum is very large in relation to the rest of the gut. It forms a spiral that occupies much of the abdominal cavity and is said to have a capacity approximately ten times that of the stomach. Its mucous membrane includes a projecting fold that spirals around the inside of the cecum as many as 25 times over its length, greatly increasing the internal surface area. The cecum terminates in a vermiform appendix with a small lumen and thick walls containing lymph follicles. The colon begins as a thin-walled sac, but soon becomes bilaterally sacculated by the presence of three longitudinal bands of muscle. Two of these disappear after a short distance, and the sacculations continue on only one side of the cecum. The third muscular band disappears approximately one-third the distance to the anus. The pika *Ochotona rubescens* appears to have paired ceca (Mitchell 1905).

Perissodactyla

This group of ungulates with an odd number of toes is presently represented only by families that contain the the equine, rhinoceros, and tapir species.

The equids are bulk-feeding grazers that spend up to 18 hours a day in the procurement of food. The rhinoceroses, which are presently represented by three genera and five species, include both grazers and browsers. The tapir family includes one living genera and four species. Most tapirs are browsers that inhabit tropical forests.

Figure 3.27 shows the gastrointestinal tract of two equine species. The stomach is simple. The cranial half is lined with nonglandular stratified squamous epithelium (Fig. 3.28). The caudal half is progressively lined with a narrow segment of cardiac mucosa and wider areas of proper gastric and pyloric glandular mucosa. Stomachs of the tapir and rhinoceros appear more elongated than that of the horse, and the tapir stomach contains a much smaller area of nonglandular epithelium.

The large intestine of horses is very complex and extremely voluminous. Although the cecum is relatively large and sacculated, the colon is much more voluminous and both sacculated and further subdivided into compartments. The cecum opens into an enlarged segment of colon (ventral colon). The next segment (dorsal colon) is, at first, narrow in diameter and then expanded once more. The terminal one-third of the colon has a lesser diameter and is referred to as the small colon. The large intestines of the rhinoceros (Fig. 3.29) and tapir (Mitchell 1905) show a somewhat similar construction.

Proboscidea

This order contains only two surviving genera and species, which inhabit Asia (*Elephas maximus*) and Africa (*Loxodonta africana*). These are the largest living land animals. Their diet consists of trees, shrubs, grasses, and aquatic plants. The gastrointestinal tract from an African elephant is illustrated in Figure 3.29. The stomach is simple and relatively narrow, with a cone-shaped cul-de-sac near its cardia. The cecum is conical and sacculated, and the colon is large in diameter and sacculated over a major portion of its length. Mitchell (1905) gave a similar description of the intestine from a young African elephant and noted that the ileum passed some distance into the wall of the hindgut, which suggested to him the presence of "primitively" paired ceca.

Sirenia

This is a group of aquatic herbivores that show some resemblance to the ungulates, elephants, and whales. It includes two genera: *Dugong*, the dugongs that inhabit a wide range of southern ocean; and *Trichechus*, the manatees of the African and American tropics. They feed on grasses and algae that nor-

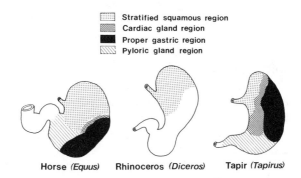

Horse *(Equus)*   Rhinoceros *(Diceros)*   Tapir *(Tapirus)*

FIGURE 3.28 Distribution of gastric epithelium in Perissodactyla. (From Bensley 1902-3.)

mally inhabit coastlines, bays, and river estuaries. A third, very large (up to 7.5 m in length) member of this order, Steller's sea cow (*Hydrodamalis gigas*) was exterminated within 30 years after its discovery in 1768 (Husar 1975).

Kenchington (1972) provided a description of the digestive system of *Dugong dugong*, which included a histological examination of gastric mucosa. The stomach was partially divided into two compartments by a ridge near its center (Fig. 3.30). A discrete glandular mass near the cardia called the cardiac gland. The remainder of the first compartment appeared to be lined with proper gastric mucosa. The second compartment, which he referred to as the pyloric region, was lined with a mucosa containing deep gastric pits. The latter showed prominent mucosal gland cells and underlying submucosal aggregations of Brunner's glands. This second compartment was separated from a third compartment by a sphincter. The third compartment appeared to be a duodenal bulb, with a pair of diverticula and no sphincter between it and the remainder of the small intestine. Examination of the dugong stomach by Marsh, Heinsohn, and Spain (1977) demonstrated that the cardiac gland contained all of the chief (pepsinogen-secreting) cells and a large portion of the parietal cells of the stomach. Some of the latter were found in the pyloric region. The duodenal diverticula were lined with a mucosa similar to that of the proximal duodenum, consisting of small villi and well-developed crypts of Lieberkuhn.

In nine specimens, the length of the small intestine averaged 5.5, and that of the large intestine 7.5 times the body length (Kenchington 1972). The colon was wider than the small intestine and demonstrated a thin muscle coat, but it was not sacculated. The cecum was described as conical in shape and about 6 inches long, but, unlike that of other mammals, very thick in its muscular coat, resembling the ventricle of the heart (Flower 1872).

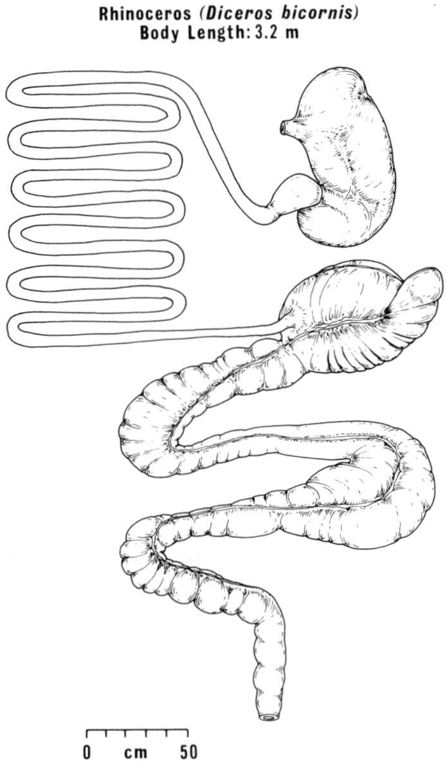

Rhinoceros *(Diceros bicornis)*
Body Length: 3.2 m

0   cm   50

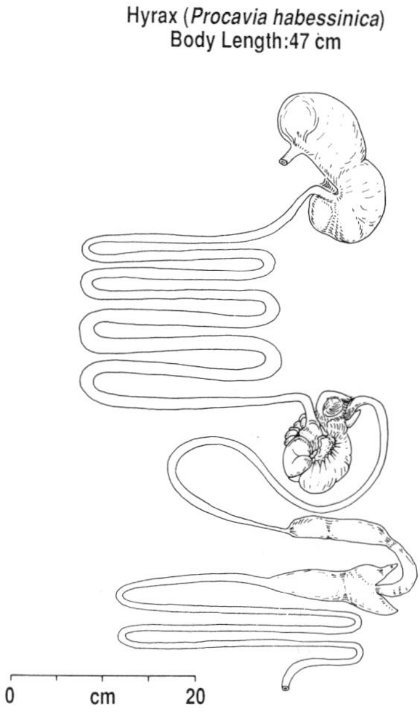

Hyrax (*Procavia habessinica*)
Body Length:47 cm

0   cm   20

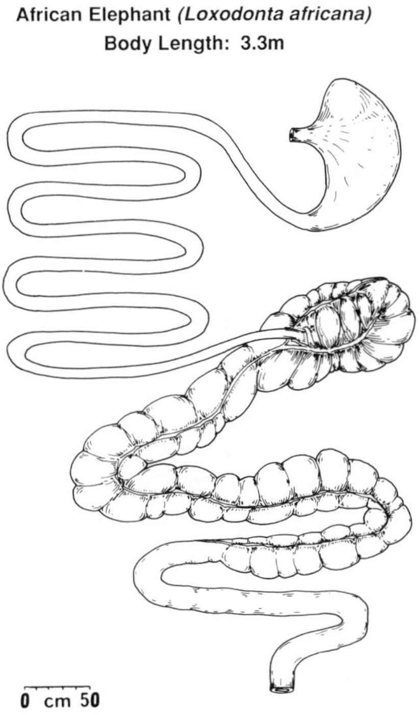

African Elephant *(Loxodonta africana)*
Body Length: 3.3m

0   cm   50

FIGURE 3.29 Gastrointestinal tracts of a rhinoceros and an elephant (from Clemens and Maloiy 1982), and a hyrax.

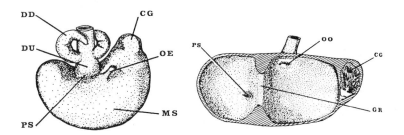

FIGURE 3.30 Stomach of a dugong. Drawing on the left shows the cardiac gland area (CG); esophagus (OE); pyloric sphincter area (PS); duodenum (DU); duodenal diverticulum (DD); and main sac (MS). Drawing to the right gives a sectional view showing the esophageal entrance (OO); cardiac gland (CG); gastric ridge (GR); and pyloric sphincter (PS). (From Kenchington 1972.)

The stomach of the manatee appears to resemble that of the dugong, in its division into two compartments with a glandular mass in the first and a pair of cecal appendages in the second. The cecum differs from that of most other mammals, in that the ileum enters a very expanded section of unsacculated proximal colon from which a pair of short, conical ceca project back along each side of the ileum. The ceca are not muscular as in the dugong. The expanded section then narrows to form a very long and coiled segment. Snipes (1984) has provided a detailed description of the anatomy of the cecum from a West Indian species of manatee.

Hyracoidea

The hyraxes are rodentlike herbivores, presently represented by only one family and 11 species, are indigenous to Africa, and are believed to be related to the elephants and sirenians. The stomach contains cranial outpocketing, or cul-de-sac, lined with nonglandular epithelium, and the first segment of intestine opens into a cecum, which is sacculated and quite capacious (Fig. 3.29). The intestine leaves the cecum, decreases to a diameter approximately that of the earlier segment of intestine, and opens into a wide segment containing an additional cecum with a pair of horns. The remainder of the intestine gradually diminishes in diameter again to continue toward the rectum. This presence of two ceca in series is unique to these animals.

Summary

The mammalian digestive tract shows a few major structural innovations as compared to lower vertebrates. These include changes in the masticatory apparatus and adaptations of the stomach and hindgut of many species. The simplest gastrointestinal tract is seen in some Insectivora and Carnivora, and a few marsupial species. This consists of a simple stomach and a relatively short intestinal tract, which may contain neither a valve nor a cecum to indicate the separation of midgut and hindgut. The hindgut in these species is neither larger than the midgut in its diameter nor haustrated. In some of these animals, this may represent a primitive condition. However, lack of a clear distinction between midgut and hindgut in some species belonging to other orders—for example, the cetaceans and edentates—probably represents regressive characteristics.

An area of stratified squamous epithelium lines the stomach of species belonging to ten of the mammalian orders, and the stomach of a number of mammals is highly developed into a capacious, compartmentalized organ. This can be seen in species belonging to the orders Chiroptera, Artiodactyla, Sirenia, Cetacia, Rodentia, Edentata, Marsupialia, and Primates. The compartments may consist of haustra, dependent on the tonus of longitudinal bands of muscle, as seen in the stomach of the marsupials and primates, or permanent sacculations. These arrangements provide for increased storage and retention of digesta and, in many cases, a site for microbial digestion. With the exception of the bats, expansion of the stomach is associated with an increase in the areas lined with stratified squamous epithelium and cardiac glandular mucosa. The evolution of homologous areas in the stomach of these animals is discussed by Langer (1987). The stomach of most of these species also contains a groove connecting the esophagus, to the region of proper glandular mucosa. Black and Sharkey (1970) concluded that this was an obligatory adaptation, to prevent dilution and fermentation of milk in the nursing young.

The hindgut shows a similar range in relative capacity and complexity. In most mammals, the midgut and hindgut can be distinguished from each other by the presence of a valve or sphincter and a cecum. The latter is paired in a few species. The cecum and colon show considerable species variation in their capacity and degree of haustration. The cecum is haustrated in many species, and this haustration often extends to the proximal colon. In a few animals such as the horse, pig, macaque, baboon, ape, and man, these haustrations extend over the entire length of the colon. The cecum is very voluminous in the lagomorphs, many rodents, and some marsupials. In other species with a well-developed large intestine, the colon provides the major site for retention of digesta. The greatest capacity and complexity of the colon is seen in the Perissodactyla and Proboscidea. The significance of the capacity, haustration, and relative length of the colon to microbial digestion and metabolism, and to recovery of electrolytes and water will be discussed in subsequent chapters.

4 Motor activity and digesta transit

The motor or muscular activities of the digestive tract are responsible for the mixing of food with secretions and for the transit of food and digesta through the alimentary tract at a rate optimal for digestion and absorption. As noted in Chapter 1, the esophagus and gastrointestinal tract of vertebrates are generally invested with an inner layer of circular muscle and an outer layer of longitudinal muscle. An additional oblique layer or sling of muscle is often present in the cranial wall of the stomach. The esophageal muscle may be striated over varying lengths, dependent upon species, but the remainder of the tract contains only smooth muscle. The circular muscle is thicker in some sections, forming sphincters or valves at various points along the digestive tract. The longitudinal layer is actually spiroform in some regions such as the esophagus, and is not evenly distributed around the circumference of the tract in other regions. For example, it can consist of only two or more longitudinal bands of muscle, distributed at intervals around the stomach, cecum, or colon of some species. The smooth muscle of the gut shows a remarkable ability to adjust its tonus in accommodation to marked changes in volume, with little or no increase in intraluminal pressure. This characteristic is most evident in the ability of the stomach muscle to undergo relaxation as it fills during a meal. However, intestinal muscle, especially that of the large bowel, demonstrates a similar capability.

The variety of mechanisms that have been adopted for the ingestion and mastication of food were mentioned in earlier chapters. The following discussion will concentrate on the motor activity of the digestive tract, and its contributions to the mixing and transit of digesta.

Motor activity

The motor activity of the digestive tract serves to mix, retain, and propel digesta. Early investigators divided these activities into different types of mixing contractions and peristalsis. The latter was defined as a moving ring of muscular activity, which consisted of a wave of relaxation followed by a wave of contraction that propelled digesta in the aboral direction along the gut. Similar propulsion in the orad direction was labeled antiperistalsis. However, Christensen (1971) stated that the esophagus appeared to be the only segment of the mammalian tract that regularly demonstrated these moving rings of contraction, and in man, even deglutition could be accomplished in about 10% of the young and in a majority of elderly individuals by a simultaneous contraction of the entire esophagus. He concluded that the motor activity can be most readily described as simply mixing and propulsive contractions. Nevertheless, we will continue to use the terms peristalsis and antiperistalsis for convenience in expression and because of their common usage.

Esophagus

Deglutition

Deglutition or swallowing involves a complicated series of mechanisms that have been described in detail for dogs and humans (Code and Schlegel 1968; Davenport 1982). Deglutition is initiated by a voluntary action consisting of food being pushed by the tongue upward and backward against the palate and into the pharynx. This stimulates receptors that initiate nerve impulses to the swallowing center in the brain stem, resulting in reflex contraction of pharyngeal muscles. The soft palate is pulled upward to close off the posterior nares. The palatopharyngeal folds pull medially to form a slit, through which food must pass. The vocal cords are pulled together, and the epiglottis is moved backward over the opening of the larynx, preventing the entrance of food into the trachea. At this same time, the upper esophageal sphincter, which includes the first 3-4 cm of the human esophagus, relaxes and the superior constrictor muscle of the pharynx then contracts, forcing food into the esophagus and initiating a peristaltic wave of contraction over the pharynx and esophagus. The entire pharyngeal stage of deglutition interrupts respiration for only 1-2 seconds, during which the respiratory center is inhibited by impulses from the deglutition center.

The circular muscle of the terminal 2-5 cm of the human esophagus serves as a lower esophageal or gastroesophageal sphincter. As peristaltic waves approach the terminal esophagus, they are preceded by a wave of reflex relaxation, transmitted in the myenteric plexus. This provides for relaxation of the gastroesophageal sphincter and a receptive relaxation of stomach muscle. Peristaltic waves originating in the pharynx pass over the length of the human esophagus in approximately 8-10 seconds. If food remains in the esophagus, distention can generate secondary waves of peristalsis at any point along its length.

Much of the research conducted on the esophagus has centered upon the mechanisms that provide for a competent gastroesophageal valve or sphincter. It is generally agreed that this sphincter functon is largely accomplished by constriction of an area of smooth circular muscle at or near the gastroesophageal junction. It has also been variously suggested that this constriction may be at least partly accomplished by 1) flaps or rosettes formed by gastric mucosal folds, 2) constrictions enforced by the diaphragmatic crura, 3) the stopcock effect of an acute angle between the terminal esophagus and cranial stomach, and/or 4) a sling formed by the gastric oblique muscle. Therefore, all of these factors need to be considered for comparisons among species.

Botha (1958, 1962) investigated the comparative anatomy of the gastroesophageal junction (GEJ) in the dogfish, sturgeon, frog, tortoise, chicken, bat, mole, hedgehog, four species of Carnivora (dog, fox, ferret, and cat), four species of rodent (rat, mouse, guinea pig, and hamster), three artiodactyls (pig, sheep, and ox), the rabbit, horse, and rhesus monkey. Functional studies were conducted on all species except the mole, hedgehog, bat, and fox. Although a sphincter or thickening of the striated, circular muscle was not seen in the distal esophagus of the dogfish, it was present in the sturgeon. He concluded that folds of gastric mucosa acted as a valve against gastroesophageal reflux in the fish. The frog and tortoise were examined under chloroform anesthesia. The frog demonstrated esophageal peristalsis, but no antiperistalsis and only an oblique angle at the GEJ. He did not mention the presence of sphincters at both ends of the esophagus, as noted earlier by Andrew (1959). No spontaneous contraction was noted in either the esophagus or stomach of the tortoise, and no constriction or thickening of muscle was seen near the GEJ. However, a well-marked cardiac angle and large folds of gastric mucosa were evident near the cardia. Compression of the tortoise stomach induced gastroesophageal reflux only after ligation of the pylorus.

Botha could find no sphincter, mucosal folds, or cardiac angle that might prevent reflux from the proventriculus and ventriculus (gizzard) of the chicken, but he suggested that fibers of gizzard muscle, passing over the terminal proventricular wall, may help serve this purpose. The sling formed by the oblique layer of stomach muscle and the highly developed gastric fundus, which contributes to the development of an acute angle at the GEJ, appeared to be characteristic only of the mammalian stomach. A long, abdominal segment of esophagus was found in the bat (*Plecotus auritus*), and, as mentioned earlier, its mucosa contained oxyntic cells. He also stated that the bat's esophageal muscle was striated over most of its length, but terminated in a highly developed, smooth muscle sphincter. Gastric folds were well developed at its cardia.

In the mammals examined by Botha, anesthesia was induced with either pentobarbital or, in the small species, tribromoethanol in amylene hydrate. Light anesthesia was then maintained with nitrous oxide. All species demonstrated primary peristalsis, and many showed a secondary peristalsis as well. Peristaltic waves passed most rapidly over the section of esophagus containing striated muscle. The smaller the species, the more rapid the rate of passage. In the monkey, the only species examined that had an extensive section of esophageal smooth muscle, the peristaltic wave slowed down at the level of the aortic arch, continuing at a slower pace until it reached the terminal esophagus. The latter then contracted on its own, forcing its contents into the stomach. The esophagus of the cat also showed an area, over its caudal 1.0-1.5 cm, that contracted more slowly and independently. In the cat, the entire lower esophagus would occasionally contract in unison and empty its contents into the stomach. Esophageal peristalsis in the rabbit and pig also was rapid until the terminal few centimeters were reached. This segment then underwent a contraction that was slower and delayed, but well-defined and unified. Although these results might be due to the distribution of smooth esophageal muscle previously noted in these species, a somewhat similar delay in esophageal emptying was witnessed in the rat, mouse, hamster, guinea pig, ferret, and dog. Contraction of the longitudinal muscle also shortened the esophagus during peristalsis, pulling the cardia forward. This was slight in some species, but marked in others, sometimes pulling the cardia into the thoracic cavity.

Each of the mammalian species in the preceding study showed a functionally competent occlusion of the terminal esophagus, which was represented by a well-defined, globe-shaped sphincter in the rabbit and a conical sphincter in the rodents. The sphincter contained both striated and smooth muscle in the rabbit, but appeared to be composed entirely of striated muscle in the rat, mouse, hamster, and mole. The lower esophageal sphincter consisted of a well-defined layer of smooth circular muscle in the bat, guinea pig, pig, and horse. In the latter species, the competence of the sphincter appeared to be excellent, and pyloric ligation was required before compression of the stomach contents would induce gastroesophageal reflux. Reflux was most easily induced in the guinea pig, but could be induced in all of these animals by either a sufficient increase in the intragastric pressure or manipulation of the GEJ or cardia. Spontaneous reflux was also occasionally noted, especially in the pig. Neither the diaphragm nor the angle between the esophagus and stomach appeared to contribute greatly to this competence, but all species demonstrated varying amounts of gastric mucosal folding and a cardiac sling of oblique muscle. The latter is especially well developed in the horse.

The terminal esophagus of the ferret, cat, dog, and monkey showed no anatomical sphincter or thickening of the circular muscle, although there was a short section of constricted esophagus above the cardia. The muscle in this section varied in tonus, but normally relaxed only during deglutition or emesis. In the cat, gastroesophageal reflux was quite easily induced by compression of gastric contents. The cat also appeared especially sensitive to manipulation of the GEJ. Reflux was hardest to induce in the dog, monkey, and ferret. In the ferret, stimulation of the pharynx induced vomiting, and this was associated with an increased tonus of gastric muscle and dilatation of the constricted portion of the terminal esophagus to the full diameter of this organ, but no local or antiperistaltic contractions of esophageal muscle were seen. Gastric folds also formed a rosette at the cardia of these animals. The effect of sectioning the diaphragmatic crura varied. In some species, it appeared to have no effect on the competence of the GEJ. It appeared to have some effect in the ferret and dog, and reflux was much easier to induce in the cat after the crura were cut. Reflux may have been partially prevented in these species by the angle of 90-100 between the terminal esophagus and stomach, especially in the ferret.

Regurgitation
Most mammals are capable of emesis or vomiting—a reflex act that serves to protect against absorption of toxic substances. The rat and horse are exceptions. The rat has highly developed senses of taste and smell, which aid in its selection of food. The horse appears to be less fortunate, and, when a horse does vomit, its stomach is often found to be ruptured. This may be due to the highly muscular lower esophageal sphincter or cardiac sling. In any event, most animals do vomit, and in those studied by Botha, regurgitation involved inhibition of the esophageal sphincters, increased tonus of the stomach muscles, and a shortening of the longitudinal fibers of the esophagus. These events, supplemented by constriction of the abdominal muscles (abdominal press), forces the gastric contents into the esophagus and mouth. There is no evidence of esophageal antiperistalsis during emesis in man, and Botha states that antiperistalsis was never seen in any of the species he studied.

Many raptors such as the hawks and owls are capable of controlling their regurgitation of digesta. This is a normal, physiological function that is associated with esophageal antiperistalsis and allows these birds to expel orally the relatively indigestible skin and skeletal parts of their prey (Duke, Jegers, Loff, and Evanson 1975; Kostuch and Duke 1975). Among mammals, the Camelidae and Ruminantia also demonstrate controlled regurgitation as part of the unique characteristic of rumination. Rumination is defined as the regurgitation, remastication, reinsalivation, and redeglutition of food. Adult cattle spend approximately one-third of their day ruminating. The act is involuntary, although it can be initiated by the presence of coarse, fibrous material in the forestomach and inhibited by any stimulus that tends to distress the animal. It has been suggested that rumination allows these species the advantage of ingesting in haste and masticating at leisure, and in greater safety. However, Gordon (1968) concluded that one of the principal advantages is the savings in energy as a result of recumbency during rumination, which he estimated to be equivalent to 10% of the caloric daily intake in sheep.

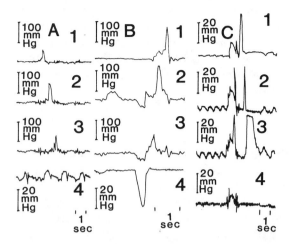

FIGURE 4.1 Pressure events recorded in the esophagus and pleural cavity during deglutition, regurgitation, and eructation. Numbers on the traces in these three series represent 1) cervical esophagus 70 cm cranial to cardia, 2) thoracic esophagus 40 cm cranial to cardia, 3) thoracic esophagus 10 cm cranial to cardia, and 4) pleural cavity. Note peristaltic deglutition wave (series A) with no associated change in respiratory cycle, as compared to regurgitation (series B), which shows the marked drop of pressure in the pleural cavity and thoracic esophagus followed by an antiperistaltic wave of esophageal contraction. Eructation (series C) is recorded in the esophagus as a rise in pressure of low amplitude and approximately 1-second duration, followed by a high-pressure wave of short duration. The first wave of pressure, which is recorded simultaneously in the pleural cavity, represents the effect of abdominal press. The second, more rapid component, is an antiperistaltic wave of contraction that is followed, on this trace, by a peristaltic wave of deglutition. (From Stevens and Sellers 1960.)

Rumination was extensively studied by French physiologists in the middle 1800s and by Dutch and German physiologists in the 1920s. Results of these and more recent studies have been reviewed by Bell (1958), Gordon (1968), and Stevens and Sellers (1968). The complete cycle of rumination is closely integrated with cyclic contractions of the ruminant forestomach, and best discussed along with that subject later in this chapter. We will concentrate here only on the regurgitation phase that directly involves the esophagus. The regurgitation phase of rumination begins with inspiration against a closed glottis, which results in a strong, negative pressure in the thoracic segment of the esophagus (Fig. 4.1B). This is accompanied by relaxation of the gastroesophageal sphincter, resulting in the aspiration of digesta from the rumen antrum into this segment of the esophagus. This digesta is then swept to the mouth by an antiperistaltic wave of esophageal contraction.

The regurgitation phase of rumination differs from emesis in both the presence of esophageal antiperistalsis and the absence of abdominal press. Emesis of abomasal contents into the reticulorumen, in association with abdominal press, is seen with blockage of the small intestine (Hammond, Dziuk, Usenik, and Stevens 1964), but trajectory emesis, with abdominal press and ejection of rumen contents from the mouth, occurs only rarely with reticulitis or ingestion of certain plant toxins.

Eructation

Eructation, the release of gas from the stomach via the esophagus, is a phenomenon that appears to be common to most species. However, it has been most extensively studied in ruminants because of its extreme importance to the survival of these animals (Sellers and Stevens 1966; Doughterty 1968, 1977). Microbial digestion in the forestomach of cattle can produce as much as a liter of gas per minute. Although much of this is CO_2, which is quite readily absorbable, approximately half of the gas is less absorbable CH_4. Furthermore, at the height of fermentation, neither gas can be absorbed from the forestomach at a rate sufficient to prevent dilatation. Therefore, ruminants are very subject to tympany or bloat—a condition seen in animals grazing on certain types of legume pasture or allowed to ingest quickly food that contains rapidly fermentable carbohydrates.

Eructation in ruminants also is intimately tied to the cyclic contractions of the forestomach and, except for its esophageal components, is therefore best discussed later with that subject. Eructation occurs at a time in the reticuloruminal cycle when the cardia is exposed to gas. The gastroesophageal sphincter opens at the same time that pressure in the rumen antrum is raised by a combination of the contraction of the dorsal sac of the rumen and abdominal press. This intraruminal pressure forces gas into the thoracic esophagus, from which it is then carried to the mouth by an antiperistaltic wave of esophageal contraction (Fig. 4.1C).

Although the ruminant esophagus demonstrates a number of unusual functional characteristics, it represents a good model for the studies of the nervous control of esophageal functions. For example, Doty (1968) referred to the studies by Roman in sheep as supplying some of the most definitive information available on feedback alterations of the course and vigor of esophageal contraction.

Stomach

Contraction of gastrointestinal smooth muscle is largely under myogenic control (Hartshorne 1981; Paul 1981), which is subject to extrinsic neurohumoral modulation (Roman and Gonella 1981). One exception to this is the ruminant forestomach, discussed below. Information on the motor activities of the dog and human stomachs, which have received the most extensive study, is reviewed by Heading (1984), Kelly (1981), and Davenport (1982). The dog and human stomachs can be divided into two distinct regions of motor activity. The proximal one-third of the stomach serves to receive and store ingested food. It undergoes receptive relaxation on the receipt of a bolus, and can distend to a large size in accommodation for food storage, with little increase in intragastric pressure. Muscle cells in this region undergo a depolarization of their membrane potentials, which can be associated with sustained tonic contractions or more rapid contractions of the muscle. However, the muscle shows no rhythmic changes in electrical potential nor peristaltic waves of contraction. The slow sustained tonic contractions serve to press fluid rapidly from the proximal segment of stomach, and to move gradually the more solid gastric contents into the distal segment.

The myoelectric activity of the distal motor region of the stomach is quite different from that of the proximal region. The electrical potentials of its muscle cell membranes undergo slow, rhythmic cycles of partial depolarization, referred to as the slow waves or basal electrical rhythm (BER). These are initiated by pacemaker muscle cells on the greater curvature of the stomach. The slow waves involve a ring of circular muscle around the entire stomach and travel varying distances in an aboral direction toward and, at times over, the pylorus and its sphincter. Although these slow waves do not initiate contraction, they sensitize the muscle to contract at a given frequency or multiples of that frequency, at which time the slow waves are associated with spike potentials and waves of muscular contraction. These contractions of distal stomach muscle mix gastric chyme with gastric secretions. They also propel the gastric chyme to the pylorus, where fluid and small particles of food are passed through the gastroduodenal junction and larger particles are retained and triturated by contractions of the antral muscles.

The frequency of slow waves and the presence of spike potentials (and contractions) are modulated by neurohumoral mechanisms (see Chapter 8). Both the slow waves and spike potentials appear to be increased by cholinergic nerve stimulation, whereas nonadrenergic inhibitory and adrenergic nerves reduce their activity. Gastrin, cholecystokinin, and motilin stimulate contraction, whereas secretin, glucagon, gastric inhibitory polypeptides, vasoactive intestinal peptide, and somatostatin have inhibitory effects. Prostaglandins also have been shown to inhibit strips of circular muscle from the antrum and pylorus of the guinea pig stomach (Milenov and Rakovska 1983).

The motor activities of the ruminant forestomach are very different from those of the human and dog stomach. The complex, cyclic contractions of the bovine forestomachs were first detailed by the classical studies of Schalk and Amadon (1921, 1928) and Wester (1926). More recent information has been reviewed by Sellers and Stevens (1966), Phillipson (1977), and Titchen, (1968). A diagramatic representation of the bovine forestomach was illustrated in Chapter 3 (Fig. 3.26). The reticulum, which is the most cranial compartment, is partially separated from the cranial sac of the rumen by a ruminoreticular fold. The cranial and dorsal sacs of the rumen are contiguous, but separated from the ventral sac of the rumen by a ring of muscular tissue, which consists of the cranial, medial, caudal, and lateral pillars of the rumen. The dorsal and ventral sacs of the rumen also are partially separated from posterior "blind" sacs by coronary pillars.

A reticuloruminal cycle of muscular contraction begins with a double (cattle) or biphasic (sheep) contraction of the reticulum. This is followed by a sequential, primary contraction of the cranial, dorsal, and ventral sacs of the rumen. The pillars also contract during this sequence, with the cranial pillar contracting first and just prior to contraction of the dorsal sac. This primary contraction of the compartments and pillars of the rumen may or may not be followed by a secondary contraction of the rumen, which involves most of these same structures, but is not preceded by the reticular contractions. The secondary contractions of the sheep rumen appear to originate in the ventral blind sac (Ruckebusch and Tomov 1973).

The reticuloruminal cycles of contraction serve to circulate and mascerate ingested food, and mix it with rumen microbes. Reticuloruminal contents consist of a ventral layer of fluid and small particulate digesta, a floating layer of fibrous plant material, and a pocket of gas that occupies the most dorsal aspect of the rumen. Immediately after ingeston, particulate food of high specific gravity such as poorly masticated kernels of corn tend to sink rapidly into the cranial sac of the rumen. The next cyclic contraction passes these forward into the reticulum, rather than over the cranial pillar into the more posterior compartments of the rumen. However, the lighter-weight, fibrous plant material is carried by reticuloruminal contractions back into the dorsal sac, and then forced ventrally by the dorsal sac contraction. Subsequent contraction of the ventral sac of the rumen forces its contents upward and, during relaxation of the cranial pillar, forward into the reticulum. This may be followed by one or more contractions of the rumen alone. Figure 4.2 shows a series of reticuloruminal cycles in cattle, consisting of double contractions of the reticulum followed by primary and, on alternate cycles, secondary contraction of the rumen.

The patterns of digesta flow during cyclic contraction of the bovine reticulorumen are shown in Figure 4.3. When the cycle includes only a primary (single) series of contractions of rumen compartments, digesta is sequentially forced by the reticular contractions into the cranial sac of the rumen and by its contraction into the dorsal sac. The subsequent contraction (raising) of the cranial pillar, and then the dorsal sac, results in the pressing of fluid and small particles into the ventral sac. Relaxation of the cranial pillar

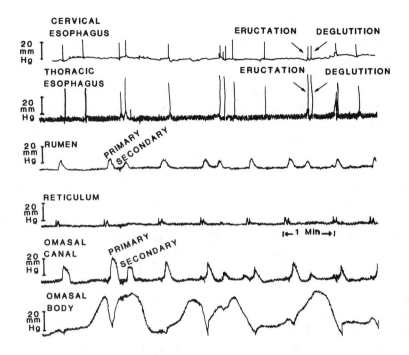

FIGURE 4.2 A 6-minute recording of reticuloruminal cycles and associated events in the esophagus and omasum. Note interruption in omasal body pressure waves during primary contraction of dorsal rumen and omasal canal. Upper traces show eructation as a biphasic wave occurring during secondary ruminal contractions and the frequent waves of deglutition associated with the swallowing of saliva. (From Sellers and Stevens 1966.)

FIGURE 4.3 Movement of digesta during the reticuloruminal cycles that involve only a primary contraction of their rumen, and those that include also a secondary contraction.

and contraction of the ventral sac then force digesta cranially; large, undigested, floating particles are pushed toward the cardia, where they may be regurgitated, and smaller particles and fluid are propelled toward the reticulo-omasal orifice. If, however, the cycle also includes secondary contractions, the cranial pillar does not fully relax at the time the ventral sac undergoes its primary contraction. Therefore, this results in a mixing of the ventral sac fluid with dorsal sac contents. The rumen then undergoes a second series of contractions during which the cranial pillar is again relaxed at the time of the ventral sac activity, and digesta follows the same course seen with the cycles that involve only primary contractions of the rumen.

The reticulo-omasal sphincter and omasal canal undergo similar cycles of contraction with each reticuloruminal cycle. These also consist of primary and secondary contractions, which occur at about the time the dorsal sac of the rumen undergoes its contractions (Fig. 4.2). The omasum serves as a two-phase pump to transfer digesta from the reticulum to the abomasum (Stevens, Sellers, and Spurrell 1960; Ehrlein and Hill 1969; Bueno and Ruckebusch 1974). Closure of the reticulo-omasal sphincter and contraction of the omasal canal together force the contents of the omasal canal into the body of the omasum between the omasal leaves. Subsequent relaxation of the sphincter and canal allows aspiration of fluid and particles from the base of the reticulum. This occurs at the height of the second reticular contraction and during the periods immediately following relaxation of the omasal canal. Transfer of digesta from the omasum into the abomasum is completed by strong and prolonged waves of contraction that pass over the omasal body (Fig. 4.2). These contractions are not seen with each cycle of contraction that involves the reticulum, rumen, omasal sphincter, and omasal canal. They are stimulated by distention of the omasal body and inhibited by abomasal distention.

Both the act of feeding and the resulting distention of the reticulorumen increase its rate of cyclic contraction and the number of secondary ruminal contractions per cycle. The coordination of contractions of the reticulo-omasal sphincter and omasal canal with those of the rumen results in an associated increase in the transit of digesta through the omasum. Outflow of digesta from the reticulorumen also appears to be affected by the osmotic activity of reticuloruminal contents in the black Bedouin goat (Shkolnik, Maltz, and Choshniak 1980). These desert animals can graze for several days without access to water—a characteristic that is necessary for survival on the sparse vegetation. As they become dehydrated, reticuloruminal outflow has been shown to decrease markedly, resulting in the more complete digestion of plant fiber. On returning to the waterhole, they can rapidly drink a volume of water equivalent to 40% of their dehydrated body weight, yet a twofold increase in reticuloruminal volume did not stimulate an immediate increase in the release of its contents into the abomasum. The fact that the drinking of isotonic saline resulted in a fivefold increase in rumen outflow within the first hour suggested that the activity of the omasal pump is inhibited by the hypotonicity of reticuloruminal contents after the imbibition of large amounts of water. This allows the forestomach of these animals to serve as a reservoir for the slow release of water to the more water-permeable segments of the gut. The rate of reticuloabomasal flow also appears to be affected by the feedback mechanism of abomasal distention on the amplitude of omasal body contractions (Stevens et al. 1960).

It was mentioned earlier that rumination and eructation are closely integrated with the cyclic contractions of the reticulorumen. Figure 4.4 shows a series of reticuloruminal cycles during a period of rumination. During rumination, the reticulum undergoes an extra contraction, just prior to the normal two. This occurs at the time of regurgitation and is believed to help by flooding the cardia with additional fluid. Immediately after the appearance of the regurgitated bolus in the mouth, the animal begins to chew its cud. Excess fluid is expressed from the bolus and immediately swallowed, but mastication continues until the remainder of the bolus is swallowed, just prior to the next cycle of rumination.

The complex series of events associated with rumination are under central regulatory control. Central regulation is indicated by the facts that fluid or balloons placed in the distal esophagus of cattle initiated secondary peristalsis rather than antiperistalsis (Sellers and Titchen 1959), and tooth abnormalities, which disturbed the mastication patterns of sheep, resulted in a correlative change in forestomach motility (Ruckebusch, Fargeas, and Dumas 1970).

Rumination has been described in both infant and adult humans (Brown 1968; Rothney 1969; Menking, Wagnitz, Burton, Coddington, and Sotos 1969; Einhorn 1977; Feldman and Fordtran 1978; Fleisher 1979). Brown noted that it was mentioned by Aristotle and Galen, and well documented as early as 1618. It appears to be an involuntary, painless act, usually seen in patients with other signs of emotional distress. Manographic and cineradiographic studies have indicated normal esophageal peristalsis, and normal activity and competence of the gastroesophageal sphincter in the absence of rumination. There is no evidence that the regurgitation is associated with antiperistaltic contractions of the esophagus, as seen in ruminants.

Figures 4.2 and 4.4 demonstrate that eructation also occurs only at a given phase of the reticuloruminal cycle. It is seen only during contraction of the dorsal sac of the rumen, more commonly on the secondary contraction in cattle. At this time, the relaxation of the reticulum and contraction of the rumen lower the level of digesta in the rumen antrum and expose the cardia to the rumen gases. Unlike regurgitation, eructation is always accompanied by abdominal press.

Cyclic contractions of the bovine forestomach occur in continuous suc-

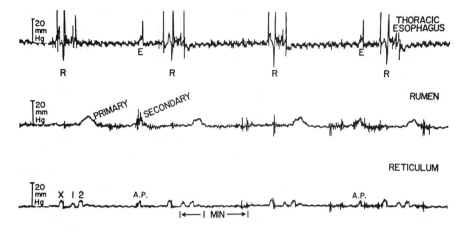

FIGURE 4.4 Three complete cycles of reticuloruminal contraction during rumination. The double reticular contraction, (1 and 2), and a primary and secondary ruminal contraction are labeled for the first cycle. The extra reticular contraction associated with regurgitation (X) is also labeled. The esophageal pressure changes during regurgitation are labeled (R) and consist of a small positive-pressure wave, followed by the negative deflection due to inspiration against a closed glottis, and then the large positive deflection due to antiperistaltic esophageal contraction (see Fig. 4.1B). Regurgitation is usually soon followed by deglutition (unlabeled) of the fluid expressed from the bolus at the beginning of mastication. A wave of deglutition, carrying the bolus to the rumen, also is seen on the esophageal trace at the end of each cycle of rumination and just before the next regurgitation. Note the close integration of the rumination and ruminoreticular cycles. Eructation (E) is labeled, and the associated increase in pressure due to abdominal press can be seen on the reticular trace (A.P.) and superimposed on the secondary wave of rumen contraction. (From Stevens and Sellers 1968.)

cession at a rate of 0.9 ± 0.1 per minute. The rate increases during feeding and decreases slightly during rumination. Ruckebusch (1975) has shown that these cycles are modified, but not entirely arrested during sleep. The cyclic motility of the ruminant forestomach is dependent on the integrity of the vagus. Vagotomy eliminates the cyclic contractions, as well as rumination and eructation. Sheep subjected to total vagotomy and sustained for two to five months by intraabomasal infusion of nutrients regained motility of the forestomach, but it was no longer synchronized into the cycles seen in the vagus-intact animal (Gregory 1982).

In contrast to the forestomach, the abomasum appears to contract in a manner similar to the stomach of humans and dogs (Ehrlein 1970). Abomasal contractions are not coordinated with those of the forestomach nor are they permanently affected by vagotomy (Duncan 1953).

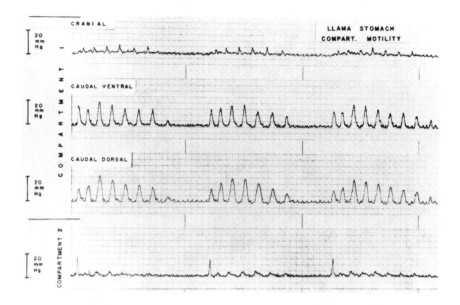

FIGURE 4.5 Pressure events simultaneously recorded from compartments 1 and 2 of the llama stomach. Note that cycles begin with a contraction of compartment 2. This is followed by a series of contractions of the caudal and cranial sacs of compartment 1. (From Vallenas and Stevens 1971a.)

Pentagastrin (see Chapter 8) has been shown to inhibit forestomach motility in sheep (Carr, McLeay, and Titchen 1970) and cattle (Ruckebusch 1971), but recent evidence suggests that this hormone may act through gastric centers in the medulla (Grovum and Chapman 1982).

The forestomach compartments of the camelid stomach also undergo cyclic, sequential contractions somewhat similar to those described for cattle and sheep (Vallenas and Stevens 1971a; Ehrlein and Engelhardt 1971). Figures 3.24 and 3.25 illustrate the llama stomach with its three major compartments and its glandular pouches. A cycle begins with a contraction of the second compartment (Fig. 4.5). This is followed by six or seven contractions of the caudal and then the cranial sacs of the first compartment. Each contraction of a compartment is associated with eversion of its glandular pouches (Fig. 4.6). During rumination, regurgitation of the bolus was associated with extra contractions of the cranial sac of the first compartment. Eructation occurred at the height of its caudal sac contraction. Cyclic contractions occurred at a rate of 0.6 ± 0.1 per minute, which is somewhat similar to the rate of reticuloruminal cycles in cattle; and the rate of llama forestomach contraction was similarly increased by feeding and decreased during rumination. Ehrlein and Engelhardt (1971) recorded more frequent

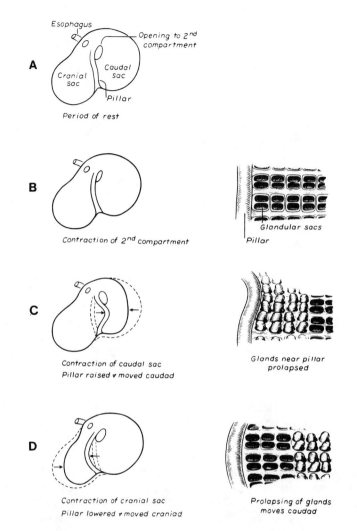

FIGURE 4.6 Basic stages in cyclic contraction of the first compartment of the llama and guanaco stomach. Drawings on left indicate contractions of pillar and sacs. In B, contraction of the second compartment can be observed only as a decrease in size of the orifice separating compartments 1 and 2. Cyclic eversion of caudal sac glandular pouches is indicated on the right by the drawings of this area during the three stages of contraction. (From Vallenas and Stevens 1971a.)

contractions of the second compartment, and a more complex cycle of forestomach motility was more recently described by Heller, Gregory, and Engelhardt (1984). The latter study indicated that the motility of the first two compartments of the llama stomach is under vagal nerve control in a manner similar to that noted in cattle and sheep.

Fluoroscopic studies of the tammar wallaby, *Macropus eugenii*, stomach showed two types of haustral contraction (Richardson and Creed 1981). One of these consisted of waves of contraction that spread in an aboral direction from one haustra to the next. The other was an intermittent narrowing of haustra. A BER rhythm of 5-6 cycles/min was recorded from the stomach muscles, but this was not seen to be associated with action potentials. Subsequent studies (Richardson and Wyburn 1983, in press) showed that haustral contractions occurred at the same frequency as the slow waves in the stomach of both the wallaby and quokka (*Setonix brachyurus*). They also confirmed the absence of action potentials in association with the slow waves of the wallaby stomach, but found that these were present in the recordings from the quokka.

Therefore, it appears that pacemaker tissue generates slow waves that govern the frequency of peristaltic contractions in the caudal region of the "simple" stomach, as well as the abomasum of ruminants. This BER also appears to control the frequency of peristalsis in the stomach of macropods, but its presence throughout the stomach and the absence of associated action potentials in the wallaby suggest that the pace may be set via the central nervous system. Cyclic contractions of the ruminant forestomach are controlled entirely by a medullary pacemaker(s) that also integrate(s) the acts of eructation and rumination with specific stages of the cycle.

Midgut

The motor activity of the small intestine is more varied and complex than that of the dog and human stomach (Weisbrodt 1981; Davenport 1982; Mathias and Sninsky 1985). The major functions of the small intestinal muscle are to mix its contents and propel digesta from the stomach to the hindgut or large intestine. Although mixing is aided by movement of mucosal folds and villi, which is believed to result from contraction of the muscularis mucosa, it is largely dependent upon rhythmic segmental contractions of the circular muscle layer. These segmental contractions are the most common type of motor activity, and their maximal frequency is determined by a BER of the muscle membrane potential, similar to that described for the stomach. In most species that have been studied, the frequency of the slow waves decrease in a stepwise fashion between the duodenum and the terminal ileum (Fig. 4.7). Since the segmental contractions are seen only in association with bursts of spike potentials superimposed on these slow waves, their maximal frequency also decreases along the small intestine. Therefore, the segmental

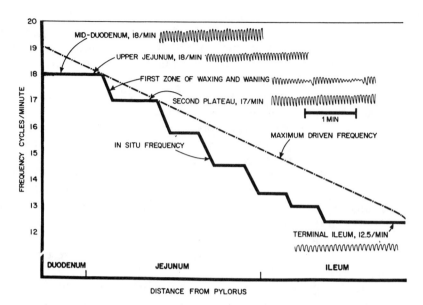

FIGURE 4.7 Frequency of the basic electrical rhythm of the small intestine of the anesthetized cat. The stepwise line shows the observed frequency—18 per minute in the duodenum and upper jejunum, and decreasing to 12.5 per minute in the terminal ileum. Segments of the electrical record from which the frequency was measured are shown. (Adapted from Davenport 1982.)

contractions may aid in the aboral propulsion of digesta. However, the steps or plateaus with a common frequency extend over considerable lengths of small intestine (Wiesbrodt 1981), and there are no gradients in the frequency of slow waves along the small intestine of the guinea pig (Calligan, Costa, and Furness 1985) or the wallaby and quokka (Richardson and Wyburn 1987). Although the frequency of BER is intrinsic to the muscle cells, the number of contractions associated with these cycles is increased by vagal nerve stimulation and cholecystokinin in at least some species.

Most of the propulsion of chyme through the small intestine is accomplished either by peristaltic contractions that move over short distances at the velocity of the electrical activity and at rates that are submultiples of the BER, or by peristaltic rushes that travel over the length of the small intestine in association with migrating myoelectric complexes (MMC). Although feeding increases the frequency of segmental contractions, it decreases propulsive activity, and peristaltic rushes are largely confined to interdigestive periods.

Myoelectric activity has been recorded in the small intestine of the chic-

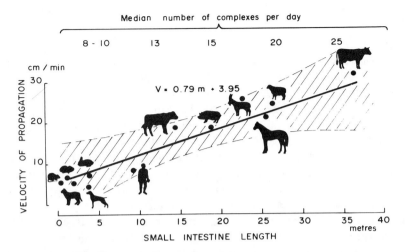

Median number of complexes per day

FIGURE 4.8 Relationship between the velocity of propagation of the myoelectric complexes and the length of the small intestine. The median daily number of jejunal complexes is high in ruminants and low in the carnivores because of the obliterating effect of feeding in the latter species. The daily number of myoelectrical complexes in the rat is far higher (36-60/day) than that indicated. The hatched area refers to the 95% confidence limits. (From Ruckebusch 1981.)

ken (Roche and Ruckebusch 1978) and reviewed for a wide range of mammals by Ruckebusch (1981). The MMC was found to pass over the upper small intestine of the dog at a rate of 6-12 cm/min, slowing down to a rate of 1-3 cm/min in the distal half of the small bowel. Six or eight sequences of MMC were measured in the small bowel of humans during an overnight fast, with an average of about 10 per day. However, both the number of complexes per day and their velocity of propagation vary with species (Fig. 4.8). Both appear to increase with the length of the intestine. In carnivores, the MMC were limited to interdigestive periods, but large, continuously feeding herbivores averaged 15-25 MMC distributed throughout the day.

Ruckebusch (1986) reviewed the comparative development of digestive motor patterns during prenatal life. He pointed out that in dog and rat pups, which are born blind, deaf, and poikilothermic, maturation of the control over gastrointestinal motility that is exerted by the central and enteric nervous systems is mainly postnatal. However, ungulates such as the sheep, calf, and foal must run with the herd within a few hours after birth, and fetal lambs demonstrate an early prenatal development of myoelectric activity in the small intestine, similar to that witnessed in premature human neonates, although the latter may have been brought about partially by extrauterine life.

Hindgut

Motility

The motor activity of the hindgut is more complex than that of the midgut and much more poorly understood. Although a number of early investigators concluded that the presence of a cecum of any length requires a mechanism for its filling, Cannon (1902) appears to have made the earliest radiographic observations of retropropulsion of contents from the proximal colon into the cecum of the cat. He also noted that the contents of more distal segments of the colon were separated into "globular masses" by segmental contractions, which could be seen to move slowly in an aboral direction. These observations were soon followed by a comparative study of large intestinal motility and digesta transit, conducted by Elliott and Barclay-Smith (1904) on the hedgehog, ferret, dog, cat, rat, guinea pig, and rabbit. The first four species were selected for the relative simplicity of their large intestine. The rat was chosen to represent an omnivore, whereas the last two species were selected to represent herbivores, with the rabbit as the extreme in cecal length and capacity. Colonic antiperistalsis was observed in all species except the dog, and resulted in a reflux of colonic digesta into the cecum of animals other than the hedgehog and ferret, which lack a cecum.

From these studies, Elliot and Barclay-Smith concluded that the colon of the rabbit could be functionally divided into three segments: 1) a proximal segment with fluid contents similar to that of the cecum and with a motor activity dominated by antiperistaltic contractions; 2) an intermediate segment of colon in which fecal nodules first appeared and peristaltic contraction predominated; and 3) a distal colon with definite fecal nodules and a motility similar to that of the intermediate colon—except that stimulation of its sacral parasympathetic nerve supply resulted in its evacuation by a massive simultaneous contraction. These functional segments of the colon were less distinct in the omnivorous rat, and least distinct in the carnivores. The entire colon of the dog appeared to them to be equivalent only to the distal segment of the rabbit colon. They concluded their discussion by pointing out that the proximal colon of mammalian herbivores tends to be characterized by marked sacculation, a condition seen throughout the large intestine of man, and they cited an earlier study of 1000 Egyptian mummies which indicated that their cecum was considerably larger than that of present-day man. Therefore, in spite of their reservation about deducing function from structure, these investigators concluded that "there can be little doubt that the human colon is rather of the herbivorous than carnivorous type."

It is somewhat surprising that the above study received relatively little attention until recent times, since it would seem to provide a reasonable working hypothesis for functional studies of the mammalian large intestine. This may be explained by the inherent distrust of experiments conducted on anesthetized animals and the fact that until quite recently, the large intestine was subjected to little study. Furthermore, it is difficult to study in conscious, intact animals, because its motility patterns tangle catheters, and its anatomical disposition almost defies radiographic examination. It also is difficult to fistulate effectively in many sites. As noted below, even its electromyographic responses show a marked difference from those of the esophagus, stomach, or small intestine.

Hukuhara and Neya (1968) described a pacemaker area in the colon of both the rat and guinea pig that was located some distance from the ileocecal junction, and could be equivalent to Elliott and Barclay-Smith's point of separation between the proximal and intermediate colon of the rabbit. This area appeared to initiate antiperistaltic waves in the direction of the cecum and peristaltic waves in the opposite direction. Hukuhara and Neya also observed occasional, strong peristaltic contractions that moved from the cecum down the colon, passing over the waves coming from the opposite direction. The inevitable collision of the antiperistaltic and peristaltic waves appeared to result in the passage of some of the digesta aborally.

Information gained from more recent studies tends to support a number of the conclusions reached by Elliott and Barclay-Smith. The proximal colon of man demonstrates retropulsion and prolonged retention of its contents, whereas the distal transverse, the descending, and the sigmoid colon tend to be divided into uniform segments, or haustra, by contracting rings that may be stationary or move slowly in an aboral direction (Christensen 1981). The transverse, descending, and sigmoid colon also demonstrate another type of movement not mentioned in the early studies. A relatively long segment undergoes simultaneous contraction, resulting in a mass, aboral movement of its contents. This process results in rapid movement of colonic digesta toward the rectum, where it can stimulate the defecation reflex, and it is most commonly seen after meals. These contractions of the colon can be stimulated by a gastrocolic reflex or the release of gastrin.

Electromyographic studies of the cat colon showed the presence of a BER consisting of slow waves similar to those found in the small intestine (Christensen, Caprilli, and Lund 1969; Christensen, Anuras, and Hauser 1974). However, they originated in the circular rather than the longitudinal muscle, and their frequency increased from the ileal-cecal-colonic junction to the midcolon, then tended to remain constant over the remaining length of the distal colon. This gradient in the frequency of the slow waves was consistent with the spread of these waves toward the cecum from pacemaker tissue located in the midcolon (Fig. 4.9). In addition to this, another independent electrical phenomenon was noted. It consisted of migrating spike bursts (MSB) that were usually directed aborally from a point near the hepatic flexure of the colon. The MSB were associated with strong contractions of circular muscle and presumed to be associated with the mass movements of

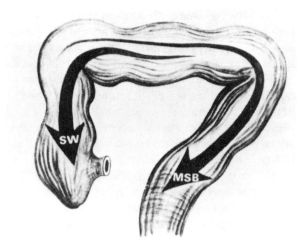

FIGURE 4.9 A proposed scheme relating digesta flow in the cat colon to the electrical slow waves (SW) and migrating spike bursts (MSB). Electrical slow waves are oriented in such a way that they appear to spread toward the cecum, away from a pacemaker whose position is highly variable about midway along the colon. Since slow waves appear to pace rhythmic contractions, such contractions should tend to produce flow with a polarity in the same direction (arrow, SW). This polarity of slow wave spread is probably not fixed. The migrating spike bursts begin at a variable position in the middle or proximal colon, and migrate toward the rectum. Since contractions accompany migrating spike bursts, such contractions should tend to produce flow with a polarity in the same direction (arrow, MSB). The migrating spike burst also has the capacity for reversal of direction. (From Christensen, Anuras, and Hauser 1974.)

digesta. A pacemaker for the rabbit colon has been located between its haustrated and nonhaustrated segments, in the fusus coli (Ruckebusch and Fioramonti 1976). Studies of the human colon (Sarna, Bardakjian, Waterfall, and Lind 1980) indicated three distinct segments of electrical control activity, with the middle segment the dominant frequency. A pacemaker area has been described also at the pelvic flexure of the equine colon, and it was concluded that this may be responsible for retropulsion of digesta in the ventral colon and cecum of these animals (Sellers et al. 1982).

A gastrocolic reflex stimulation of colonic motility as a result of gastric distension has been noted in the dog (Tansy, Kendall, and Murphy 1972) and horse (Sellers, Lowe, and Brondum 1979), and a ruminocecal reflex has been described in sheep (Fioramonti and Ruckebusch 1979). A gastrocecal reflex also has been described in the rabbit, horse (Ruckebusch and Vigroux 1974), and pig (Fioramonti and Bueno 1977).

The appearance of a distinct hindgut in vertebrates correlates with the evolution of terrestrial species, characterized by their greater need to conserve water. It provided a site for the retention of digesta, the additional removal of inorganic electrolytes and water of dietary origin, and the retrieval of electrolytes and water secreted into the upper digestive tract (see Chapter7). In lower vertebrates, it also provided a means of retrieving electrolytes and water of urinary origin. The kidneys of amphibians, reptiles, and most birds are limited in their ability to concentrate urine, as compared to most terrestrial mammals. Minnich (1970) concluded that much of the urinary sodium and water excreted by the kidneys of the desert iguana were reabsorbed by the cloaca. However, Junqueira, Malnic, and Monge (1966) concluded that the colon also was responsible for absorption of urinary ions and water in snakes (*Xenodon, Philodryas*, and *Crotalus* sp.). Antiperistaltic contractions have been observed in the colon of the terrapin *Chinemys reevsii* (Hukuhara, Naitoh, and Kameyama, 1975). There also have been numerous demonstrations that urine can be refluxed from the cloaca through the colon and into the ceca of chickens (Browne 1922; Koike and McFarland 1966; Akester, Anderson, Hill, and Osboldiston 1967; Skadhauge 1968, 1973), Japanese quail (Akester et al. 1967; Fenna and Boag 1974), roadrunners, *Geococcyx californianus* (Ohmart, McFarland, and Morgan 1970), and turkeys (Dziuk 1971). Bjornhag and Sperber (1977) estimated that 20-24% of the urine flow is transported to the ceca of turkeys. This does not seem to hold for the African ostrich (Skadhauge et al. 1984). Electromyographic studies of the hindgut of the opossum, a mammal that has retained the cloaca, showed a progressive increase in the frequency of slow waves from the ileocecal junction to the rectum (Anuras and Christensen 1975).

Therefore, the pacemaker that initiates antiperistaltic waves of contraction may have originated in the cloaca of lower vertebrates, suggesting that the motor activity of their entire hindgut is equivalent to that described by Elliott and Barclay-Smith (1904) for the proximal segment of the rabbit colon. However, with the loss of the cloaca, the pacemaker appears to have migrated to a more proximal position in the colon of most mammals.

Digesta transit

The rate of digesta transit through the gastrointestinal tract can be estimated by the use of "markers"—substances that are not normally secreted, digested, or absorbed by the gut. Passage of fluid digesta can be estimated by administering soluble markers and the passage of particles by the use of insoluble, particulate markers. Although the rate of digesta transit has been measured in a variety of species, useful species comparisons are often limited by differences in experimental procedure and design. For example, there can be large differences in the normal diet and dietary regimen of various species, and the rate of passage varies with the diet and feeding intervals. Experiments comparing the rate at which particulate digesta move through the gastrointestinal tract of different species are especially difficult to design, as the rate of transit is dependent on particle size and specific gravity, as well as the structural and functional characteristics of the gastrointestinal tract.

The advantages and limitations of various substances used as markers for the liquid and particulate fractions of the digesta have been reviewed by Faichney (1975), Warner (1981a), Van Soest (1982), and Clemens (1982). A good marker must be recoverable. It must flow with the portion of the digesta (fluid or particulate) that is being investigated, and it must not be digested, absorbed, or adsorbed in the process. Chromium-labeled ethylenediaminetetraacetic acid (Cr-EDTA) and polyethylene glycol (PEG) generally satisfy the above criteria for measurement of fluid transit. They are normally absorbed in only small quantities, and can provide sensitive and accurate measurements, especially when isotopically labeled. Particulate markers present a greater problem. Plastic particles are indigestible and retain their dimensions and specific gravity if they escape mastication or trituration by a gizzard. They are a good measure of the transit of food particles with identical characteristics, but they do not necessarily match the specific gravity of normal digesta particles. Feed particles stained with a dye or labeled with isotopes represent the true characteristics of a particulate ingesta. However, digestion of these particles can change their dimensions during transit and release the marker to the fluid or to other particles in the digesta. Therefore, chromium or rare earth elements fixed to food particles by a mordant that prevents their digestion are most suitable for measuring the effect of a particles size on its rate of transit.

Warner (1981) provided a very comprehensive review of studies of digesta transit through the gut of 61 species of mammal and four species of birds, determined with a variety of procedures. He concluded that the mean transit time, the time when one-half of the marker had been excreted, was the best single measure of digesta passage. Table 4.1 gives the mean transit times for 34 species of mammal and four species of bird. It shows that the shortest mean retention times were recorded by small mammalian carnivores (shrews and mink) and voles on green feed. The birds also registered moderately short retention times. The longest retention times were recorded by arboreal folivores, the koala and the sloths (not shown in table). Warner concluded that the retention time not only varied with the diet, frequency of feeding, ambient temperature, pregnancy, exercise, and age, but it also could show a substantial coefficient of variation between and even within animals under highly controlled conditions.

Figures 4.10, 4.11, and 4.12 represent a series of studies of the transit of fluid (Cr-EDTA or PEG) markers through the digestive tract of the raccoon, dog, bush baby, vervet monkey, pig, pony, rabbit, and hyrax. Most of the fluid

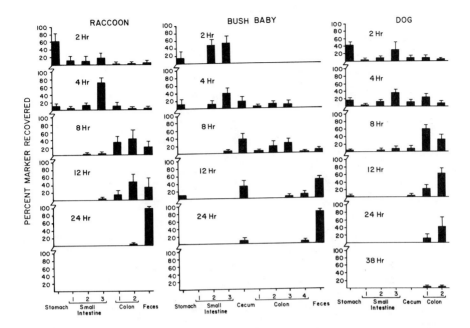

FIGURE 4.10 Percentage of digesta fluid marker recovered from the gastrointestinal tract of the raccoon, bush baby, and dog at the specified times following administration. Raccoons were fed a commercially prepared, pelleted, low-concentrate, high-fiber diet. Bush babys were fed a commercially prepared primate diet, and dogs were fed a commercial cereal-based diet. Animals were fed at 12-hour intervals for at least two weeks prior to the study. Digesta fluid markers were orally administered at the time of feeding. Animals were sacrificed in groups of three at the times designated following the meal, and sections of the gut were immediately separated by ligatures for recovery of markers (Modified from Banta, Clemens, Krinsky, and Sheffy 1979; Clemens and Stevens 1979; Clemens 1980.)

marker left the stomach of the bush baby, pony, and rabbit within 2 hours after intragastric administration, and the stomach of the raccoon, dog, vervet monkey, and pig within 4 hours after it was administered. Approximately 80% of the marker was retained by the hyrax stomach at the end of 4 hours. Much of the marker had reached the large intestine of the pony and rabbit within 2 hours, the large intestine of the vervet monkey within 4 hours, and the large intestine of the remaining species within 8 hours after it was given with the meal. The greatest variation in retention time was seen in the large intestine. Most of the fluid marker was excreted in the feces of the raccoon and bush baby within 24 hours, and the feces of the dog within 38 hours of administration. However, after 38 hours, the vervet monkey, hyrax, and pig had excreted only 50%, 40%, and 20%, respectively, of the fluid marker that

TABLE 4.1 Mean retention times of digesta in the gut of mammals and birds

Species	Conditions[a]	Mean retention time (hr)	Reference
MAMMALIA			
MARSUPIALIA			
Possum, *Trichosurus vulpecula*	—	46 ± 20	Gilmore (1970)
Koala, *Phascolarctos cinereus*	Particles	130	Cork, Warner, and Harrop (1977)
	Solutes	200	Ibid.
Quokka, *Setonix brachyurus*	Mixed diet 1	28	Calaby (1958)
	Lucerne	26	Ibid.
	Mixed diet 4	36	Ibid.
Red kangaroo, *Megaleia rufa*	Lucerne	34.8 ± 3.7	Foot and Romberg (1965)
	Oat straw	44.2 ± 7.3	Ibid.
	Lucerne, ad lib.	41.1 ± 4.4	McIntosh (1966)
	Lucerne, restricted	56.9 ± 9.2	Ibid.
	Oat chaff	47.9 ± 7.4	Ibid.
	Lucerne hay	28.4	Forbes and Tribe (1970)
	Oat straw	42.9 ± 9.5	Ibid.
	Lucerne hay	38.6 ± 15.1	Ibid.
	Oat straw	44.5 ± 11.2	Hume and Dellow (1980a)
	One meal/d; particles	29.3 ± 4.1	Warner (1981b)
	One meal/d; solutes	15.7 ± 3.2	Ibid.
	Continuous feeding; particles	19.1 ± 3.7	Ibid.
	Continuous feeding; solutes	11.3 ± 1.2	Ibid.
INSECTIVORA			
Water shrew, *Neomys fodiens*	Mealworms	1.26 ± 0.33	Kostelecka-Myrcha and Myrcha (1964c)
	Wheat in mealworms	2.33 ± 0.42	Kostelecka-Myrcha and Myrcha (1965)
PRIMATES			
Rhesus, *Macaca mulatta*	Rubber, knots	20 ± 3	Hoelzel (1930)
Man, *Homo sapiens*	Mixed meals	45.6 ± 11.1	Luckey et al. (1979)
LAGOMORPHA			
Rabbit, *Oryctolagus cuniculus*	Particles	15.2 ± 4.3	Piekarz (1963)
			Laplace and Lebas (1975)

	Feed/Marker	Value	Reference
RODENTIA			
Bank vole, *Clethrionomys glareolus*	Green feed	3.10 ± 0.86	Kostelecka-Myrcha and Myrcha (1964a)
	Wheat grain	11.18 ± 2.19	Ibid.
	Acorns	9.63 ± 2.37	Ibid.
	Wheat in mixture	12.24 ± 4.05	Kostelecka-Myrcha and Myrcha (1964b)
Field vole, *Microtus agrestis*	Green feed	2.77 ± 0.96	Kostelecka-Myrcha and Myrcha (1964a)
	Wheat	12.29 ± 1.88	Ibid.
	Green feed in mixture	2.65 ± 0.59	Kostelecka-Myrcha and Myrcha (1964b)
	Green feed marker wheat diet	6.80 ± 2.12	Ibid.
Common vole, *Microtus arvalis*	Green feed	3.19 ± 0.45	Kostelecka-Myrcha and Myrcha (1964a)
	Wheat	13.45 ± 1.51	Ibid.
Snow vole, *Microtus nivalis*	Green feed	3.94 ± 0.93	Ibid.
	Wheat	12.83 ± 1.35	Ibid.
Rat, *Rattus norvegicus*	Mixed feed	28.8 ± 10.9	Thompson and Hollis (1958); Varga (1976)
Nutria, *Myocastor coypus*			
Male	—	12.7 ± 0.7	Gill and Bieguszewski (1960)
Female	—	7.8 ± 0.6	Ibid.
CARNIVORA			
Dog, *Canis familiaris*	Mixed feed	22.6 ± 2.2	Banta, Clemens, Krinsky, and Sheffy (1979)
Raccoon, *Procyon lotor*	—	14.7 ± 1.3	Clemens and Stevens (1979)
Mink, *Mustela vison*	—	4.2 ± 0.9	Sibbald, Sinclair, Evans, and Smith (1962)
Cat, *Felis catus*	Rubber	3.3	Hansen (1978)
		13	Hoelzel (1930)
PROBOSCIDEA			
Indian elephant, *Elephas maximus*	—	32.7 ± 2.2	Gill (1960)
HYRACOIDEA			
Rock hyrax, *Procavia habessinica*	1-cm particles	53	Clements (1977)
	5-mm particles	42	Ibid.
	2-mm particles	36	Ibid.
	Solutes	41	Ibid.

(continued)

TABLE 4.1 Continued

Species	Conditions[a]	Mean retention time (hr)	Reference
PERISSODACTYLA			
Horse, *Equus caballus*	Nonpelleted	33	Hintz and Loy (1966)
	Pelleted	28	Ibid.
	Several diets	37.9 ± 5.3	Vander Noot, Symons, Lydman, and Fonnesbeck (1967)
	Long hay	35.5 ± 8.4	Wolter, Durix, and Letourneau (1974)
	Ground hay	25.9 ± 4.5	Ibid.
	Pelleted hay	31.2 ± 3.7	Ibid.
	Several diets	28.8 ± 5.1	Wolter, Durix, and Letourneau (1976)
ARTIODACTYLA			
Pig, *Sus scrofa*	Hay, grain	43.3 ± 6.2	Clemens, Stevens, and Southworth (1975a)
Red deer, *Cervus elaphus*	Several diets	33.1 ± 8.9	Milne, MacRae, Spence, and Wilson (1978)
White-tailed deer, *Odocoileus virginianus*	Hay, concentrates	22.7 ± 9.0	Mautz and Petrides (1971)
	Natural feed	22.5 ± 4.4	Ibid.
Mule deer, *Odocoileus hemionus*	*Agropyron spicatum*	55.5 ± 13.8	Milchunas, Dyer, Wallmo, and Johnson (1978)
	Epilobium angustifoloum	63.2 ± 10.2	Ibid.
	Vaccinium sp.	45.5 ± 18.5	Ibid.
Moose, *Alces alces*		72	Gill (1959)
Buffalo, *Bubalus bubalis*	80% of maintenance	94.8 ± 3.3	Ponnappa, Uddin, and Raghavan (1971)
	Maintenance	68.0 ± 15.5	Kuman and Raghavan (1974)
	12% of maintenance	61.0 ± 7.9	Ibid.
Cattle, *Bos taurus*	Several diets	65.1 ± 10.4	Ibid.
		68.8 ± 28.2	Mäkelä (1956); Brandt and Thacker (1958); Garner, Jones, and Ekman (1960); Shellenberger and Kesler (1961); Miller, Perry, Chandler, and

112

Zebu, *Bos indicus*	—	63.4 ± 7.2	Cragle (1967); Miller, Swanson, Lyke, Moss, and Byrne (1974)
Yak, *Bos grunniens*	—	78.2	Phillips (1961)
Bison, *Bison bison*	—	78.8	Schaefer, Young, and Chimwano (1978)
Suni, *Nesotragus moschatus*	Lucern leaves	16.8 ± 5.9	Hoppe and Gwynne (1978)
	Lucern stems	20.5 ± 2.1	Ibid.
Sheep, *Ovis aries*	Several diets	47.4 ± 26.5	Rogerson (1958); Hungate (1966); Faichney and Black (1974); Thewis, Francois, and Thill (1975); Thewis, Francois, Debouche, and Thielemans (1976); Westra and Christopherson (1976); Kennedy, Young, and Christopherson (1977)
Goat, *Capra hircus*	—	38.1 ± 3.3	Castle (1956a)
Age: 4 weeks	—	56.8 ± 12.2	Castle (1956b)
Age: 7 weeks	—	44.9 ± 3.2	Ibid.
Age: 10 weeks	—	41.5 ± 2.0	Ibid.
Age: 3-15 months	—	43.4 ± 2.3	Ibid.
AVES			
ANSERIFORMES			
Goose, *Anser anser*	Pelleted	5.7	Clemens, Stevens, and Southworth (1975b)
GALLIFORMES			
Chicken, *Gallus gallus*	Mixed diet	6.6 ± 2.1	Hurwitz and Bar (1966); Sklan, Dubrov, Eisner, and Hurwitz (1975)
Ptarmigan, *Lagopus mutus*	Particles	3.2	Gasaway, Holleman, and White (1975)
	Solutes	12.3	Ibid.
Turkey, *Meleagris gallopavo*	Solute	10.0	Björnhag and Sperber (1977)
PASSERIFORMES			
Whistlers, *Pachycephala* spp.	—	2.9 ± 0.6	Keast and Walsh (1979)

[a]Dash indicates that diet or makers were not specified.
Source: Reduced from Warner (1981a).

113

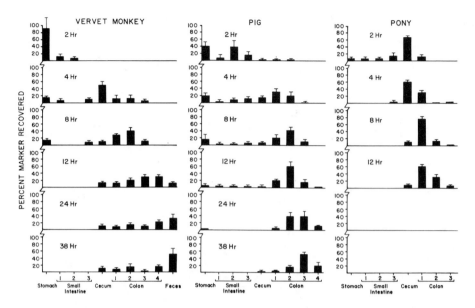

FIGURE 4.11 Percentage of digesta fluid marker recovered from the gastrointestinal tract of the vervet monkey, pig, and pony. The diet of the vervet monkey was the same as that fed the bush baby, and pigs and ponies were fed the same diet as the raccoons in the studies described in Figure 4.10. The dietary regimen and administration of markers also were the same, except that results for the recovery of marker from the pony large intestine were obtained after the administration of marker directly into the cecum via a fistula. (Modified from Argenzio, Lowe, Pickard, and Stevens 1974a; Clemens, Stevens, and Southworth 1975a; Clemens 1980.)

had been administered to each of these animals. The pony excreted only 5% of the fluid marker 12 hours after its administration into the stomach, and 30% of the fluid marker within 48 hours after its administration directly into the cecum (Argenzio, Lowe, Pickard, and Stevens 1974a). The rabbit study was terminated at the end of 24 hours, but at that time, its rate of excretion matched that of the hyrax. The cecum was the principal site of fluid marker retention in the intestines of the rabbit and hyrax, but the colon was the principal site for the retention of the marker by the large intestine of the other species.

The transit of particulate markers through the gastrointestinal tracts of the above animals was measured by the simultaneous intragastric administration of sections of polyethylene tubing, 2 mm in diameter and 2 or 5 mm in length. These particles were retained within the stomach of each species for a longer period than the fluid marker. The percentage excreted in the feces within the first 24 hours was as follows: raccoon, 95%; bush baby, 80%; dog

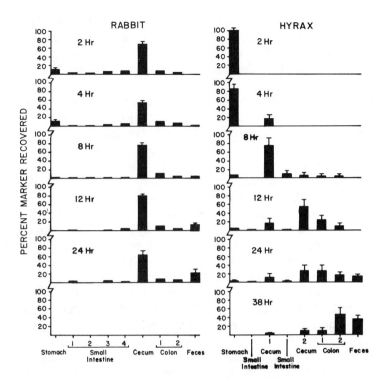

FIGURE 4.12 Percentage of digesta fluid marker recovered from the gastrointestinal tract of the rabbit and hyrax. Rabbits were fed a commercially prepared, pelleted rabbit diet, and hyrax were preconditioned to a diet of lucern leaves. Other conditions of the experiment were identical to those of animals described in Figures 4.10 and 4.11. (Modified from Pickard and Stevens 1972; Clemens 1977.)

and rabbit, 30%; pig, 12%; and hyrax and pony, 5%. Simultaneous administration of fluid and particulate markers directly into the cecum of rabbits, via cecal fistula, resulted in cecal retention of the fluid marker and rapid excretion of particulate markers in the feces. However, when these markers were infused into the pony cecum, the particulate markers were retained for substantially longer periods of time, principally in the proximal (dorsal and ventral) colon (Argenzio et al. 1974). Uden (1978) measured the transit time for fluid (Co-EDTA) and particulate (Cr-mordanted food particles) in horses, cattle, goats, and sheep. The horses excreted both the fluid and the particulate markers at a considerably more rapid rate than the ponies in the above study, and there was much less difference in the relative rate of fluid and particle transit. This could be due to differences in the nature of the particles that were used or the dietary regimen (see Chapter 6). Ponies were given a ground (pelleted) feed at 12-hour intervals, and the horses were fed a mixture of hay and concentrate. Uden found that the retention times for both the fluid and particulate markers were considerably longer in ruminants, as compared to the horse, and the particles were retained much longer than fluid.

There is considerable evidence that fluid and particles can pass through the digestive tract of chickens, turkeys, and many other birds at a very rapid rate (Sturkie 1965, 1970; Duke, Dziuk, and Hawkins 1969; McBee 1977). However, Leopold (1953) found that the feces of gallinaceous birds contained two types of digesta and that one of these had a composition similar to that of cecal contents. Administration of fluid and particulate markers to the Alaskan rock ptarmigan (*Lagopus mutus*) resulted in a prolonged accumulation of fluid marker in the ceca and rapid excretion of particles in the feces (Gasaway, Holleman, and White 1975). However, "cecal defecation" occurred at intervals of approximately 8.6 hours, with an average discharge of 56% of the cecal contents. Therefore, digesta transit may be more complicated in the bird than previously considered.

The times required for passage of food through the digestive tract of various reptiles are listed in Table 4.2. Guard (1980) measured transit time for digesta markers in two carnivorous and two herbivorous reptiles (Table 4.3). Fluid marker passed more quickly than particles through the gut of all four species. Radiological examination of the caiman and terrapin showed the stomach to be the major site of particle retention. However, the cecum and proximal colon also were major sites for the retention of markers by the herbivorous lizard and tortoise. Karasov, Petrossian, Rosenberg, and Diamond (1986) listed mean retention times obtained from studies of 11 species of reptile, but the variety of markers used in those studies make comparisons difficult. Interspecies comparisons are further complicated in poikilothermic animals by variation in ambient temperatures. This also limits the comparison of these results to those obtained with birds and mammals. Little comparable information appears to be available on the rates of digesta transit through the gastrointestinal tract of amphibians and fish.

Selective retention of digesta fluid or particles

Fluid appears to leave the stomach of all animals more rapidly than particulate matter of any size. This is especially true for the forestomach of sheep and cattle, which have received a great deal of study with respect to their mechanism for the release of fluid and particulate digesta (Hyden 1961; Balch and Campling 1965; Grovum and Williams 1973; Faichney 1975; and Van Soest 1982). Because of the ventral location of the reticulo-omasal orifice and the filtering arrangement of the omasum, particulate digesta can-

TABLE 4.2 Time required for food passage through the digestive tract

Species	Ambient temperature (°C)	Food passage time[b]
Tortoise, Geochelone elephantopus	—[a]	7–20
Lizards		
Lacerta muralis	—	33–40 hr
Lacerta sicula	—	33–40 hr
Lacerta viridis	—	32–45 hr
Snakes		
Eunectes murinus	—	35
Python molurus	—	38
Coluber constrictor	—	4
Diadophis punctatus	35	1.5
	33	2–3
	25	3–5
	18–20	7–8
	15	14
Elaphe schrenckii		5–8
Lampropeltis getulus	23	7
Nerodia sipedon	21	3.5
Natrix natrix	25	6–7
	30	4–4
Thamnophis sauritus	21–26	4–5
Vipera berus	21	7
Agkistrodon piscivorus	—	3–12
Crotalus mitchellii	—	5–6
Crotalus tortugensis	—	9

[a]Data not available.   [b]In days, except where hours are shown.
Source: Skoczylas (1978).

TABLE 4.3 Mean transit time (hours) for digesta markers through the entire gastrointestinal tract

Species	Liquid marker	Particulate markers		
		2 mm long	5 mm long	10 mm long
Caiman crocodilus	41	162	162	162
Chrysemys picta belli	35	56	57	60
Geochelone carbonaria	<48	270	285	363
Iguana iguana	<48	207	221	386

Note: Values represent two trials on four specimens of each species except C. crocodilus (three specimens). Liquid marker was polyethylene glycol or $BaSO_4$. Particulate markers were cut from 2.2-mm-diameter polyethylene tubing.
Source: Guard (1980).

not escape the reticulorumen until it is reduced to a given size and gains a specific gravity that is neither too light nor too heavy to allow its release from the reticulum. King and Moore (1957) found that plastic particles with a volume of 20-30 mm3 and a specific gravity of approximately 1.2 showed the highest rate of transport out of the reticulorumen of cattle. However, the rate at which both fluid and particles leave the reticulorumen is affected markedly also by the diet and feeding intervals.

The forestomach of macropod marsupials is the major site for digesta retention in these animals. However, fluid and particles pass more rapidly through the forestomach of kangaroos than through that of sheep and cattle, and with a greater degree of separation in the transit of fluid versus particles (Fig. 4.13). This apparently is due to a lesser degree of mixing in the kangaroo forestomach. The longest digesta transit times have been recorded in sloths. Honigmann (1936) reported a minimum transit time of 96 hours and maximum time of 1128 hours for the two-toed sloth, and Montgomery and Sunquist (1978) recorded 5% excretion of glass beads 3 mm in diameter in 60 hours and 95% excretion in 50 days. Much of the delay was due to retention of fecal pellets in the rectum, for up to 7 or 8 days, but the major site of retention was in the stomach (see Fig. 3.12).

Studies of coprophagy in the rabbit provide some insight relative to mechanisms by which the large intestine can selectively retain digesta. In 1882, Morot, a French veterinarian, reported that the domestic rabbit excreted hard, dry, fecal pellets during the day and soft, fluid, fecal pellets at night. Only the soft feces, which tend to attach to the surface around the anus, were reingested. Southern (1940, 1942) later demonstrated that wild rabbits produce their soft feces during the day, after the return to their burrow from early morning foraging, suggesting that this diurnal variation in fecal composition was related to feeding habits. Morot had concluded that the hard feces represented material that had passed through the digestive tract twice, but Madsen (1939) concluded that the difference between hard and soft feces was due to a difference in rhythmic activity of the intestine.

Eden (1940) analyzed the two types of fecal pellets and found that the soft feces contained approximately 30% protein and 15% crude fiber, as opposed to 10% protein and 30% fiber in the hard fecal pellets. From this finding, plus analysis of ash content and the relative digestibility of protein in soft feces and cecal contents, he concluded that the soft feces had a composition quite similar to that of cecal digesta. This suggested that the differences in fecal composition might be explained by a direct passage of some digesta from the ileum into the colon to form the dry pellets, and a periodic, massive evacuation of cecal contents to form the softer version. Thacker and Brandt (1955) reached a similar conclusion, postulating that the soft feces were produced by contraction of the spiral musculature of the cecum and rapid passage through the colon. However, Yoshida, Pleasants, Reddy, and Wostmann

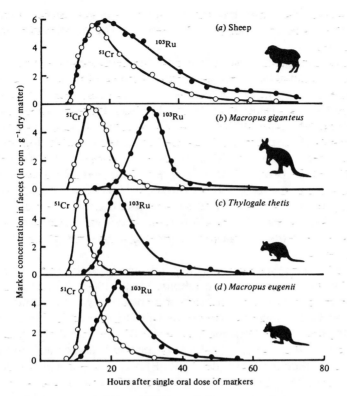

FIGURE 4.13 The pattern of appearance of the fluid marker [^{51}Cr]EDTA and the particulate marker [^{103}Ru]phenanthroline in the feces after a single oral dose in sheep, Western gray kangaroo (*Macropus giganteus*), red-necked pademelon (*Thylogale thetis*), and tammar wallaby (*Macropus eugenii*) fed chopped lucern hay ad libitum. Note the faster elimination of both markers in the macropodine marsupials, and the greater separation of the two markers in the macropodines than in the sheep. (From Hume 1982a; after Dellow 1979.)

(1968) proposed that the lower protein content of hard feces was due to absorption of protein, or the by-products of its breakdown, during passage through the colon.

Subsequent studies (Sperber 1968; Pickard and Stevens 1972; Björnhag 1972) showed that the fluid and fine particles were separated from larger particles of digesta in the proximal colon of the rabbit, and that this was accomplished by mechanical expression rather than by the absorption of fluid. The fluid and fine particles, including microorganisms, were retained in the proximal colon and cecum by cyclic retropropulsive digesta flow, and the larger particles were evacuated more rapidly as hard feces. Björnhag (1981) observed that the retrograde transit of digesta occurred along the wall of the haustra, leaving the lumen free for aboral transit of gas and other contents, and that this process was periodically interrupted by the formation and transit of soft feces.

Fioramonti and Ruckebusch (1976) noted a decrease in the motility of the cecum and proximal colon and an increase in the motility of the distal colon during the production of soft pellets, and that these passed through the colon 1.5-2.5 times faster than the hard pellets. Both the electrical (Ruckebusch and Hörnicke 1977) and motor (Ehrlein, Reich, and Schwinger 1983) activity of the proximal, haustrated colon increased during the formation of hard feces. Formation of soft feces was accompanied by a decrease in the motility of the proximal colon and cecal base, and increased motility of the colon distal to the fusus coli pacemaker area. Subsequent studies (Pairet, Bouyssou, and Ruckebusch 1986) showed that the formation of soft feces was accompanied by a 30% reduction in the myoelectric activity of the distal colon and indicated that endogenous prostaglandins may play a role in soft feces production.

The ability to retain selectively fluid and small particles in the cecum and proximal colon is not limited to rabbits. Small (300-500 mm) plastic particles administered into the stomach of guinea pigs were accumulated and retained by their cecum in a manner similar to that noted for fine particles in the rabbit (Jilge 1980). Cork and Warner (1983) measured digesta transit in the koala with the use of ^{51}CrEDTA as the fluid marker and ruthenium-labeled tris-ruthenium chloride (^{103}Ru-P) as an adsorbant to particles. Digesta transit was measured by collection of feces in one series of experiments, and by analysis of the contents of gut segments in a separate series. Fluid marker was retained longer than particulate markers, and both were retained longer than that reported for animals other than the sloth. They concluded that the principal site of digesta retention was the cecum/proximal colon, which appeared to act as a single mixing compartment, and that the ^{103}Ru-P in this compartment was probably associated with small particles of digesta, because of the tendency of this marker to transfer from labeled to unlabeled particles (Faichney and Griffiths 1978). Cork and Warner pointed out the similarities between the koala and rabbit with respect to sites of retention and separation of fluid from large particles, in spite of the fact that the koala lacks haustra in this area of the large intestine and does not practice coprophagy. The brushtail possum, *Trichosurus vulpecula*, also has demonstrated prolonged mean retention times of 64 hours for fluid marker and 71 hours for ^{103}Ru-P (Wellard and Hume 1981). However, there was no significant difference between the transit times of these two markers, and part of the delay was due to prolonged retention in the rectum.

A mechanism for the selective retention of small food particles and microorganisms would be a valuable asset to small herbivores in which the cecum

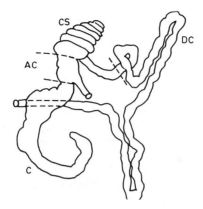

FIGURE 4.14 Cecum and colon of Scandinavian lemming. Ampulla coli (AC); cecum (C); colonic spiral (CS); distal colon (DC). (From Sperber et al. 1983.)

is the major site of digesta retention. However, separation of digesta fluid and particles has been proposed also for the equine large intestine. The right dorsal colon of donkeys and ponies was found to contain higher concentrations of nitrogen and small particles, and a higher water content than the distal (small) colon, suggesting preferential retention of small nitrogen-rich particles, especially microorganisms, and fluid by the dorsal colon (Björnhag, Sperber, and Holtenius 1984).

The proximal colon of the lemming has been shown to have a highly developed mechanism for separation of fluid and small particles, particularly microorganisms, from the larger particles of digesta (Sperber 1968; Sperber, Björnhag, and Ridderstrale 1983). The proximal colon of these animals can be divided into an ampulla coli (a short, wide segment extending from the entry of the ileum to the base of the colonic spiral), an inner spiral of about four windings around an axial structure, and an outer spiral that runs down to the base and leaves it to join the distal colon (Fig. 4.14). The mucosal surface of the proximal colon has a complicated pattern of folds, which permanently project into the lumen (Fig. 4.15). The largest of these starts at the ampulla coli and runs as a longitudinal fold for about two windings of the inner spiral. The free edge of this fold comes close to the axial attachment of the wall, thus dividing the lumen of the inner spiral into a narrow channel and main channel. The narrow channel opens into the ampulla coli by a narrow slit and extends, at its distal end, as a longitudinal groove connecting to oblique furrows in the colonic wall. The contents of the cecum, ampulla coli, and main channel of the inner spiral consist of a similar mixture of food residues and bacteria. However, the narrow channel contains dense masses of bacteria and mucus with little food residue in evidence. The main source

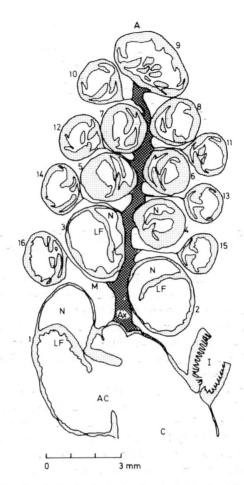

FIGURE 4.15 Section through the colonic spiral. Composite from sections obtained from a 21-day-old lemming. Apex (A); ampulla coli (AC); axial structure (Ax); cecum (C); ileum (I); longitudinal fold (LF); mesentery (M); narrow channel (N). Numbers 1-8 refer to sections through inner spiral; 10-16, to sections through outer spiral. (From Sperber et al. 1983.)

of the mucus appears to be a thickened area of inner spiral mucosa, distal to the narrow channel. It was concluded that cecal digesta are mixed with mucus secreted by the middle part of the inner spiral, resulting in an aggregation of bacteria with mucus, which is transported by antiperistaltic contractions into the narrow channel and back to the ampulla and cecum. This mechanism causes microorganisms to be retained in the cecum and proximal colons, and would explain the low microbial counts and nitrogen concentration of the feces produced by these animals.

A somewhat similar mechanism for the separation of microorganisms from fluid and particles was demonstrated in the colon of guinea pigs and chinchilla (Holtenius and Björnhag 1985). The proximal colon contains a furrow, formed by two mucosal folds about 20 cm long in the guinea pig and twice this length in the chinchilla. Contents of the furrow contained significantly higher concentrations of nitrogen and adenosine triphosphate (ATP), indicating that most of the nitrogen was present in microorganisms. Labeled bacteria, infused into the proximal colon, were transported into the cecum, and their concentration in the most proximal segment of the furrow was twice that measured in the lumen.

Sperber et al. (1983) also examined the rat for the presence of mechanisms similar to those noted in the lemming. The ampulla coli of the rat is small and difficult to distinguish. However, the main part of the proximal colon contained two rows of oblique folds, with a mucosa that was thick and composed of long, tubular glands in the luminal half of the fold. Mucus-containing masses of long, fusiform bacteria were often found close to the mucosa and between the folds. Rats fed on a fiber-rich diet produced two types of fecal pellets. One of these had a composition similar to that of cecal contents, whereas the other had a much lower protein and nitrogen content. These investigators concluded that the mucus-bacteria mixture was normally returned to the cecum, and cited evidence (see Chapter 6) of abnormal accumulation of mucus in the cecum of germ-free rats. Periodic aboral transit of the mucus-bacteria mixture would account for the production of the protein-rich pellets. Björnhag (1987) concluded that animals less than 1 kg in body weight such as the vole, lemming, rat, guinea pig, and chinchilla adopted a colonic separation mechanism that separates bacteria from the remainder of the digesta, in contrast to that of the lagomorph, marsupial, and equine species, which selectively retains fluid, small particles, and bacteria.

The evidence mentioned earlier for cecal retention of fluid digesta by birds suggests that the avian hindgut also contains a mechanism for the selective retention of fluid and small particles. However, from studies of turkeys, geese, and guinea fowl, Björnhag and Sperber (1977) concluded that such retention could be explained by the retrograde propulsion of urine.

Summary

The sequential and episodic events of digestion and absorption are highly dependent upon the motor activities of the digestive tract. Food is rapidly transported by the esophagus from the mouth to the stomach, present in most vertebrates, where it is retained for the processes of gastric digestion. As food particles are reduced in size, they are released into the midgut. This reduction in particle size is aided by regurgitation of less digestible components of prey in raptors and by regurgitation and remastication of plant fiber in ruminants. Retention of particulate digesta in the stomach varies from a few hours, in most species, to extremely long periods of time in carnivores that ingest their prey in one piece, and in herbivores with a complex voluminous stomach. The transit of digesta through the midgut of a species can vary with prevalence of different types of motor activity, structural variations such as the spiral valve and pyloric ceca of some fish, and the relative length of this segment of the gut. Digesta transit is delayed once more in the hindgut of most vertebrates by a variety of mechanisms.

Retention of digesta in the hindgut of reptiles and birds can be aided by antiperistaltic contractions, which originate at or near the cloaca and can reflux urinary excretions and digesta the length of the colon and into the cecum (if present). The advent of a cecum, or ceca, which provided an additional site for digesta retention, also required a mechanism for their reflux filling. An intervening sphincter or valve prevented the further reflux of hindgut contents into the midgut, where they could interfere with its digestive and absorptive functions. Separation of the exits from the urinary and digestive tracts in most mammals removed the opportunity for the hindgut to serve as a site for the recovery of urinary electrolytes. This was associated with a lengthening of the colon and displacement of the pacemaker that initiates retropulsive contractions to a more proximal segment of the colon. Delay in digesta transit increases the opportunity for the multiplication of indigenous microorganisms and, as we will see in Chapters 6 and 7, introduces other problems with respect to the buffering of their end products and retention of water within the lumen. Therefore, specialization of the cecum and proximal colon for retention of fluid and small particles, and the distal colon for recovery of electrolytes and water, would be advantageous to terrestrial vertebrates and, especially, the herbivorous species.

5

Carbohydrate, fat, and protein digestion by endogenous enzymes, and absorption of end products

The principal organic components of plants and animals are carbohydrates, lipids, proteins, and nucleic acids. Most of these are originally produced in plants by photosynthesis and obtained by animals through consumption of plants or other animals. Although some organic compounds are ingested in a form that can be readily absorbed, most of them require conversion into a limited number of simpler compounds prior to their absorption. This is accomplished by enzymes produced by the digestive system of the host animal or by microbes indigenous to its digestive tract. This chapter will concentrate on the contributions of enzymes generated by the digestive system of vertebrates.

Table 5.1 lists the principal forms of organic compounds found in the diet of vertebrates, the major endogenous enzymes that act on them, and the end products. All of the enzymes are proteins that catalyze the hydrolysis of their substrates. This often involves the sequential hydrolysis of large molecules by a series of digestive enzymes. The end products consist of a limited number of monosaccharides, fatty acids, alcohols, peptides, amino acids, and other compounds suitable for assimilation. This list does not include all of the enzymes that may be present, and some of those that are listed are absent in many species. Furthermore, the complement and concentrations of digestive enzymes can change markedly during ontogenesis, especially in animals that undergo metamorphosis from larval to adult forms and in those such as the mammals that demonstrate significant changes in their diet during early development.

The demonstration of endogenous digestive enzymes is complicated by the presence of enzymes in the diet, desquamated cells, and indigenous microbes. Even the isolation of enzymes from homogenates of tissue from the digestive tract or its accessory glands does not prove that they are involved in normal digestive processes. For example, some disaccharidases and peptidases found in the lumen membrane or cytosol of intestinal cells hydrolyze substrate absorbed from the lumen, but many intracellular enzymes are involved only with cellular metabolism. The source of a digestive enzyme can be equally difficult to determine, as its presence in the lumen at a given site along the tract does not necessarily indicate the site of release.

125

TABLE 5.1 Principal food components and endogenous digestive enzymes of vertebrates

Substrate	Extracellular enzymes	Intestinal mucosal enzymes	End products
Carbohydrates			
Amylose			
Amylopectin	α-Amylase		
Glycogen	→ Maltose, Isomaltose, α-1,4 Oligosaccharides →	Maltase / Isomaltase →	Glucose
Chitin	Chitinase → Chitobiose →	Chitobiase →	Glucosamine
Sucrose		Sucrase →	Glucose / Fructose
Lactose		Lactase →	Glucose / Galactose
Trehalose		Trehalase →	Glucose
Lipids			
Triglycerides	Lipase / Colipase →		β-Monoglyceride / 2 Fatty acids
Phospholipids	Phospholipase →	Phosphatase →	Alcohol / Fatty acids / Phosphate
Cholesterol esters	Cholesterol esterase →		Cholesterol / Fatty acid

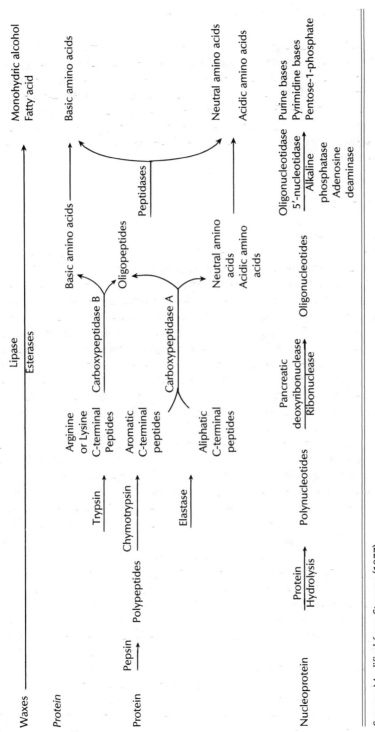

Source: Modified from Stevens (1977).

These problems help explain why the presence or source of many digestive enzymes has been clearly demonstrated in only a relatively few species.

This chapter describes the principal organic compounds found in the diet of vertebrates and how they are digested and absorbed by various species. The text of Vonk and Western (1984) serves as a major source for comparative information on enzymatic digestion in vertebrates. It also provides additional information on invertebrates and the structural characteristics of individual enzymes.

Carbohydrates

Components of food

Carbohydrates consist of individual monosaccharides or monosaccharides linked by glycosidic bonds to form chains that are two (disaccharides), three to ten (oligosaccharides), or more (polysaccharides) units in length. The chemical structures for monosaccharides commonly found in plants and animals are given in Figure 5.1. Only glucose and fructose are found as free monosaccharides in any important quantities in plants, and glucose is the only free monosaccharide present to any degree in animals. The disaccharide most commonly found in plants is sucrose (glucose-fructose), which serves as the principal form by which carbohydrate is transported. The major disaccharides found in animals are trehalose (glucose-glucose), the blood transport carbohydrate of insects; and lactose (glucose-galactose), the carbohydrate found in mammalian milk.

Most of the carbohydrate in plants and animals is present in the form of polysaccharides that serve either for the structural support of cell walls or for reserve storage. Figure 5.2 shows the monosaccharides and linkages for a number of common structural polysaccharides. The principal structural carbohydrates of terrestrial plants are cellulose, hemicellulose, and pectins (Van Soest 1982). Cellulose is the major constituent of cell walls in higher plants, providing 20-40% of the dry matter. It consists of straight chains of glucose with a β-1,4 linkage. Hemicellulose is a mixture of polysaccharides of varying composition. It can include xylose, galactose, arabinose, mannose, and other monosaccharides (Kronfeld and Van Soest 1976), but often with a common main core consisting of a xylose chain with a β-1,4 linkage. Pectin, the cementing substance of plant cell walls, consists primarily of an α-1,4-linked galacturonic acid chain. The cell walls of algae also can contain hemicellulose and pectin along with other structural carbohydrates (Vonk and Western 1984). Chitin is a structural polysaccharide found in the cell walls of bacteria, fungi, and many invertebrates (Table 5.2). It consists of an unbranched polymer of β-1,4-linked N-acetyl-D-glucosamine.

The major storage reserve carbohydrates of terrestrial plants are amylose and amylopectin, the two polysaccharides that make up plant starch.

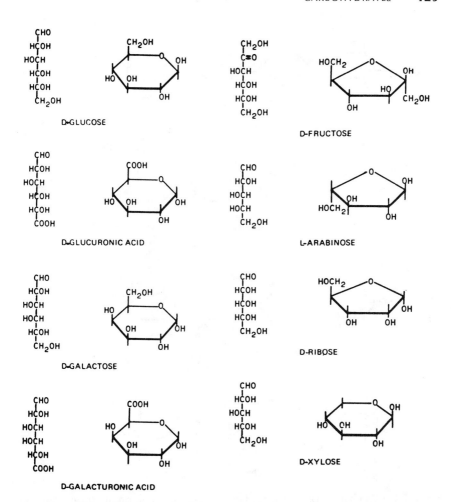

FIGURE 5.1 Structures of the more common sugars in forages. Open-chain formulae and the corresponding ring forms are shown. Ring closure results in the formation of a new asymmetric center on carbon-1 (carbon-2 in the case of fructose), which may be α (up) or β (down). Open-chain formulae number carbons consecutively from the top. Ring formulae are numbered clockwise, beginning with the extreme right (see Figure 5.3). Configurations shown are those most commonly found in glycosidic linkage. However, variation occurs in the forms of glucose and galactose. Only glucose and fructose are encountered in any important amount in the free form. Quantitatively less important sugars (not shown in the figure) include mannose, rhamnose, and 2-deoxyribose, which occur in hemicellulose, pectins, and deoxyribonucleic acid, respectively. A variety of other sugars occur in small amounts. The projection of the ring forms shown here is illustrated according to the Haworth convention, which does not reveal actual ring geometry. (From Van Soest 1982.)

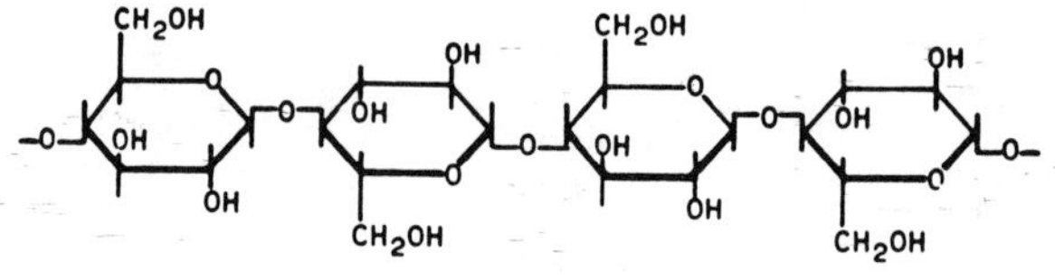

CELLULOSE

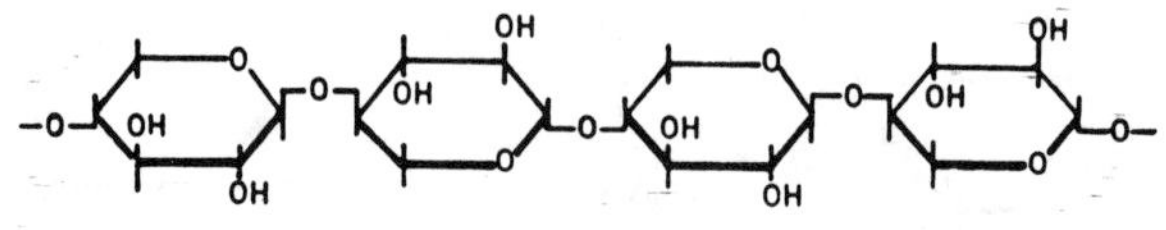

XYLAN (HEMICELLULOSE)

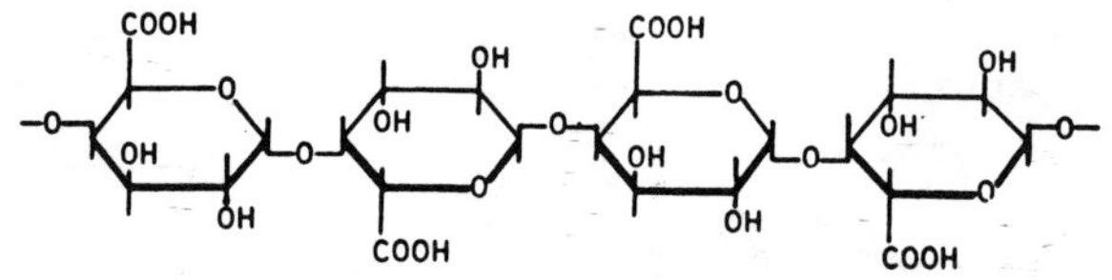

POLYGALACTURONIC ACID (PECTIN)

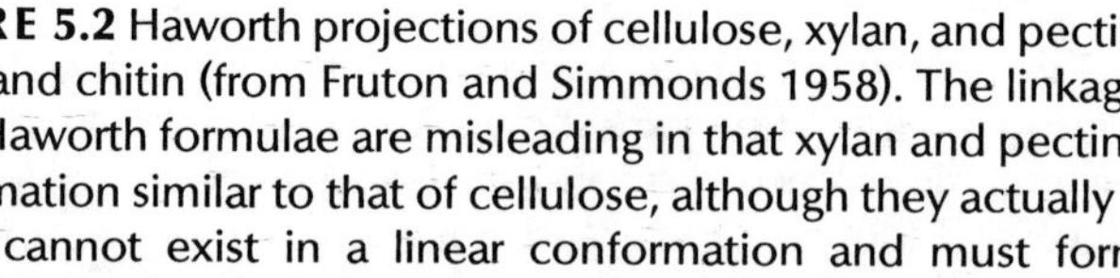

Chitin

FIGURE 5.2 Haworth projections of cellulose, xylan, and pectin (from Van Soest 1982); and chitin (from Fruton and Simmonds 1958). The linkage in pectic acid is α-1,4. Haworth formulae are misleading in that xylan and pectin appear to have a conformation similar to that of cellulose, although they actually differ. Pectin, like starch, cannot exist in a linear conformation and must form kinks or coils. However, the axial position of carbon-4 in galacturonic acid results in a different configuration, as compared to starch.

Amylose consists almost entirely of α-1,4-linked glucose units. Amylopectin consists of the same chains with branches, formed by α-1,6 linkages, usually at intervals of 24 glucose units. Vonk and Western (1984) list other storage carbohydrates found in algae such as Floridean starch, which is similar to amylopectin, and laminaran, a polymer of β-1,3-linked glucose molecules. Glycogen, the principal storage reserve carbohydrate of animals, also is

TABLE 5.2 The distribution of chitin in living organisms

Phylum	Class	Structure
Fungi	Chytriodiomycetes	Cell walls
	Zygomycetes	Cell walls
	Ascomycetes	Cell walls of filamentous types
		Cell wall component of yeasts
	Basidiomycetes	Cell walls
Algae	Bacillariophyceae	Chitin fibrils of centric diatoms
Coelenterata	Hydrozoa	Perisarc of polyps, pneumatophore of some siphonophores
Aschelminthes	Nematoda	Egg shells
Phoronida		Tubes
Bryozoa		Exoskeleton, operculum
Brachiopoda		Pedicle, shell
Mollusca	Gastropoda	Radula, jaws
	Cephalopoda	Shell, dorsal shield
Annelida	Polychaeta	Jaws, chaetae, gut lining
	Oligochaeta	Chaetae
Arthropoda		Exoskeleton, lining of foregut and hindgut, apodemes, egg shells
Chaetognatha		Hooks
Pogonophora		Tubes, setae

Source: Vonk and Western (1984).

similar to amylopectin in structure, except for shorter chains (10-14 units) between α-1,6 branches of the glucose chain. It can comprise up to 2% of the fresh weight of muscle and 10% of the fresh weight of liver (Lehninger 1975).

Digestion

Structural polysaccharides
Although endogenous enzymes of vertebrates can hydrolyze the α-1,4 and α-1,6 linkages in starch and glycogen, they cannot hydrolyze the β-1,4 linkages of cellulose and hemicellulose, or the α-1,4 linkages in pectins and galactans (Van Soest 1982). Failure to hydrolyze these linkages is due partially to an inability to attack β-1,4 linkages. However, because of its axial position in the molecule the α-1,4 linkage of pectin is highly resistant also to the attack of α-amylase (Van Soest 1982). Microbial fermentation of these carbohydrates will be discussed in the following chapter.

Chitin is digested by a wide range of vertebrates. Chitinase is an intracellular enzyme in many microorganisms. Therefore, its presence in the digestive tract of vertebrates does not necessarily indicate endogenous enzyme production. However, Jeuniaux and co-workers have demonstrated chitinase in the gastric mucosa and pancreas of many mammals, birds, reptiles, and adult amphibians (Table 5.3). It also was found in the pancreas of the mole, hedgehog, pig, starling, blackbird, amphibians, and most of the reptiles examined. Chitinase was absent from the digestive tract of teleosts that lacked a stomach and pyloric ceca, but it has been demonstrated in the gastric mucosa of many telosts (Jeuniaux 1961, 1963; Dandrifosse 1963; Okutani 1966; Colin 1972; Micha, Dandrifosse, and Jeuniaux 1973b; Fänge, Lundblad, and Lind 1976; Fänge, Lundblad, Lind, and Slettengren 1979). It also was found in homogenates of their pyloric ceca, probably in association with disseminated pancreatic tissue. Chitinase was present in the gastric mucosa of sharks (*Squalus* and *Scyliorhinus*) and rays (*Raja* spp.), but none was found in their pancreas (Micha et al. 1973b). Information on chimaeras and cyclostomes appears to be limited to the demonstration of pancreatic chitinase in chimaeras and low levels of intestinal chitinase activity in hagfish (Fänge et al. 1976, 1979).

Storage polysaccharides
Amylose, amylopectin, and glycogen are hydrolyzed by α-amylase to form maltose, isomaltose, maltotriose, and other α-1,4 oligosaccharides (Fig. 5.3). This enzyme is secreted by the pancreas or pancreatic tissue of all vertebrates and by the salivary glands of echidna, possum (*Trichosurus*), dog, rats, mice, guinea pig, squirrel, vole, grasshopper mouse (*Onychomys*), rabbit, artiodactyls (pigs, cattle, sheep, goats, deer), and primates (Vonk and Western 1984). Pancreatic amylase is evident in prenatal humans, horses, cattle, and sheep (Koldovský 1970). It increases postnatally, most rapidly after weaning.

Salivary amylase has been reported in the chicken, turkey, goose, and pigeon, but there does not appear to be conclusive evidence for this (Ziswiler and Farner 1972). Amylase activity also has been demonstrated in the crop (Bolton 1962, 1965), gizzard (Hewitt and Schelkopf 1955), and bile (Farner 1943) of chickens, but its origin is uncertain. High levels of amylase were found in the stomachs of the turtles *Chelonia* and *Dermochelys* by Tanaka (1942), and in the stomach of a snake (*Agkistrodon*) by McGeachin and Bryan (1964), but these may represent regurgitation of pancreatic secretion. The pancreas appears to be the main source of amylase in amphibians. The same holds true for fish, although the diffuse distribution of pancreatic tissue in many species makes the site of amylase secretion difficult to trace.

Pancreatic and salivary enzymes are α-amylases, and are activated by chloride. Molecular weights and activation energies are known for some of these, and the amino acid composition has been determined for those of the chicken and a few mammals (Vonk and Western 1984).

An intestinal γ-amylase, which does not require chloride activation and produces a stepwise release of terminal glucose from starch, glycogen, and

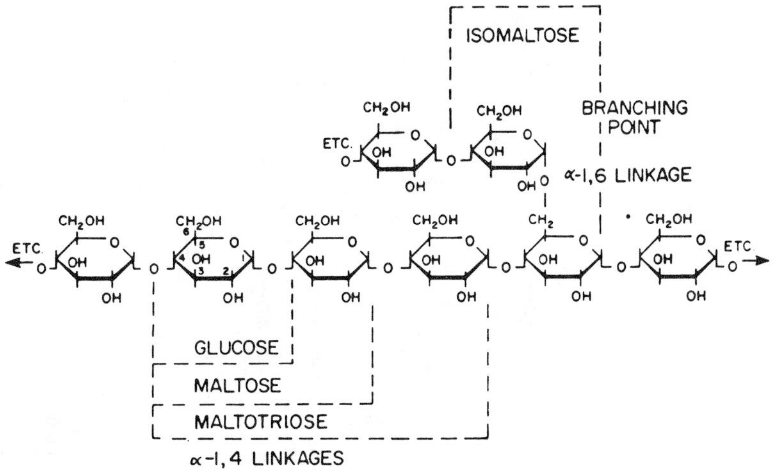

FIGURE 5.3 The structure of starch. Hydrolysis catalyzed by pancreatic amylase occurs at the α-1,4 linkage, and the products of hydrolysis are straight-chain oligosaccharides. Because pancreatic amylase does not catalyze hydrolysis of the α-1,6 linkages, isomaltose is also a product of hydrolysis. Further hydrolysis is catalyzed by the maltases and the isomaltase of the brush border of intestinal epithelial cells. (From Davenport 1982.)

oligosaccharides, also has been described (Alpers and Solin 1970; Dahlqvist and Thompson 1963; Eggermont 1969). Vonk and Western (1984), however, attributed this to a lysosomal enzyme.

Disaccharides
Disaccharides, with the exception of chitobiose, are hydrolyzed to monosaccharides by a set of enzymes located in the brush border of intestinal cells. These can be divided into seven enzymes or enzyme groups: maltase, isomaltase (oligo-1,6-α-glucosidase), sucrase (invertase), trehalase, lactase, cellobiase, and chitobiase. There may be more than one enzyme per group, and enzymes from different groups may act on the same substrate.

Tables 5.4 and 5.5 list intestinal disaccharidases found in mammals. Maltase, isomaltase, and sucrase have been found in most mammals that have been examined, although sucrase was absent from the intestine of the echidna, short-nosed bandicoot, kangaroo, pinnipeds, cattle, and sheep. When present, sucrase levels tended to be lower than those of the other two enzymes. Trehalase, which would be most useful to insectivores, was found in many of the mammals examined. However, it was absent from the domestic cat, panther, and pinnipeds, and only traces were found in the koala intestine.

TABLE 5.3 Chitinase activity in gastric mucosa and pancreas of birds, reptiles, and amphibians

Species	Gastric mucosa	Pancreas	Reference
MAMMALS			
Bat (Rhinolophus ferrumequinum)	+	0	Jeuniaux (1962a)
Mole (Talpa europaea)	+	+	Jeuniaux (1962a), Jeuniaux and Cornelius (1978)
Hedgehog (Erinaceus europaeus)	+	+	Jeuniaux (1962a)
Dog (Canis familiaris)	+	0	Cornelius, Dandrifosse, and Jeuniaux (1975)
Fox (Vulpes vulpes)	+	0	Cornelius, Dandrifosse, and Jeuniaux (1975)
Marten (Martes)	0	0	Cornelius, Dandrifosse, and Jeuniaux (1975)
Cat (Felis)	0	0	Cornelius, Dandrifosse, and Jeuniaux (1975)
Sloth (Choloepus)	0	0	Jeuniaux (1962b)
Mouse (Mus musculus)	+	-	Frankignoul and Jeuniaux (1965)
Rat (Rattus norvegicus)	+	0	Frankignoul and Jeuniaux (1965)
Hamster (Cricetus cricetus)	+	0	Frankignoul and Jeuniaux (1965)
Prosimian (Perodicticus potto)	+	0	Beerten-Joly, Piavaux, and Goffart (1974)
Monkey (Cebus capucinus)	+	0	Jeuniaux and Cornelius (1978)
Human (Homo sapiens)	0	0	Jeuniaux and Cornelius (1978)
Pig (Sus scrofa)	+	+	Jeuniaux (1962a)
Sheep (Ovis aries)	0	0	Jeuniaux (1962a)
Rabbit (Oryctolagus)	0	0	Jeuniaux (1962a)
BIRDS			
Sparrow (Passer domesticus)	+	0	Jeuniaux (1962a); Jeuniaux and Cornelius (1978)
Robin (Erithacus rubecula)	+	0	Jeuniaux (1962a); Jeuniaux and Cornelius (1978)

Japanese nightingale (*Liothrix lutea*)	+	0	Jeuniaux and Cornelius (1978)
Starling (*Sturnus vulgaris*)	+	+	Jeuniaux and Cornelius (1978)
Blackbird (*Turdus merula*)	+	+	Jeuniaux (1962a)
Carrion crow (*Corvus corone*)	+	0	Jeuniaux and Cornelius (1978)
Chicken (*Gallus gallus*)	+	0	Jeuniaux (1962a)
Barn owl (*Tyto alba*)	+	0	Jeuniaux and Cornelius (1978)
Pigeon (*Columba palumbis*)	0	0	Jeuniaux (1962a)
Parrot (*Psittacus erithacus*)	0	0	Jeuniaux and Cornelius (1978)
REPTILES			
Terrapin (*Emys*)	+	+	Jeuniaux (1962a)
(*Clemmys*)	+	+	Jeuniaux (1962a)
Tortoise (*Testudo hermanni*)	+	+	Jeuniaux (1962a)
Lizard (*Lacerta*)	+	+	Jeuniaux (1962a)
(*Uromastyx*)	+	+	Jeuniaux (1962a)
(*Anolis*)	+	+	Jeuniaux (1962a)
(*Chamaelo*)	+	+	Jeuniaux (1962a)
(*Augus*)	+	0	Jeuniaux (1962a)
Snake (*Natrix natrix*)	0	0	Micha, Dandrifosse, and Jeuniaux (1973a)
AMPHIBIANS			
Adult frog (*Rana temporaria*)	+	+	Jeuniaux (1963)
Adult toad (*Bufo marinus*)	+	+	Micha, Dandrifosse, and Jeuniaux (1973b)
Salamander (*Triturus alpestris*)	+	+	Micha, Dandrifosse, and Jeuniaux (1973b)
(*Salamandra salamandra*)	+	+	Micha, Dandrifosse, and Jeuniaux (1973b)

TABLE 5.4 Diet and intestinal disaccharidase activities of prototherian and methatherian mammals

	Food	Maltase	Isomaltase	Sucrase	Lactase	Trehalase	Cellobiase
PROTOTHERIA							
Tachyglossus aculeatus							
Echidna	Ants	5.5	4.4	0	0.07	2.6	0.02
METATHERIA							
Antechinus stuartii[a]							
Marsupial mouse	Insects	42.6	—	10.2	0.06	8.1	0.06
Dasyurus maculatus							
Tiger cat	Carnivorous	69.4	38.4	4.9	1.3	23.7	0.14
Perameles nasuta							
Long-nosed bandicoot	Insects	18.9	12.0	5.0	0.68	11.1	0.26
Isoodon obesulus							
Short-nosed bandicoot	Insects	—	—	0	—	4.0	0.22
Trichosurus vulpecula							
Brushtail possum	Mainly plants	41.2	22.9	6.8	0.47	7.2	0.20
Pseudocheirus peregrinus							
Ringtailed possum	Herbivorous	0.22	0.11	0.06	11.2	—	0.10
Phascolarctos cinereus							
Koala	*Eucalyptus* leaves	12.0	5.2	1.9	0.68	0.006	1.43
Macropus giganteus							
Gray kangaroo	Grass	0.3	0.12	0	0.05	0.12	0

Note: All data are taken from adult specimens. Data are expressed in units (1 μmol substrate/min) per gram of mucosa (net wt.).
[a]Activities are expressed as units per gram of intestine (net wt.).
Source: Modified from Kerry (1969) by Vonk and Western (1984).

Lactase and cellobiase activities have been reported in the intestine of many mammals. Unlike most other carbohydrases, the activity of these enzymes is highest at birth and decreases, or may disappear, in adult animals (Koldovský 1970). Lactase is absent or present in only trace amounts in the intestine of newborn pinnipeds, which receive little lactose in their milk. Its absence from the intestine of the adult echidna, short-nosed bandicoots, ringtail possum, and elephant may be due to loss after weaning. The presence and function of cellobiase are more difficult to understand. Cellobiase is a β-glucosidase that can hydrolyze the β-1,4 linkage of cellobiose, the end product of cellulose hydrolysis by C_1 and C_x cellulytic enzymes (Vonk and Western 1984). However, the latter are not endogenous enzymes of vertebrates, and microbial digestion of cellulose leads primarily to the release of fatty acids, CO_2, and CH_4 (Van Soest 1982). Vonk and Western (1984) point out, however, that lactase has been shown to have both β-galactosidase and β-glucosidase activity in the pig and calf, and some of the cellobiase may be of bacterial or lysosomal origin.

Activities of some of the above disaccharidases were examined in the intestinal mucosa of the European crane and Chinese quail (Zoppi and Schmerling 1969), chicken (Brown and Moog 1967), and five species of sea birds (Kerry 1969). Maltase, isomaltase, and sucrase levels in the crane and chicken were similar to those of mammals. The quail showed lower levels of these enzymes. The sea birds demonstrated the lowest levels of maltase and sucrase, similar to those of the quail. Low levels of lactase also were found, but trehalase activity was not clearly demonstrated in any of these birds.

Zoppi and Schmerling (1969) demonstrated the presence of maltase, isomaltase, sucrase, and trehalase in the intestine of the herbivorous tortoise *Testudo hermanni*, and Hourdry, Chabot, Menard, and Hugon (1979) found maltase and trehalase (no sucrase) in the intestinal brush border of the frog *Rana catesbeiana*. Intestinal homogenates were examined for disaccharidases in the teleosts *Salmo gairdneri* (Kitamikado and Tachino, 1960), *Oncorhynchus nerka* (Ushiyama, Fujimori, Shibata, and Yoshimura 1965), *Plecoglossus* and *Pagrus* (Kawai and Ikeda 1971), and *Cyprinus carpio* (Vonk and Western 1984). Maltase activity was found in the intestine of all of these species, and the carp intestine showed additional sucrase and cellobiase activities. Disaccharidase levels do not appear to have been examined in the intestine of the more primitive classes of fish.

Chitobiase attacks the end products of chitin digestion, releasing glucosamine. Chitobiase activity was demonstrated in gastric, pancreatic, and intestinal tissue of the insectivorous horseshoe bat, *Rhinolophus* (Jeuniaux 1962a), the gastric mucosa of a number of teleosts (Jeuniaux 1963; Fänge et al. 1979), the pancreas of chimaeras, and the intestinal mucosa of hagfish (Fänge et al. 1979). However, the chitobiase activity of gut contents was higher than that of tissues in these species, and chitobiase was present in the

TABLE 5.5 Distribution of intestinal disaccharidase activities in eutherian mammals

Order	Examples	Maltase	Isomaltase	Sucrase	Lactase	Trehalase	Cellobiase
Chiroptera	Eidolon helvum	+		+	−		
Primates	Tupaia, Nycticebus	+	+	+	+	+	+
	Perodicticus, Oedipomidas	+	+	+	+	+	+
	Tamarinus, Aotus, Saimiri	+	+	+	+	+	+
	Cebus, Macaca, Papio, Man	+	+	+	+	+	+
Lagomorpha	Oryctolagus	+	+	+	+	+	(+)
Rodentia	Sciurus	+		+	(+)	+	(+)
	Cricetus cricetus	+	+	+	+	+	
	Rat	+		+	(+)	+	(+)
	Cavia porcellus	+		+	(+)	+	(+)
Carnivora	Dog	+		+	+	+	+
	Ursus maritimus	+		+	Trace	Trace	
	Mustela	+		+	+	+	
	Cat	+		+		0	
	Panthera leo	+		+		0	

Pinnepedia					
Zalophus calfornianus[a]	+	0	0	0	0
Eumetopias jubatus[a]	+	0	0	0	0
Arctocephalus spp.	+	+	Trace	0	0
Odobenus r. divergens	+	+	Trace	Trace	
Mirounga leonina	+	+	Trace	0	
Phoca vitulina[a]	+	0	0	0	0
Perissodactyla					
Horse	+	+	+	+	0
Proboscidea					
Elephas maximus	+	+	0	+	+
Artiodactyla					
Tayassu tajacu	+	+	(+)	Trace	
Pig	+	+	+	+	+
Cow	+	(0)	+	+	+
Sheep	+	0	+	+	+
Goat	+	+	(+)	+	(+)

Note: Enzyme activity is designated + (= present), trace, or 0 (= absent). Results in brackets indicate that an alternative substrate was used. These include palatinose (for isomaltose), o-nitrophenyl-β-D-glucoside (For cellobiase), the two latter substrates revealing β-galactosidase and β-glucosidase activities, respectively.

[a]Suckling juveniles examined.

Source: Vonk and Western (1984), with addition of Perissodactyla data from Roberts (1975).

gut contents of many species that lack chitinase. These findings, plus the fact that this enzyme does not appear to be a component of the intestinal brush border, led Vonk and Western (1984) to conclude that it may be of dietary or microbial origin.

Absorption

Absorption of carbohydrate is limited to the active or passive transport of monosaccharides across intestinal mucosa. Glucose is actively transported (i.e., against its concentration gradient) into the intestinal cells, and a major fraction of this is linked to the absorption of sodium (see Chapter 7). This system transports D-glucose, D-galatose, and D-xylose. Fructose is absorbed by facilitated diffusion, and mannose, L-glucose, and other monosaccharides are absorbed more slowly by passive diffusion. These basic mechanisms of monosaccharide absorption seem to apply to all vertebrates. However, the intestine of the frog *Rana catesbeiana* does not exhibit active transport of D-xylose (Lawrence 1963), and Alliot (1967) reported that the intestine of the cat shark *Scyliorhinus* absorbed N-acetyl-D-glucosamine more rapidly than glucose.

Lipids

Components of food

Lipids are organic compounds that are insoluble in water and soluble in organic solvents such as ether and chloroform. They are called fats if solid and oils if liquid at normal environmental temperatures, and they serve as a major source of energy reserve in many plants and animals. Those of greatest interest as components of food are the triglycerides, phospholipids, glyco-lipids, and waxes. Each of these contains esters of alcohol and fatty acid. Triglycerides are neutral esters of glycerol and three fatty acids, and the major form in which lipid is stored by animals and the seeds of plants. Phospholipids consist of a monohydric alcohol, fatty acid, phosphoric acid, and a base such as choline. The phospholipids are major components of the outer cellular and cell organelle membranes and, therefore, an important fraction of the lipids present in many animals. Lecithin is the principal phospholipid of animals. Glycolipids are found mainly in photosynthetic tissue and provide much of the lipid content of pasture grasses and clover. They consist of amino alcohol, sugar (glucose, galactose), fatty acid, and sometimes other organic acids. Waxes are esters of complex, monohydric alcohols and a long-chain fatty acid. They are found in the cuticle of many plants and the hive of bees, but their greatest importance derives from their presence in marine invertebrates, fish, and the spermaceti of whales (Schmidt-Nielsen 1983).

Benson, Lee, and Nevenzel (1972) found that waxes were stored or utilized by all organisms examined that were taken from depths below 1000 m. Wax esters constituted 20% of the lipids found in copepods (small crustaceans) from cold, deep water (Lee et al. 1972) and Antarctic species of planktonic crustaceans at depths of 0-300 m (Bottino 1975). Because of the importance of planktonic crustaceans as a link in the marine food chain, Benson and Lee (1975) concluded that 50% of the organic material synthesized by phytoplankton is stored temporarily by marine animals as wax.

The fatty acids that have been identified in lipids have a terminal carboxyl group, and since they are derived from acetate, most of them have an even number of carbon atoms. Although up to 9% of those formed in mammalian milk have a length of only 4-14 carbons (Davenport 1982), the fatty acids in most lipids have 14-22 carbons (Vonk and Western 1984). A large percentage of those found in animals and higher plants are C_{16} and C_{18} fatty acids. Planktonic algae contain high percentages of these and also C_{20} and C_{22} acids.

Digestion

Lipids are hydrolyzed by lipase, which can act only at a lipid-water interface, and by a variety of esterases that can hydrolyze only those lipids that are in solution (Table 5.1). However, even the more water-soluble lipids form micelles as they approach saturation. Therefore, the same substrate can be attacked by lipases or esterases, depending on its concentration. The lipases described in most vertebrates hydrolyze triglycerides at their C_1 and C_3 ester bonds, releasing two fatty acids and β-monoglyceride, although a nonspecific lipase that also attacks the C_2 position has been described in rats (Mattson and Volpenheim 1968; Patton 1975), dogs, the Northern fur seal *Callorhinus ursinus,* and teleosts (Patton, Nevenzel, and Benson 1975), and the leopard shark *Triakis semifasciata* (Patton 1975). Lipases also hydrolyze monoesters, but at a much slower rate. Esterases hydrolyze monoesters such as lecithin and cholesterol ester to release lysolecithin, cholesterol, and their respective fatty acids.

Lipase hydrolyzes lipids that contain long-chain fatty acids more rapidly than do the esterases, whose specificity appears to be more dependent on the type of alcohol than the structure of the fatty acid. Although lipases have been found in salivary, pharyngeal, and gastric secretions of various species, those secreted by the pancreas are the most important to digestion. Esterases also are secreted by the pancreas and other tissues, but their origin and function are less clearly understood.

Much of the dietary lipid is in the form of fats or oils, which must be emulsified to provide the surface area required for efficient digestion. Biliary secretions provide the principal endogenous source of emulsifying agents. They contain the bile salts and phospholipids (principally lecithin) that aid in the emulsification and absorption of dietary fat, and also serve as a major route for excretion of cholesterol and the end products of hemoglobin catabolism.

Information on the structure and evolution of vertebrate bile salts has been reviewed by Haslewood (1964, 1967). All bile salts appear to be derived from cholesterol. They consist primarily of sulfated alcohols in fish and amphibians, and taurine or glycine conjugates of bile acids in vertebrates above this level.

$$R \cdot OH + HO \cdot SO_3^- \rightarrow R \cdot O \cdot SO_3^- + H_2O$$
bile alcohol $\qquad\qquad$ sulfate ester

$$R \cdot COOH + H_2N \cdot CH_2 \cdot CH_2 \cdot SO_3H \rightarrow R \cdot CO \cdot NH \cdot CH_2 \cdot CH_2 \cdot SO_3H + H_2O$$
bile acid $\qquad\qquad$ taurine $\qquad\qquad\qquad\qquad$ taurine conjugate

$$R \cdot COOH + H_2N \cdot CH_2 \cdot COOH \rightarrow R \cdot CO \cdot NH \cdot CH_2 \cdot COOH + H_2O$$
bile acid $\qquad\qquad$ glycine $\qquad\qquad\qquad$ glycine conjugate

Glycine conjugates appear to be restricted to eutherian mammals. Carnivores have taurine conjugates, herbivores have mostly glycine conjugates, and omnivores have a mixture of both. Bovidae, which have a mixture of the two conjugates, are an exception to this general rule. Because of the hydrophilic and lipophilic nature of different parts of their molecules, bile salts act as detergents for the emulsification of fat. Bile salts and lecithins are responsible for the formation of micelles, which contain the biliary lipids and bile salts that are released into the intestine.

Lipase activity is inhibited by bile salts, but this is counteracted by the presence of colipase, a small protein secreted by the pancreas. Once the emulsification process has been initiated, lysolecithin, released by hydrolysis of lecithin, and the end products of triglyceride hydrolysis also act as strong detergents. Micelles, with a diameter of 0.5-1.0 μm and a surface made up of bile salts, accumluate the long-chain fatty acids, monoglycerides, phospholipids, and fat soluble vitamins released by the processes of lipid digestion.

Although the diet of most adult vertebrates contains relatively low amounts of fat, mammalian milk can contain levels of from 1.9% in the horse (Hamosh 1979) to 49% in the Tasmanian fur seal, *Arctocephalus pusillus* (Kerry and Messer 1968). Lipase secreted by the lingual salivary glands has been shown to be important to the digestion of milk triglycerides in newborn humans, cattle, and rats (Hamosh 1979). Gastric lipases have been reported in a number of species, and the lipase found in the gastric contents of humans has been shown to differ from that of the pancreas, but a clear demonstration that lipase is secreted by the stomach, versus lingual or other sources, appears to be lacking (Vonk and Western 1984).

Pancreatic lipases appear to be the most important enzymes for the digestion of fat in most vertebrates that have been studied. They have been demonstrated in the pancreas of young chicks (Laws and Moore 1963), the western rattlesnake (Patton 1975), and the frog *Rana esculenta* (Scapin and Lambert-Gardini 1979), and, along with colipase, in the interpyloric cecal tissue of the rainbow trout (Léger 1979). It also has been reported in the pancreas of the barndoor skate, *Raja radiata* (Brockerhoff and Hoyle 1965), and the leopard shark *Triakis semifasciata*, blue shark *Prionace glauca*, dogfish *Squalus suckleyi*, and stingray *Urolophus halleri* (Patton 1975). A high degree of lipase activity was demonstrated also by the intestinal epithelium of the hagfish *Myxine glutinosa* (Adam 1963).

Mammalian pancreatic enzymes include phospholipase A_2 and cholesterol esterase (Vonk and Western 1984). Phospholipase A_2, which has been demonstrated in the pancreatic juice of humans, rats, and pigs, hydrolyzes lecithin to form lysolecithin (Kidder and Manners 1978), which is then largely reabsorbed. This enzyme also can substitute for colipase (Bläckberg, Hernell, and Olivecrona 1981). Cholesterol esterase has been obtained from pancreatic secretions of rats and pigs. It hydrolyzes the ester to cholesterol and fatty acid, and may release vitamins A, D, and E from their esters (Brockerhoff and Jensen 1974).

Brockerhoff and Jensen (1974) described a variety of other lipolytic esterases in the digestive tract of animals, but the specificity and origin of those found in gut contents are unclear. The fact that some of these are involved in cellular metabolism, including re-formation of triglycerides, questions their function as digestive enzymes in gut contents. Snake venoms contain a number of esterases, including phospholipases (Elliot 1978), but their primary function appears to be to aid in the penetration of toxins.

The importance of wax esters in the food chain of marine animals was mentioned earlier. Benson, Lee, and Nevenzel (1972) examined the digestion of wax esters and triglyceride by pancreatic lipase of the pig, and by homogenates of pyloric ceca from the chum salmon *Oncorhynchus keta* and the anchovy *Engraulis mordax*. Porcine lipase hydrolyzed triglyceride 3.5 times more rapidly than wax esters, and the tissue homogenates from the salmon hydrolyzed triglycerides 12 times more rapidly. However, the material from anchovies, which normally feed on copepods, hydrolyzed wax esters at twice the rate of triglycerides. A study by Patton and Benson (1975) of seven marine teleosts showed that wax esters were hydrolyzed most rapidly by those with a stomach and pyloric ceca, and that the released alcohol was converted to acid and then acyl lipids. An additional study of five marine teleosts (Patton et al. 1975) found that triglyceride was hydrolyzed more rapidly than wax esters and that both substances were hydrolyzed by a nonspecific lipase that also split β-monoglyceride.

Absorption

Glycerol, short- and medium-chain fatty acids, triglycerides of these fatty acids, and some phospholipids such as lecithin can be absorbed directly from the lumen contents without incorporation into micelles (Davenport 1982). However, most products of lipid digestion are absorbed from the micelles when they come in contact with intestinal mucosal cells. Triglycerides and phospholipids are then resynthesized within the cell and assembled into chylomicrons, which are small spheres, 0.1-3.5 nm in diameter that are coated with a mixture of protein, cholesterol, triglyceride, and phospholipid. The chylomicrons pass through the cell and are released into intercellular space for removal by the lymphatic system. Short-chain fatty acids and some glycerol are absorbed directly into the blood. Conjugated and ionized bile salts are rapidly absorbed by active transport from the terminal ileum.

Protein

Dietary and endogenous protein

Plant and animal proteins consist of chains of L-amino (or imino) acids linked together with peptide $(-NH \cdot CO-)$ bonds. The contributory amino acids are listed in Table 5.6. A given protein may contain up to 20 different amino acids. The number of amino acids and the differences in amino acid sequence, cross-linkages and spatial relations allow a wide variety of dietary proteins. The gut contents also contain endogenous protein derived from secretions, desquamated cells, and the escape of plasma protein. It has been estimated that the human digestive tract receives 10-30 g of protein from secretions and 10 g from desquamated cells, plus 1.9 g of plasma albumin excreted into the stomach each day (Davenport 1982).

Digestion

With the exception of some newborn mammals, protein is sequentially hydrolyzed by extracellular endopeptidases that attack peptide bonds along the protein chain, and exopeptidases, which split off terminal amino acids (Fig. 5.1). The action of these enzymes releases oligopeptides, dipeptides, and amino acids. Oligo- and dipeptides are further hydrolyzed by enzymes in the brush border or contents of intestinal cells with the subsequent passage of amino acids and small amounts of dipeptides into the blood.

Endopeptidases

The principal endopeptidases found in mammals are pepsin, chymosin (rennin), trypsin, chymotrypsin, elastase, and collagenase. Each of these enzymes is secreted as an inactive zymogen, which prevents it from attacking host tissue prior to its release, and is activated within the lumen of the digestive tract.

Pepsin. Secretion of pepsinogen and HCl by proper gastric glands was discussed in Chapter 1. Pepsinogen is activated to pepsin in an acid media, and pepsin then serves as an autocatalyst for the release of additional enzyme. The enzyme is most active at pH 4 and 2. Pepsin favors the hydrolysis of peptide bonds to which an aromatic amino acid provides the amino group, especially where this aromatic amino acid group is present on both sides of the bond (Fruton and Simmonds 1958). The end products of pepsin digestion are large polypeptides, oligopeptides, and some amino acids. Vonk and Western (1984) described the activities of a variety of pepsins that have been isolated from the stomachs of humans, swine, cattle, dogs, and whales. Pepsinogens also are secreted by tissues other than proper gastric glands.

Two or three pepsinogens have been isolated from the proventriculus of chickens (Donta and Van Vunakis 1970a,b), and Herpol (1964) found that this was supplemented by pepsinogen produced by the gizzard in birds of prey (*Falco, Athene, Buteo*). Pepsin has been found in the stomach of numerous reptiles (Dandrifosse 1974) and adult amphibians, but appears to be absent in the larvae of frogs and toads (Forte, Limlomwongse, and Forte 1969). Pepsins have been isolated from the stomach of many teleosts, including salmon, pike, perch, tuna, bowfin, albacore, cod, and hake (Vonk and Western 1984), as well as the sleeper shark *Squalus acanthius* (Vonk 1927). Four pepsins have been described in the stomach of the dogfish *Mustelus canis* (Bar-Eli, White, and Van Vunakis 1966; Merrett, Bar-Eli, and Van Vunakis 1969). Secretion of pepsinogen by the esophagus of some species of fish, adult amphibians, reptiles, and mammals was discussed in Chapter 1.

Pepsin is not essential for the digestion of protein, as evidenced by the absence of pepsinogen and HCl secretion by the stomach of the spiny anteater *Tachyglossus aculeatus* (Griffiths 1965) and some larval amphibians, and their absence from cyclostomes, chimaeras, lungfish, and families of teleosts that lack a stomach. Humans with achlorhydria, the inability to secrete pepsinogen and HCl, can retain their nitrogen balance (Davenport 1982), and the same is true after surgical removal of the stomach from humans and a number of other species.

Chymosin. Chymosin, or rennin, was first described in the abomasum of young domestic cattle as the principal agent responsible for the clotting of milk. Its proteolytic activity is similar to but weaker than pepsin, and milk is clotted by both pepsin and trypsin as well. However, Foltmann (1981) suggested that the appearance of weakly proteolytic chymosin in the neonate prior to the peptic digestion could serve to clot milk without damage to immunoglobulins. Chymosin is absent from the human stomach, but immunological reactions of gastric mucosal extracts from the dog, cat, rat, porcupine, kangaroo, horse, and zebra (Foltmann 1981; Foltmann et al. 1981) suggest a wide distribution among mammals.

TABLE 5.6 Classification of amino (and imino) acids derived from proteins

Amino acid (Symbol)	Molecular weight (M)	R group[a]	Other characteristics[b]
Aliphatic			
Monoamino-monocarboxylic acids			
Glycine (Gly)	75.1	UP	
Alanine (Ala)	89.1	NP	
Serine (Ser)	105.1	UP	Alcoholic
Cysteine (Cys)	121.1	UP	Contains sulfur, cross-links with another Cys by oxidation to a covalent disulfide bond
Threonine (Thr)	119.1	UP	Has 2 asymmetric carbon atoms
Methionine (Met)	149.2	NP	Contains sulfur
Valine (Val)	117.1	NP	
Leucine (Leu)	131.2	NP	
Isoleucine (Ile)	131.2	NP	Has 2 asymmetric carbon atoms
Monoamino-dicarboxylic acids			
Aspartic acid (Asp)	133.1	Acidic	b
Asparagine (Asn)	132.1	UP	Amide
Glutamic acid (Glu)	147.1	Acidic	b
Glutamine (Gln)	146.1	UP	Amide

	MW	Classification[a]	Remarks
Diamino-monocarboxylic acids			
Lysine (Lys)	146.2	Basic	[b]
Hydroxylysine (Hlys)	162.2		Has 2 asymmetric carbon atoms; only in collagen and gelatin
Arginine (Arg)	174.2	Basic	[b]
Diamino-dicarboxylic acids			
Cystine (CyS-SCy)	240.3	UP	Contains sulfur
Aromatic			
Monoamino-monocarboxylic acids			
Phenylalanine (Phe)	165.2	NP	
Tyrosine (Tyr)	181.2	UP	Alcoholic
Heterocyclic			
Monoamino-monocarboxylic acids			
Tryptophan (Try or Trp)	204.2	NP	Potential metal ligand[b]
Histidine (His)	155.2	Basic	Imino acid; disrupts α-helicity of protein chain
Proline (Pro)	115.1	NP	
Hydroxyproline (Hyp)	131.1		Imino acid; has 2 asymmetric carbon atoms; only in collagen and gelatin

[a]UP, acid with uncharged polar R group; NP, acid with nonpolar (hydrophobic) R group.
[b]These polar amino acids are ionized over a wide range and can form ionic bonds in the protein structure.
Source: Vonk and Western (1984).

Trypsin. Trypsinogen is secreted by the pancreas and converted to trypsin by enterokinase, an enzyme secreted by intestinal mucosal cells. Small amounts of trypsin then catalyze release of the remaining enzyme, the other pancreatic endopeptidases (chymotrypsin, elastase, and collagenase), and the exopeptidases (carbohypeptidases). Trypsin acts chiefly on peptide bonds in which the carbonyl portion is provided by basic amino acids, especially arginine. However, its function as an activator of other pancreatic proteinases makes it very important to the digestion of proteins. The structures of trypsinogens and trypsins have been described for a number of species (Vonk and Western 1984).

Pancreatic trypsin has been isolated from chickens (Whiteside and Prescott 1962) and turkeys (Ryan 1965). Zendzian and Barnard (1967a,b) identified trypsin in the pancreas of reptiles (*Podocnemis, Pseudemys, Chrysemys,* and *Chelydra*) and the amphibians *Amphiuma, Necturus, Rana pipiens,* and *Rana catesbeiana.* Extracts from the pancreas of the lungfish *Protopterus* were found to contain three trypsinogens (Reeck, Winter, and Neurath 1970), and trypsin was isolated from the pancreas of the carp (*Cyprinus carpio* (Cohen, Gertler, and Birk 1981a,b). Demonstration of trypsin activity in teleosts that do not have a distinct pancreas has been difficult. However, trypsin has been found in extracts of the hepatopancreas of goldfish and pyloric ceca of salmon, tuna, groupers, barracuda, bass, and mackerel (Vonk and Western 1984).

Trypsin has been found in the pancreas of dogfish (Prahl and Neurath 1966), nurse shark, and stingray (Zendzian and Barnard (1967a,b), as well as the pancreas of *Chimaera monstrosa* and the intestinal mucosa of hagfish (Nilsson and Fänge 1969, 1970). The distribution and content of total proteinase, trypsin, and chymotrypsin in pancreatic tissue of various vertebrates are listed in Table 5.7.

Chymotrypsin. Chymotrypsin is a pancreatic endopeptidase that preferentially releases peptides with a C-terminal hydrophobic and, more specifically, aromatic amino acid. Because these are more commonly present than the two basic amino acids, chymotrypsin digests a larger number of bonds than trypsin. The structure and activity of bovine chymotrypsin have received the most extensive study. In addition to the nonmammalian species listed in Table 5.7, chymotrypsin has been found in the lungfish *Protopterus,* the hepatopancreas of goldfish, and pyloric ceca of bass, mullet, mackerel, and bowfin (Vonk and Western 1984).

Elastase. Elastase is another endopeptidase found in the mammalian pancreatic secretions. It attacks elastin, a fibrous protein of arteries and ligaments that is resistant to other proteinases, and is capable of hydrolyzing insulin, ribonuclease, and the protein of some bacteria. Elastase splits bonds adjacent to uncharged, nonaromatic amino acids, with a preference for serine and alanine. Porcine elastase has been purified and crystallized, and its amino acid sequence elaborated (Hartley and Shotton 1971). Elastase activity has been found in the pancreas of chickens, lungfish, stingrays, and chimaeras, and the pyloric ceca of tuna, bass, and mullet (Vonk and Western 1984).

Collagenase. This enzyme hydrolyzes collagen, another fibrous protein of vertebrates that is a chief constituent of the connective tissue associated with skin, tendons, and bones. Collagenase appears to have been demonstrated only in the pancreatic juice of dogs (Takahashi and Seifter 1974) and *Chimaera monstrosa* (Nilsson and Fange 1969).

Exopeptidases

Exopeptidases confine their activity to the removal of terminal amino acids from the peptide chain. These include C-terminal peptidases (carboxypeptidases), which attack peptides with a free carboxyl group, and N-terminal peptidases (aminopeptidases), which attack substrates with a free amino group.

Carboxypeptidases. Carboxypeptidases A and B are secreted by the pancreas. Carboxypeptidase A is especially active on the peptides with C-terminal aromatic or branched aliphatic residues that are end products of chymotrypsin digestion. The bovine enzyme is secreted in two forms of zymogen that are activated and then converted by trypsin to four different enzymes with essentially the same specificity. Carboxypeptidase B acts on the polypeptides with a C-terminal lysine or arginine that are end products of trypsin digestion.

Carboxypeptidase A has been demonstrated in pancreatic tissue of humans, cattle, pigs, and reptiles (turtles and terrapins), as well as the pancreas, hepatopancreas, or pyloric ceca of teleosts, elasmobranches, and cyclostomes (Vonk and Western 1984). There is also some evidence for its presence in the pancreas of chickens and adult amphibians. Carboxypeptidase B has been demonstrated in the pancreas of humans, cattle, pigs, dogs, rats, lungfish, and dogfish, and the pyloric ceca of cod, bass, mullet, and carp (Vonk and Western 1984).

Intestinal exopeptidases. Recent evidence indicates that the brush border enzymes of the rat small intestine can digest considerable amounts of intact protein, even in the absence of pancreatic enzymes (Guan et al. in press). However, the enzymes in the brush border and fluid contents (cytosol) of the enterocyte are believed to be responsible principally for the hydrolysis of peptides. This stage of protein digestion has been reviewed by Kim and

TABLE 5.7 Distribution and content of proteinases in the pancreas of vertebrates

	Total proteinase (casein)	Trypsin (BAEE)	Chymotrypsin (BTEE)	RNase group
Group A: 20-60 mg proteinase per gram pancreatic tissue				
Terrapins, *Chrysemys picta*	58	9.1	12.0	C
Pseudemys elegans	53	7.6	16.0	B
Chelydra serpentina	33	6.2	11.0	B
Chrysemys picta (fasted at 5°C)	26	2.2	4.4	—
Frog, *Rana catesbeiana* (fed)	22	5.2	5.5	C
Group B: 10-20 mg proteinase per gram				
Terrapin, *Podocnemis unifilis*	19	7.0	3.2	B
Congo eel, *Amphiuma* sp.	19	6.0	2.8	C
Horse	14	6.0	2.8	B
Frog, *Rana pipiens*	14	1.9	1.2	C
Dog	12	2.2	2.0	C

Frog, *Rana catesbeiana* (fasted)	12	1.9	1.9	C
Cow	10.3	3.2	6.4	A
Mud puppy, *Necturus maculosus*	10	2.5	2.7	C
Sting ray, *Dasyatis americana*	10	1.8	1.4	C
Group C: 0–10 mg proteinase per gram				
Turkey	8.8	3.5	1.3	B
Goat	8.7	2.9	4.1	A
Barracuda,[a] *Sphyraena barracuda*	7.3	0.5	0.4	C
Chicken	6.0	0.6	1.0	B
Wallaby, *Macropus eugenii*	5.5	1.7	0.5	A
Nurse shark, *Ginglymostoma cirratum*	3.4	0.3	0.1	C
Dogfish, *Squalus suckleyi*	3.0	0.5	1.3	C
Tuna,[a] *Thunnus secundodorsalis*	1.6	0.06	0.18	C
Rabbit	1.4	0.4	0.2	C

Note: Activities of the substrates shown are expressed as the equivalent amount of bovine trypsin (BAEE and casein) or chymotrypsin (BTEE) giving the indicated activity under the same conditions.

[a]Pyloric ceca were used so the pancreatic fraction would be small; this may account for their low ranking.

Source: After Zendzian and Barnard (1967a), from Vonk and Western (1984).

Erickson (1985) and Silk, Grimble, and Rees (1985). The final stages of peptide hydrolysis are performed by aminopeptidases, tripeptidases, and dipeptidases found in the brush border and cytosol of intestinal absorptive cells. Brush border enzymes have greater activity against tetrapeptides and larger oligopeptides, but most of the tri- and dipeptidase activity is located in the cell contents. The concentration of these enzymes has been shown to increase from duodenum to mid-ileum, and then decrease in the remainder of the small intestine of the rat (Robinson 1960), pig (Josefsson and Lindberg 1965), and sheep (Symons and Jones 1966).

Four brush border dipeptidases (Tobey et al. 1985) and seven cytosol enzymes that hydrolyze small peptides (Rapley, Lewis, and Harris 1971) have been reported in human enterocytes. Vonk and Western (1984) listed seven exopeptidases found in the intestinal cells of mammals. Three aminopeptidases—leucine aminopeptidase, aminotripeptidase, and aminopeptidase A—respectively split bonds of the C-terminal residues of leucylpeptides, aminotripeptidases, and L-series acidic residues that carry a free amino group. The dipeptidases consisted of L-glycylglycine and L-glycylleucine dipeptidases, prolinase, and prolidase. Prolinase is specific for the amino group of L-prolylglycine and L-hydroxyprolylglycine. Prolidase was specific for L-glycylproline and L-glycylhydroxyproline.

Leucine aminopeptidase activity has been demonstrated in the intestinal cells of reptiles (*Testudo, Lacerta*, and *Natrix*), the intestinal microvilli of teleosts (*Perca, Cottus, Enophrys*, and *Misgurnus*), and intestinal mucosal extracts of chimaeras and hagfish (Vonk and Western 1984). Hydrolysis of L-glyclglycine and/or other glycine dipeptides has been demonstrated with intestinal mucosal extracts from teleosts, chimaeras, and hagfish.

Absorption

Amino acids and peptides
Amino acids are transported into intestinal cells by at least four carrier-mediated systems that transport sodium in a manner similar to the system that transports glucose, as well as several sodium-independent carrier-mediated systems (Hopfer 1987). These systems favor the transport of L-amino acids. One carrier transports neutral amino acids that have a free carboxyl group, a free α-hydrogen, and a neutral side chain. A second carrier transports basic amino acids including L-arginine, L-lysine, D- and L-ornithine, and L-cysteine. A third carrier transports L-proline, L-hydroxyproline, sarcosine, dimethylglycine, and betaine. The dicarboxylic amino acids, aspartic and glutamic, are not served by these carriers, but it appears that they are absorbed into the cell and transformed into alanine (Schultz, Yu-tu, Alvarez, and Curran 1970). A sodium-independent transport system has been described for the efflux of neutral amino acids from the cell to the blood (Mircheff, van Os, and Wright 1980).

Tetrapeptides are degraded by brush border enzymes before entering the cell contents, but tripeptides and dipeptides can be absorbed from the lumen faster than their constituent amino acids. There is substantial evidence for the active transport of dipeptides into the intestinal cell of mammals and the frog (*Rana pipiens*) by a carrier-mediated process that shows competitive inhibition between peptides, but no competition between the peptide and its constituent free amino acids (Silk et al. 1985). There is conflicting evidence on whether this process is sodium dependent, and Vadivel and Leibach (1985) proposed that it is energized instead by a proton (H) gradient maintained by Na-H exchange at the luminal membrane. Estimates of the amounts of peptide reaching the blood vary from significant quantities of some peptides, reported in the rat (Gardner 1982), to very large amounts reported in calves (Webb, 1986). The nutritional significance of dipeptide absorption is generally believed to be small. However, as pointed out by Silk et al. (1985), the possible presence of 400 dipeptides and 8000 tripeptides makes this difficult to assess.

Intact protein

Protein can be absorbed from the intestine into the lymphatic system during the first 24-36 hours following the birth of cats, pigs, goats, sheep, cattle, and horses (Koldovský, 1970). It may be absorbed over a longer period by dogs, up to three weeks by mice and rats, and as long as six weeks following the birth of hedgehogs. Brambell (1970) demonstrated an inverse relationship between the ability of a species to transfer immunity (γ globulins) by the placental versus intestinal routes (Table 5.8).

Absorption of IgG antibodies by the neonatal rat intestine involves their selective binding to receptors on the brush border of intestinal cells (Davenport 1982). Rat globulins are absorbed much more rapidly than albumen or the globulins of other species. Human infants also are capable of absorbing γ globulins and other macromolecules, but these are absorbed by a nonspecific process of pinocytosis and largely digested within the cell. Nevertheless, intact protein can be absorbed by infants and adult humans in quantities sufficient to produce allergic reactions.

The transfer of immunity from dam to progeny is limited by not only the ability of the intestines to absorb these proteins, but also by the degradation of globulin by the acid and enzymes of the neonatal digestive tract (Koldovský 1970). For example, gastric secretion of HCl is well developed in the guinea pig prior to birth, and although gastric contents of humans are usually neutral at birth, the pH decreases rapidly the first day. The pH of gastric contents in rats remains high (5.0-6.5) for a few days after birth. Both pancreatic and intestinal proteinases increase during gestation and continue to increase throughout the suckling period. Trypsin activity increases the first week after birth in human infants and calves, the first five weeks after birth in pigs, and,

TABLE 5.8 Transmission of passive immunity

Species	Prenatal[a]	Postnatal[a]
Horse	0	+++ (24 hr)
Pig	0	+++ (24–36 hr)
Ox, goat, sheep	0	+++ (24 hr)
Quokka (*Setonix*)	0	+++ (180 d)
Dog, cat	+	++ (1–2 d)
Fowl	++	++ (<5 d)
Hedgehog	+	++ (40 d)
Mouse	+	++ (16 d)
Rat	+	++ (20 d)
Guinea pig	+++	0
Rabbit	+++	0
Man, monkey	+++	0

[a]0, no absorption or transfer; + to +++, degrees of absorption or transfer.
Source: Brambell (1970).

most rapidly, between the fourth and sixth week in rats. However, a trypsin inhibitor is present in the colostrum of calves.

Nucleic acids

All cells contain ribonucleic acid (RNA) and deoxyribonucleic acid (DNA), which consist of the pentose sugars (ribose or deoxyribose), phosphate, and a purine or pyrimidine base. Ribonucleic acid is found in the cytosol and some nuclei. It aids in the control of cellular chemical activities. Deoxyribonucleic acid is found in the nucleus of plant and animal cells, or packaged separately in the cell contents of prokaryote organisms, and serves for the transmission of genetic information. Most of the RNA and (except for bacteria and virus) DNA are bound to protein. Hydrolysis of nucleoproteins by gastric and pancreatic proteinases releases these nucleic acids (Table 5.1), which are hydrolyzed by pancreatic ribo- and deoxyribonucleases to form polynucleotides. The polynucleotides are then further hydrolyzed by intestinal nucleotidases, deaminases, and phosphorylases to form end products capable of being passively or actively transported into intestinal cells (Davenport 1982).

Barnard (1969a,b) examined the distribution of pancreatic ribonucleases among vertebrates. Although ribonucleases have been recorded in all classes examined, Table 5.9 shows the considerable amount of species variation in their level of activity. These levels appeared to be constant for most species and, unlike other pancreatic enzymes, independent of the diet or feeding state. The species with the highest ribonuclease levels were ruminants and macropod marsupials, which have large numbers of microbes entering the intestine from the stomach. Smith and McAllan (1971) estimated that 20% of the microbial nitrogen leaves the forestomach of cattle in the form of polynucleotides, and 75-80% of the microbial RNA and DNA is digested in the duodenum. The polynucleotide nitrogen of bacteria may be largely converted to urea in these animals, but the phosphate released by their hydrolysis can provide an important source for both the animal and its microbes.

Summary

Many of the endogenous, extracellular digestive enzymes that hydrolyze substrates in the gastrointestinal cavity are common to all vertebrates. Some exceptions to this are the pepsins (absent from some species of fish, larval amphibians, and a few mammals), phospholipase A_2 and cholesterol esterase (demonstrated only in mammals), and elastase and carboxypeptidase B (not demonstrated in reptiles, amphibians, or cyclostomes). Other extracellular enzymes such as chitinase are found in members of each vertebrate class, but not in all species.

Endogenous digestive enzymes that are present in the brush border or contents of intestinal cell have been examined in few mammals and very few lower vertebrates. Maltase appears to be common to all vertebrates. Sucrase and trehalase are absent from the intestine of some mammals and other vertebrates. Lactase is absent from the intestinal mucosa of some neonate and many adult mammals, and it appears to be absent from lower vertebrates. Less information is available for the comparison of intestinal peptidases.

The levels of enzyme activity can vary with species, diet, or the stage of an animal's development. Carnivores tend to have higher levels of proteinases than omnivores or herbivores, and animals that have large numbers of indigenous microbes in their foregut tend to have higher levels of pancreatic nuclease. Major differences in the diet of larval amphibians or neonate mammals, as compared to adult animals, are reflected by marked differences in the levels of their various digestive enzymes. There is evidence that enzyme levels can also adjust to dietary changes following maturation in some species.

Phagocytosis or pinocytosis appears to be confined mostly to newborn animals of some species. Lipid-soluble substances can be absorbed by passive diffusion. Absorption of water-soluble organic compounds is mainly limited to passive diffusion of a few monosaccharides and amino acids, and to the active transport of others such as glucose, galactose, xylose, most amino acids, and some dipeptides. It now appears that major quantities of the amino acids that reach the blood are constituents of di- and tripeptides that are actively taken up from the lumen and hydrolyzed by enzymes of the enterocyte.

TABLE 5.9 Distribution and content of ribonuclease in the pancreas of vertebrates

Group A: 200-1200 µg RNase per gram pancreatic tissue

Cow	1200	Wallaby, *Macropus eugenii*	515
Bison, *Bison bison*	1180	Mouse	395
Sheep	1080	Lizard, *Iguana iguana*	380
Goat	1000	Uganda kob, *Kobus kob*	270
Kangaroo, *Macropus rufus*	600	Golden hamster	260
Elk, *Cervus canadensis*	550	Rat	260
Kangaroo, *Macropus giganteus*	530	Guinea pig	240

Group B: 20-100 µg RNase per gram

Terrapin, *Podocnemis unifillis*	90	Armadillo, *Dasypus novemcintus*	30
Pig	80	Horse	25
Terrapin, *Pseudemys elegans*	65	Turkey	25
Hippopotamus, *H. amphibius*	62	Chicken	20
Terrapin, *Chelydra serpentina*	60	Opossum, *Didelphis marsupialis*	20
Caiman, *C. crocodilus*	53		

Group C: 0-20 μg RNase per gram

18	Whale, *Eschrichtius robustus*	
18	Dogfish, *Squalus suckleyi*	
17	Frog, *Rana catesbeiana*	
12	Snake, *Bungarus fasciatus*	
9	Terrapin, *Chrysemys picta*	
8	Mud puppy, *Necturus maculosus*	
8	Grouper, *Epinephelus striatus*	
7	Frog, *Rana pipiens*	
7	Lungfish, *Protopterus aethiopicus*	
5	Dolphin, *Tursiops truncatus*	
5	Stingray, *Dasyatis americana*	
4	Toad, *Bufo marinus*	
3	Pigeon	
2	Monkey, *Macaca mulatta*	
2	Tuna, *Thunnus secundodorsalis*	
2	Barracuda, *Sphyraena barracuda*	
2	Shark, *Ginglymostoma cirratum*	
1	Man	
0.7	Elephant, *Loxodonta africana*	
0.5	Dog	
0.5	Cat	
0.5	Rabbit	
0.5	Congo eel, *Amphiuma* sp.	
<1	Toad, *Bufo americanus*	
<1	Snake, *Natrix taxi*	

Note: Enzyme activity was determined from the rate of RNA hydrolysis at pH 7.4. Whole pyloric ceca (i.e., not pure pancreatic tissue) from tuna and barracuda were used.

Source: After Barnard (1969a,b) from Vonk and Western (1984).

Therefore, the assimilation of nutrients by vertebrates is largely dependent on the hydrolysis of food by endogenous enzymes. However, indigenous microbes also can play an important role in the provision of nutrients, and their contribution will be discussed in the following chapter.

Despite similarities in the enzymes and absorptive mechanisms, the efficiency of these processes also depends on the mucosal surface area of the midgut, which can be greatly increased by pyloric ceca, ridges, villi, and microvilli or by an increase in intestinal length. Karasov, Solberg, and Diamond (1985) examined the rate of glucose and proline uptake by segments of small intestine from three reptiles and three mammals of equal size: the desert iguana (*Dipsosaurus dorsalis*), chuckwalla (*Sauromalus obesus*), box terrapin (*Terrapene carolina*), wood rat (*Neotoma lepida*), hamster, and guinea pig. Although all of these animals showed similar mechanisms for the active transport of D-glucose and L-proline, the total capacity for glucose and proline absorption was seven times higher in the mammals. This was found to be due mainly to a 4 to 5.5 times greater area of mucosal surface in the mammals, but also to the fact that the normalized mucosal weight of the reptiles was one-half that of the mammals and the reptile intestine operated at lower temperatures during the night.

6

Microbial fermentation and synthesis of nutrients, and absorption of end products

When one considers all of the arrangements and mechanisms that have evolved in the digestive system, those involving a symbiotic relationship between an animal and its gastrointestinal protozoa, algae, fungi, and bacteria are in many respects the most interesting and least understood. This relationship is a widespread occurrence among both invertebrates and vertebrates. It is critical to growth or even the survival of some species, with the symbiont supplying oxygen, utilizable energy, a mechanism for the synthesis of essential amino acids and vitamins, or a means for the recycling of waste products.

The gastrointestinal tract of mammals is colonized by microorganisms shortly after birth. They become restricted principally to the stomach and large intestine, presumably as a result of to the slower rate of digesta transit through these parts of the tract. The small intestine contains relatively few bacteria although significant numbers of organisms can be found in the ileum of many species. This may be due to a slower rate of passage through the terminal small intestine or occasional regurgitation of large intestinal contents into the ileum.

Dubos (1966) divided the flora found in the gut of a given species into two categories. "Allochthonus" are transient organisms that may be frequently found. "Autochthonous" microorganisms, which are symbionts present in a given species under a wide range of conditions, include lactobacilli, bacteroid, and fusiform bacteria. The type and number of microorganisms present in the digestive tract after weaning are affected by a number of conditions. Microbial populations are affected by marked changes in diet, especially if the change is too rapid. However, the normal balance of gastrointestinal microorganisms can be affected more by starvation or overfeeding than by dietary changes. Gut microorganisms also can be affected by marked changes in digesta pH and the administration of antibiotics or other antimicrobial drugs.

The microorganisms indigenous to the gut serve a number of beneficial purposes. For example, studies of "gnotobiotic" (germ-free) animals show that the microflora normally present protect the gut from disease, both by stimulation of immune mechanisms and by direct competition with pathogenic organisms. Gastrointestinal microorganisms are capable of reduction,

159

dehydration, ring fusion, and aromatization of a wide range of compounds. They can digest mucus, bilirubin, gastrointestinal enzymes, urea, protein, and other nitrogenous compounds, plus a wide range of drugs. They can produce organic acids, CO_2, CH_4, H_2, NH_4, and toxic amines.

Gut microorganisms are capable of converting plant material of little direct nutritional value into readily utilizable nutrients. This occurs in the stomach of numerous mammals, and in the large intestine of a wide range of species. It provides a means of fermenting structural carbohydrate such as cellulose and hemicellulose into organic acids, which can be readily absorbed and utilized as a source of energy. These microorganisms are capable also of synthesizing B vitamins and high-quality protein, in their own cell bodies, from nonprotein nitrogenous sources. The processes of microbial fermentation and synthesis have some basic requirements that help explain many of the structural and functional adaptations seen in the gastrointestinal tract of vertebrates. The primary requirements are 1) anaerobic conditions, 2) a means of maintaining a relatively neutral pH in the face of substantial organic acid production, and 3) provision for the more prolonged contact of microorganisms and digesta needed for this relatively slow process. The latter requires a site or sites in the gastrointestinal tract that can selectively retain the microorganisms and their substrate until this process is complete.

It is obvious that microbial fermentation and nutrient synthesis is most useful to herbivores confined to a diet high in plant fiber and low in readily available carbohydrate, protein, and vitamins. The corollary to this is that the low nutritional value of this type of diet requires the ingestion and processing of a large volume of food. This requirement, plus the need to retain digesta within the gut for longer periods of time, accounts for the greater volume capacity of the herbivore digestive system. Although the ability to contain and retain larger volumes of digesta is well developed in the hindgut of some reptiles and birds, and many mammals, the earliest and the most extensive studies of the contributions of gut microorganisms to nutrition were conducted on the forestomach of domestic sheep and cattle. Therefore, the information gained from these studies provides the basis for most of the following discussion.

Carbohydrate fermentation

Stomach

Ruminants

The ruminant stomach (see Fig. 3.26) consists of a forestomach (reticulo-rumen and omasum) that is lined with nonglandular stratified squamous epithelium, and a glandular compartment (abomasum), which is analogous

TABLE 6.1 Fermentative properties of ruminal bacteria

Species	Function[a]	Products[b]
Bacteroides succinogenes	C,A	F,A,S
Ruminococcus albus	C,X	F,A,E,H,C
Ruminococcus flavefaciens	C,X	F,A,S,H
Butyrivibrio fibrisolvens	C,X,PR	F,A,L,B,E,H,C
Clostridium lochheadii	C,PR	F,A,B,E,H,C
Streptococcus bovis	A,SS,PR	L,A,F
Bacteroides amylophilus	A,P,PR	F,A,S
Bacteroides ruminicola	A,X,P,PR	F,A,P,S
Succinimonas amylolytica	A,D	A,S
Selenomonas ruminantium	A,SS,GU,LU,PR	A,L,P,H,C
Lachnospira multiparus	P,PR,A	F,A,E,L,H,S
Succinivibrio dextrinosolvens	P,D	F,A,L,S
Methanobrevibacter ruminantium	M,HU	M
Methanosarcina barken	M,HU	M,C
Spirochete sp.	P,SS	F,A,L,S,E
Megasphaera elsderii	SS,LU	A,P,B,V,CP,H,C
Lactobacillus sp.	SS	L
Anaerovibrio lipolytica	L,GU	A,P,S
Eubacterium ruminantium	SS	F,A,B,C

[a]C = cellulolytic, X = xylanolytic, A = amylolytic, D = dextrinolytic, P = pectinolytic, PR = proteolytic, L = lipolytic, M = methanogenic, GU = glycerol-utilizing, LU = lactate-utilizing, SS = major soluble sugar-fermenting, HU = hydrogen-utilizing.
[b]F = formate, A = acetate, E = ethanol, P = propionate, L = lactate, B = butyrate, S = succinate, V = valerate, CP = caproate, H = hydrogen, C = carbon dioxide, M = methane.
Source: Modified by Allison (1984) from Hespell (1981).

to the stomach of the human and dog. The reticuloruminal contents of sheep and cattle can provide 10-15% of the animal's body weight. They become colonized, predominantly with *Escherichia coli aerogenes* and streptococci, during the first week after birth (Eadie and Mann 1970). These are later joined by lactobacilli, which persist, along with the streptococci, in the nursing animal. Weaning is followed by the development of the extremely complex group of flora and fauna that are characteristic of the adult animal. The microbiology of the rumen has been reviewed by Hungate (1966), Bryant (1977), Wolin (1979), and Allison (1984).

Table 6.1 lists some of the most important bacteria found in the rumen of sheep and cattle, along with their fermentative properties. Culture counts give estimates of 10-50 billion bacteria per gram of fluid rumen contents. Microscopic counts, which include organisms that are dead or require specific culture media, can give values two to three times higher than this. Although protozoa are less numerous than the bacteria, they can occupy an almost equal volume of the rumen contents. The most important protozoa are anaerobic ciliates that belong to the families *Isotrichidae* and *Orphryo-*

scolecidae (Ogimoto and Imai 1981). Rumen protozoa are capable of fermenting carbohydrate, digesting protein, and hydrogenating fatty acids, but they are not essential to the ruminant digestive system and their contributions remain unclear (Clarke 1977; Coleman 1980; Hobson and Wallace 1982). In addition to the bacteria and protozoa, anaerobic mycoplasmas can be found at concentrations of 10^5-10^7 organisms per gram. Their significance is unknown.

Microbial fermentation of carbohydrate and synthesis of protein and vitamins are the subjects of a number of early reviews (Hungate 1968; Bryant 1977; Phillipson 1977). Rumen microorganisms ferment sugars, starches, and structural carbohydrates such as cellulose, hemicelluloses, and pectins to short-chain organic acids, CO_2, CH_4, and H_2. The principal organic acids, regardless of carbohydrate substrate, are acetic, propionic, and butyric acid. Because these three fatty acids can be readily separated from the low concentrations of longer-chain organic acids that are present in rumen contents by steam distillation, they became generally referred to as the volatile fatty acids (VFA). The total concentration of VFA in rumen contents varies between 60 and 120 mmol/L, depending on diet and time after feeding. When animals are fed a diet of hay or other roughage, the proportions consist of 60-70% acetate, 15-20% propionate, and 10-15% butyrate. Addition of grain or other readily fermentable starches or sugars to the diet increases the rate of fermentation and decreases the acetate/propionate ratio. Because the VFA are formed as free acids with a pK of 4.8, the pH of rumen contents varies inversely with the rate of VFA production. However, the pH is normally maintained between 5.5 and 7.0 by the addition of HCO_3 and PO_4 from the saliva, secretion of HCO_3 by rumen epithelium, and absorption of VFA by rumen epithelium. Engorgement of grain, fruit, or other foods containing rapidly fermentable carbohydrate can depress the pH of rumen contents to lower levels that result in destruction of normal microorganisms, rapid multiplication of lactobacilli, ulceration of forestomach epithelium, and systemic acidosis and dehydration (Dirksen 1970).

Rumen gases vary in their rate of production and their composition with time after feeding (Fig. 6.1). Carbon dioxide is derived from fermentation of carbohydrate and the neutralization of VFA with HCO_3. Methane is believed to be almost totally derived from the reduction of CO_2 by formate, succinate, and H_2. The latter reaction is believed to account for the low concentrations of H_2 in the rumen except for the first few days of a fasting period. Nitrogen and O_2 are added from swallowed air, although N_2 also can diffuse into the rumen from the blood. Oxygen is rapidly assimilated by rumen microorganisms and some of the CO_2 is directly absorbed from the forestomach, but most of the gases produced in the rumen are removed by eructation, as described in Chapter 4.

The low pH and proteolytic activity of abomasal contents inhibit further

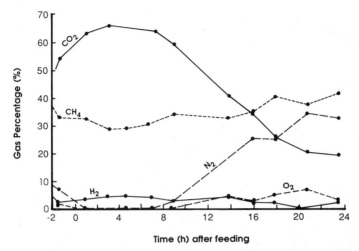

FIGURE 6.1 Composition of rumen gases in a dairy cow on a ration of alfalfa hay and grain. (From Washburn and Brody 1937.)

production of VFA by rumen microbes, and this, plus the further absorption of VFA and the addition of abomasal secretions, results in relatively low levels of these fatty acids in abomasal contents. Cattle fed a hay diet demonstrated mean VFA concentrations of 6 mEq/L in the abomasum, versus 60 mEq/L in the rumen (Svendsen 1969). Gas was released from the abomasal contents of these animals and escaped back into the rumen at a rate of approximately 0.8 L/hr. When the same animals were fed a high-grain diet, VFA concentrations of both abomasal and rumen contents doubled and the volume of gas released from the abomasum contents more than doubled. The gases released from abomasal contents prior to feeding were composed of approximately 17% CO_2, 65% CH_4, and 17% N_2. These levels remained relatively constant in animals on a hay diet, but when animals were fed a high-grain diet the percentage of CO_2 increased to approximately 25%, with a reduction of CH_4 and N_2 to approximately 60% and 8%, respectively. The composition of these gases differs from that noted in the rumen. The N_2 presumably arises from the atmosphere via diffusion from blood and the swallowing of air, and the CO_2 could be readily explained by the acidification of forestomach digesta in the abomasum. The CH_4 could have been derived from reduction of CO_2. However, this would require relatively strong proton donors plus enzymes, suggesting the presence of methanogenic organisms even at the relatively low pH of abomasal contents.

Although microbial fermentation is an inefficient means of utilizing sugar and starch, it provides the only means of digesting the structural carbohydrate of plants. It has been estimated that VFA provides 70% or more of the energy required by adult cattle and that a major portion of this is derived from forestomach fermentation. The protozoa of the ruminant forestomach are believed to contribute relatively little to VFA production, but they do contribute to the production and storage of starch and protein for later digestion during passage through the gastrointestinal tract.

Other species

Microbial fermentation similar to that seen in the forestomach of Ruminantia has been demonstrated in the voluminous, compartmentalized stomachs of other Artiodactyla (camelids, peccaries, babirusa, and hippopotamus), macropod marsupials, colobus, and langur monkeys, and sloths. The stomach contents of camels, which constitute 10-17% of their body weight, contained bacteria (Hungate et al. 1959; Williams 1963) and protozoa (Lubinsky 1957) similar to those seen in the rumen of sheep and cattle. The pH and both the total concentration and relative composition of VFA in the first two major compartments of the camelid stomach (see Fig. 3.24) were similar to those recorded in the rumen (Williams 1963; Vallenas and Stevens 1971b), and the same was found to be true for the rates of VFA, CO_2, and CH_4 production (Hungate et al. 1959). The forestomach compartments of the hippopotamus stomach contained 75% of the total weight of gut contents and large numbers of bacteria and ciliated protozoa along with total concentrations and relative proportions of VFA similar to those recorded in the domestic ruminants (Thurston, Noirot-Timothée, and Arman 1968; Clemens and Maloiy 1982).

The stomach contents of macropod marsupials, which can constitute 15% of the body weight, demonstrate many similarities to the rumen contents of sheep and cattle (Hume 1984). Moir, Somers, and Waring (1956) found that microflora in the stomach contents of the quokka (*Setonix brachyurus*) were similar to those of the sheep rumen and capable of converting starch and cellulose to VFA. The populations of bacteria in the forestomach of the red-necked pademelon (*Thylogale thetis*) were similar in number to those of sheep on the same diet (Dellow 1979). Ciliated protozoa have been found in the forestomach of most macropods that have been examined (Moir 1965; Harrop and Barker 1972; Lintern-Moore 1973; Dellow 1979). Dellow (1979) found fungal sporangia like those of the rumen in the forestomach of the Eastern gray kangaroo (*M. giganteus*), the red-necked wallaby (*M. rufogriseus*), the wallaroo (*M. robustus robustus*), and the swamp wallaby (*Wallabia bicolor*).

The stomach of the red-necked pandemelon and red-necked wallaby contained VFA at concentrations similar to those found in the rumen of sheep fed the same diet, but the kangaroos showed a higher rate of VFA production and a lower ratio of acetate to propionate, suggesting less digestion of structural carbohydrates (Hume 1977). The stomach of the quokka and tammar wallaby (*M. eugenii*) showed a higher H_2 content and a lower production of CH_4 as compared to ruminants (Engelhardt, Wolter, Laurenz, and Hemsley 1978), and only negligible amounts of CH_4 were found in the stomach of the gray kangaroo (Kempton, Murray, and Leng 1976). Hume (1984) suggested that the lower rate of CH_4 production may be due to the more rapid transit of digesta (see Chapter 4), which would tend to inhibit the establishment of the slow-growing methanogenic bacteria. The concentration of VFA is highest in the initial sacciform segments of the kangaroo stomach and progressively decreases as digesta passes through the tubiform and into the hindstomach, which secretes pepsinogen and HCl (Hume and Dellow 1980b; Dellow and Hume 1982; Dellow, Nolan, and Hume 1983), supporting the early findings of Barker (1961) that indicated direct absorption of VFA from the macropod stomach.

The colobus and langur monkeys are arboreal folivores with a voluminous, partially sacculated stomach. Gastric contents of colobus monkeys can constitute 10.5-20.6% of the total body weight (Kuhn 1964; Ohwaki et al. 1974), and those of the langur monkey were found to equal 17% of its body weight (Bauchop and Martucci 1968). Stomach contents from both groups of monkeys demonstrated a range of pH and bacterial flora similar to those of the sheep and bovine rumen, but an absence of protozoa (Kuhn 1964; Bauchop and Martucci 1968; Ohwaki et al. 1974). Concentrations and proportions of VFA similar to those of rumen contents were found in the stomach contents of the colobus (Drawert, Kuhn, and Rapp 1962; Ohwaki et al. 1974) and langur monkeys (Bauchop and Martucci 1968), and CH_4 and CO_2 were the principal gaseous end products of microbial fermentation.

The two-toed sloths (*Choloepus*) and three-toed sloths (*Bradypus*) are arboreal herbivores that feed on leaves, young shoots, and twigs. The stomach of a three-toed sloth comprised 20-30% of the animal's body weight (Britton 1941). Studies of the gastric contents of *Choloepus hoffmanni* showed a rich bacterial population and a range of pH and total concentration of VFA similar to those found in the rumen of sheep and cattle, but ciliated protozoa were absent (Denis, Jeuniaux, Gerebtzoff, and Goffart 1967).

Although the stomachs of dugongs and manatees are relatively complex and voluminous, the dugong stomach constituted only 5% of the body weight and contained only low levels of VFA as compared to the large intestine, which comprised 10% of the body weight and contained high levels of these fatty acids (Murray, Marsh, Heinsohn, and Spain 1977). Therefore, it was concluded that the large intestine plays a more important role in the fermentation of carbohydrate by these animals.

There has been a great deal of speculation over the functional significance of the voluminous and highly compartmentalized stomachs of the toothed and baleen whales. It has been suggested that this allows for storage of large quantities of undigested prey. However, VFA have been reported in the forestomach of small, toothed whales (Morii 1972, 1979; Morii and Kanazu 1972) and, more recently, that of the bowhead (*Balaena mysticetus*) and gray baleen (*Eschrichtius robustus*) whales (Herwig, Staley, Nerini, and Braham 1984). The total concentration and composition of VFA estimated for the stomach contents of the baleen whales were similar to those recorded in domestic ruminants. Although kelp has been found in the stomach of gray whales, Herwig et al. (1984) suggested that the large quantities of chitin that were present may serve as the substrate for fermentation.

Elsden, Hitchcock, Marshall, and Phillipson (1946) and Phillipson (1947) were the first to note that VFA were present in the "simple" stomachs of the rat, rabbit, pig, horse, and dog. Figure 6.2 gives the mean concentration of VFA at different sites along the gastrointestinal tract of the raccoon, dog, pig, bush baby, vervet monkey, and pony. It shows that the stomach of each species contained measurable quantities of VFA. Those of the pig, bush baby, vervet monkey, and pony contained mean concentrations of 20-40 mmol/L. A similar range of VFA concentrations has been demonstrated in the stomach of the hyrax (Clemens 1977), baboon and Sykes monkey (Clemens and Phillips 1980), koala (Cork and Hume 1983), and elephant (Clemens and Maloiy 1982).

Substantial concentrations of lactic acid also were present in the stomachs of the dog, pig, and (especially) the pony. Although lactic acid may be released from gastric mucosa, the higher concentrations demonstrated in the stomach of pigs and ponies by Alexander and Davies (1963) suggest that it was produced largely by indigenous microbes. Microorganisms capable of producing lactic acid could be favored over those producing VFA by a lower pH or high concentrations of rapidly fermentable carbohydrate. There was no evidence, however, of an inverse correlation between the cyclic changes in the mean pH and lactic acid concentrations of the gastric contents in these species (see Figs. 7.9, 7.10, and 7.11).

Concentrations of VFA and lactic acid remained relatively low throughout the small intestine of each of these animals. The number of viable bacteria in the contents of the human small intestine is relatively low (10^4-10^6/g), in contrast to populations of 10^{10} or more found in the rumen or in the large intestine of man and other mammals that have been studied.

Hindgut
Bacteria indigenous to the hindgut have been described in taxonomical terms for chickens (Ochi, Mitsuoka, and Sega 1964; Smith 1965; Salanitro et al. 1978) and a few mammals (Savage 1977; Hoogkamp-Korstanje et al. 1979; Robinson, Allison, and Bucklin 1981; Finegold, Sutter, and Mathisen 1983; Allison 1984; Mikel'Saar, Tjuri, Väljaots, and Lencner 1984; Norin, Gus-

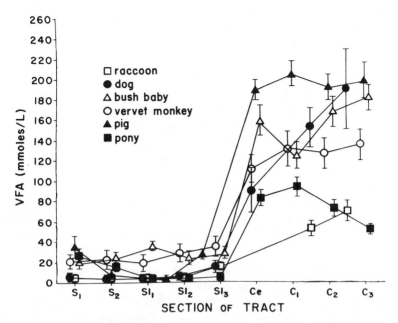

FIGURE 6.2 Mean (± SE) values for concentrations of volatile fatty acids (VFA) along the gastrointestinal tract of raccoon (*Procyon lotor*), dog (*Canis familiaris*), bush baby (*Galago crassicaudatus*), vervet monkey (*Cercopithecus pygerythrus*), pig (*Sus scrofa*), and pony (*Equus caballus*). All animals were fed at 12-hour intervals during a 3- to 4-week period before the study. Raccoons, pigs, and ponies were given the same pelleted diet. Each value represents the mean from twelve animals, killed in groups of three at 2, 4, 8, and 12 hours after feeding. Sections of tract were oral (S_1) and aboral (S_2) halves of the stomach; two or three equal segments of small intestine (SI_1, SI_2, SI_3); cecum (Ce); and two or three segments of colon (C_1, C_2, C_3). (Illustration modified from Argenzio, Southworth, and Stevens 1974; Clemens, Stevens, and Southworth 1975a; Banta, Clemens, Krinsky, and Sheffy 1979; Clemens and Stevens 1979; Clemens 1980.)

tafsson, Lindblad, and Midtvedt 1985). Microflora found in the large intestine of various species are quite similar to one another (Clarke 1977) and demonstrate a number of similarities with those found in the rumen (Wolin 1981; Allison 1984). At least 400 species, representing 40 genera of bacteria have been isolated from human feces (Savage 1986). However, the populations associated with epithelial tissue differ from those found in the lumen, and fecal populations do not necessarily represent the lumen contents of the entire large bowel (Allison 1984). Anaerobic protozoa have been demonstrated in the large intestine of horses, rhinoceroses, tapirs, elephants, chimpanzees, gorillas, and a few species of rodents.

Barcroft, McAnally, and Phillipson (1944) were the first to note that the composition and total concentrations of VFA in the cecum of the pig and colon of sheep were similar to those in the rumen. Further studies (Elsden et al. 1946; Phillipson 1947) demonstrated that the similarity in total VFA concentration extended to the cecum and colon of the ox, deer, rat, rabbit, and dog. The above measurements were made under a variety of conditions with respect to diet, time after feeding, and time after death. However, the results of controlled studies, illustrated in Figure 6.2, show that the mean daily concentrations of VFA in the large intestine of the dog, raccoon, bush baby, vervet monkey, pig, and pony were equal to or greater than those found in the rumen of sheep and cattle. Lactic acid concentrations were low in the large intestine of the raccoon and dog, and extremely low in the large intestine of the pig and pony.

Volatile fatty acid levels in the large intestine of the baboon and Sykes monkey (Clemens and Phillips 1980) were similar to those in the vervet monkey. The same was true for the dugong (Murray et al. 1977). Volatile fatty acids levels in the rhinoceros large intestine were similar to those seen in the pony, but higher concentrations (122-148 mmol/L) were found in the large bowel of the elephant (Clemens and Maloiy 1982). Concentrations of VFA in the cecum of the rabbit (Hoover and Heitmann 1972) and greater glider *Petauroides volans* (Cork and Hume 1978) and in the koala hindgut (Cork and Hume 1983) were considerably lower than those in the other species listed above.

A change in the ratio of starch to fiber in the diet fed the dog, pig, and pony resulted in only minor changes in the pH and in the concentrations of VFA and lactic acid in either the stomach or large intestine. However, an increase in dietary fiber approximately doubled the volume and, therefore, the VFA content of the large intestine of the dog and pony. Measurements of individual VFA showed lower acetate:propionate ratios in the large intestine of pigs fed a high starch diet (Argenzio and Southworth 1974; Imoto and Namioka 1978). Thus, it appears that microbial digestion of carbohydrate in the large intestine produces VFA in proportions and concentrations similar to those seen in the ruminant forestomach, and that an increase in dietary starch levels has a similar effect on the acetate:propionate ratio.

The structural carbohydrates of plants (cellulose, hemicelluloses, and pectins) can provide a major source of the VFA found in the large intestine of herbivores. It was estimated that the large intestine accounted for 63-73% of the neutral detergent fiber digested by ponies (Hintz et al. 1971) and 12% of the cellulose digested by sheep (Goodall and Kay 1965). However, significant quantities of starch granules reached the large intestine of rats, mice, hamsters, guinea pigs, rabbits, pigs, and human subjects fed potato starch (Baker, Nasr, Morrice, and Bruce 1950). These amounts were decreased markedly by boiling or grinding of the starch before feeding or by substitution of cornstarch for potato starch. Keys and DeBarthe (1974) demonstrated that although less than 8% of the starch in the diet passed the terminal ileum of pigs fed corn, wheat, or milo, this increased to 21% when these animals were given barley. Studies of ponies (Hintz et al. 1971), cattle (Karr, Little, and Mitchell 1966), and sheep (Orskov, Fraser, and McDonald 1971) fed a diet containing high levels of corn revealed that up to 29%, 15%, and 6%, respectively, of the starch escaped to the large intestine.

Endogenous polysaccharides provide an additional source of substrate. Vercellotti, Salyers, and Wilkins (1978) suggested that mucus, which is 80% polysaccharide, is a major source of the carbohydrate presented to the large intestine of humans. This would allow the recovery of the constituents of mucus secreted by the gastrointestinal tract. Guinea pigs and rats maintained under germ-free conditions show a five- to tenfold increase in cecal volume in association with diarrhea and impaired motility of the intestine (Gordon and Bruckner 1984). The increased cecal volume appeared to be due to a marked increase in the colloid osmotic pressure of cecal contents as a result of the accumulation of excess mucinous material, particularly hexosamines and hexuronic acid. Amino acids of dietary or endogenous origin may serve as substrates for VFA production, as demonstrated in the ruminant forestomach (el-Shazly 1952a,b). Ehle, Robertson, and Van Soest (1982) concluded that most of the VFA produced in the human large intestine was derived from endogenous carbohydrate.

Calloway (1968) reviewed the information available on the composition of large-intestinal gases in the dog, rat, pig, ox, horse, and human. As in the rumen, these consisted primarily of CO_2, CH_4, H_2, N_2, and O_2. There was considerable variation among species and, especially, with changes in diet. It has been estimated that the intestinal contents provided approximately 13% of the CH_4 produced in the gastrointestinal tract of sheep (Murray, Bryant, and Leng 1976). In a study of 11 normal human subjects, Levitt and Bond (1970) reported a composition of $14 \pm 7\%$ CO_2, $9 \pm 9\%$ CH_4, $19 \pm 16\%$ H_2, $0.7 \pm 0.5\%$ O_2, and $64 \pm 21\%$ N_2 in the large intestine, which would indicate that the intestinal gases of humans have a much lower percentage of CO_2 and CH_4 and a much higher percentage of H_2 and N_2 than gases found in the rumen. In fact, CH_4 is said to be absent in about two-thirds of the human population (Levitt and Bond 1970). On the basis of studies of rumen gas composition, part of the variability noted in the above studies may result from differences in diet, time after eating, and duration of digesta retention.

Relatively little is known about the microbes indigenous to the digestive tract of lower vertebrates and their contribution to nutrition. Indigenous microorganisms have been described in the intestine of herbivorous fish such as the surgeon fish, *Acanthurus* species (Fishelson, Montgomery, and Myrberg 1985), but their function is unknown. There is evidence for cellulose digestion in the carp intestine (Shcherbina, Mochul'skaya, and Erman 1970; Shcherbina and Kazlauskene 1971), and Stickney and Shumway

(1974) found cellulase activity in the gut of 16 species of marine detritus feeders and the freshwater catfish (*Ictalurus punctatus*). Yet no cellulase activity could be found in the gut of the herbivorous algae feeder *Tilapia esqulenta* (Fish 1951), or the grass carp *Ctenopharyngodon idella* (Cross 1969), which consumes its weight in vegetation each day. There appears to be no information on microbial fermentation by herbivorous amphibian larvae. Bacteria in the large intestine of the adult leopard frog (*Rana pipiens*) showed a number of similarities to those of the mammalian large intestine, including the presence of acetogenic and butyricogenic organisms (Gossling, Loesche, and Nace 1982). However, the frog large intestine contained lower counts (10^{10}/g) and a greater proportion of facultative anaerobes.

Szarski (1962) suggested that the paucity of present-day reptilian herbivores was due to their inability to grind plant material efficiently and maintain a constant body temperature. Ostrom (1963) concluded that it was due to the construction and mobile articulation of the mandible, which provides a less efficient plant-grinding mechanism than that of the Mesozoic reptiles, chelonians, and other herbivorous tetrapods. Sokol (1965) noted that the herbivorous species of reptile tended to be larger and concluded that a minimal body size was required for efficient tearing and grinding of plant material. Pough (1973) found that all species of lizards within the families containing herbivores (Agamidae, Gerrhosauridae, Iguanidae, and Scincidae) that weighed more than 300 g as adults were herbivores, whereas species that weighed less than 50-100 g were carnivores. The small juveniles of herbivorous species also were carnivores or omnivores. This was attributed to the inability of the larger animals to satisfy their energy requirements on a diet of insects and, in the absence of larger alternative prey, a reliance on vegetation. A number of tortoises and some marine turtles also are true herbivores. Guard (1980) demonstrated VFA in the hindgut of the carnivorous caiman (*Caiman crocodilus*) and an omnivorous terrapin (*Chrysemys picta belli*) and relatively high concentrations of VFA in the cecum of an herbivorous lizard (*Iguana iguana*) and tortoise (*Geochelone carbonaria*).

Bacteria colonize the intestinal tract of chicks within the first few days after hatching (Ochi, Mitsuoka, and Sega 1964; Smith 1965). Within the first week, this becomes stabilized to a predominance of lactobacilli in the small intestine and *Bacteroides, Clostridium, E. coli,* and anaerobic lactobacilli in the cecum (Salanitro et al. 1978). It has been estimated that 11% of the energy requirements of the chicken could come from acetate production in the ceca (Annison, Hill, and Kenworthy 1968), and VFA levels of 60-70 mmol/L were measured in the cecum and colon of geese (Clemens, Stevens, and Southworth 1975b) under conditions similar to those described for the mammals in Figure 6.2. However, the relatively small size of the chicken and goose ceca suggests that microbial fermentation would contribute little to the energy requirements of these birds.

Most of the information on microbial fermentation in birds is derived from studies of various species of grouse (McBee 1977). Fermentation of carbohydrate and the production of VFA in concentrations and composition similar to those of the mammalian large intestine have been demonstrated in the willow ptarmigan (McBee and West 1969), an Alaskan grouse that winters primarily on buds and twigs of the willow, and in the rock ptarmigan (Gasaway 1976) and red grouse (Moss and Parkinson 1972). It was estimated that the VFA produced in the cecum may provide 4-11% of the energy requirements of these birds.

Absorption of volatile fatty acids

Early studies showed that most of the VFA produced in the rumen of sheep were directly absorbed (Barcroft et al. 1944). Volatile fatty acids are absorbed at similar rates from the omasum (Engelhardt and Hauffe 1975). Although concentrations of VFA are much lower in the abomasum, they are rapidly absorbed at the lumen pH of abomasal contents. The observation that the rate of VFA absorption increased with either a decrease in the pH of rumen contents or an increase in the chain length of the individual fatty acids (Ac<Prop<But) (Danielli, Hitchcock, Marshall, and Phillipson 1945) suggested that they were absorbed in their more lipid-soluble, undissociated form. However, the concentration of the three fatty acids in the venous return from the rumen actually decreased with chain length (Masson and Phillipson 1951), and it was found that the different rates of absorption and transport to the blood could be readily explained by differences in the rate that acetate, propionate and butyrate were metabolized by rumen epithelium (Pennington 1952; Pennington and Sutherland 1956; Stevens and Stettler 1966a). Absorption of VFA in only the undissociated form also failed to account for their relatively rapid uptake at a pH of 6.8 or above, where less than 1% of the fatty acid would be present in this form.

Ash and Dobson (1963) found that the absorption of acetate from the rumen of sheep was accompanied by an increase in HCO_3 and a decrease in the P_{CO_2} of lumen contents. They concluded that the hydration of CO_2 to H_2CO_3 in the rumen generated a continuous source of protons for the absorption of undissociated acetate and HCO_3 for release into the lumen contents. As this would account for only one-half of the acetate absorbed, they proposed that the remainder was absorbed by passive diffusion or active transport of the anion. However, in vitro studies indicated that rumen epithelium was impermeable to the transmural passive diffusion of acetate anion (Stevens and Stettler 1966b). Further studies showed that rumen epithelium could transport acetate against a transepithelial electrochemical gradient, but in the direction of blood to lumen—the wrong direction to account for the apparent absorption of anion (Stevens and Stettler 1967).

Subsequent experiments demonstrated that rumen epithelium transported other weak organic acids to the lumen bath and weak organic bases in the opposite direction against their respective electrochemical gradients (Stevens, Dobson, and Mammano 1969). A model was proposed that explained these results on the basis of 1) the electrical and pH gradients between a tissue compartment and the bathing solutions, and 2) a difference in the membranes separating this compartment from the opposing bathing solutions with respect to their permeability to the dissociated and undissociated forms of the fatty acid. It was suggested that epithelial cells provided the compartment, in which case the transport of these organic electrolytes would be driven by the electrochemical gradient of protons between cell compartments and bathing solutions. A similar model was proposed for the transport of weak organic electrolytes across the small intestine of rats, but it was suggested that this was due to an intercellular or submucosal compartment at a pH higher than that of the opposing bathing solutions (Jackson 1973; Jackson, Tai, and Steane 1981).

Table 6.2 compares in vitro studies of VFA transport by the gastric and/or large intestinal epithelium of the pig, pony, and dog, to results obtained with rumen epithelium. Stratified squamous epithelium from the pony stomach showed no evidence of VFA uptake or transport, but each of the remaining tissues transported VFA at rates similar to that of the rumen epithelium. Mucosa from the pony large intestine showed additional similarities to rumen epithelium in the metabolism of VFA, impermeability to passive diffusion of acetate anion, and tendency to transport acetate to the lumen.

Perfusion of the colon of various species with solutions similar to those normally present showed two general patterns of absorption and secretion (Table 6.3). Absorption of VFA from the proximal colon of the pony and the colon of the pig and human was accompanied by an increase in the HCO_3 and decrease in the P_{CO_2} levels of lumen contents, in a manner similar to that described for the reticulorumen. This was associated with the absorption of Na at a rate much slower than that of VFA. The goat and dog colon absorbed Na and VFA at an equivalent rate with no net appearance of HCO_3 in the lumen. Although the colon of the pig and dog absorbed VFA at approximately the same rate per unit area of mucosal surface, the dog's colonic mucosa absorbed Na and water at twice the rate noted for the pig. In vitro studies showed also that although VFA was absorbed at similar rates by the proximal and distal colon of the pony, the distal colon absorbed Na six times more rapidly. Therefore, differences in the relative rates of VFA and Na absorption appear to result largely from different rates of Na absorption.

Microbial fermentation would be favored by the release of HCO_3 and retention of water in the lumen of the forestomach or large intestine. A more rapid rate of Na absorption is advantageous to the conservation of both Na and water (see Chapter 7). Table 6.3 includes an estimate of the contribution of the absorbed VFA to the metabolic requirements of each species.

TABLE 6.2 Transport of VFA across isolated, short-circuited gastrointestinal epithelium

Tissue	Ox		Pig		Pony		Dog	
	Loss lumen side	Gain blood side	Loss lumen side	Gain blood side	Loss lumen side	Gain blood side	Loss lumen side	Gain blood side
Stratified squamous epithelium	10.5 ± 1.4[a]	2.4 ± 0.4	6.1 ± 2.4	0.4 ± 0.1	1.4 ± 1.4	0.002 ± 0.02	—	—
Cardiac mucosa	—	—	8.3 ± 1.8	1.1 ± 0.3	—	—	—	—
Proper gastric mucosa	—	—	5.0 ± 1.0	0.3 ± 0.04	9.4 ± 3.8	0.2 ± 0.1	—	—
Pyloric mucosa	—	—	5.8 ± 1.1	0.5 ± 0.02	5.5 ± 1.4	0.4 ± 0.1	—	—
Cecal mucosa	—	—	10.3 ± 2.9	4.3 ± 0.6	8.2 ± 0.8	1.6 ± 0.2	—	—
Proximal colon mucosa	—	—	8.0 ± 0.8	3.7 ± 0.6	9.8 ± 1.4	1.8 ± 0.2	8.8 ± 0.3	4.0 ± 0.2
Distal colon mucosa	—	—	9.8 ± 1.0	3.1 ± 0.2	6.5 ± 1.5	1.5 ± 0.2	8.8 ± 0.3	4.0 ± 0.2

Note: Values are means ± SE for the rate of VFA transport (lumen to blood) obtained during 2.5-hour experimental periods. Ringer's solution, containing a 90 mM equimolar mixture of acetate, propionate, and butyrate, was used to bathe the lumen surface of tissue, and normal Ringer's was used to bathe blood surface; both solutions were buffered at pH 7.4 with bicarbonate. The transmucosal electrical potential difference (PD) was clamped at zero.

[a]Volatile fatty acids in $\mu mol/cm^2 \cdot hr$.

Source: Modified from Stevens and Stettler (1966b), Argenzio, Southworth, and Stevens (1974), Argenzio and Southworth (1974), and Herschel, Argenzio, Southworth, and Stevens (1981).

TABLE 6.3 In vivo absorption or net appearance of volatile fatty acids (VFA), Na, HCO$_3$, and water by the isolated reticulorumen of sheep, and the large intestine of pony, pig, human, goat, and dog

Species	Segment	pH	VFA (mmol/L)	Sodium (mmol/L)	Bicarbonate (mmol/L)	Chloride (mmol/L)
				Bathing solution		
Sheep	Reticulorumen	7.0	134	149	9.1	29
Pony	Ventral colon	6.1	100	140	20	20
Pig	Proximal colon					
	Distal colon and rectum	6.4	107	122	15	15
Human	Cecum, colon, and rectum	7.4	90	120	20	40
Goat	Colon and rectum	6.0	70	100	20	30
Dog	Colon and rectum	6.4	90	122	27	15

Note: Bathing solutions were isotonic to plasma of each species but varied in initial composition. In sheep, pony, pig, and human studies, the VFA consisted only of acetate. The solution used in the goat study contained 60 mM acetate and 10 mM propionate. That used to perfuse the dog colon was an equimolar mixture of acetate, propionate, and butyrate. Positive values designate net absorption, and negative values net appearance within the lumen. Appearance of HCO$_3$ in the goat and dog colon was not significant (NS). A steady-state perfusion technique was used on all species except the sheep and pony, in which a

Even bearing in mind the differences in pH of bathing solutions and evidence that VFA production may be the rate-limiting factor in its absorption from the pig colon (Argenzio and Whipp 1979), it would appear that VFA could provide substantial portions of the energy required by each species.

Perfusion studies also showed an interdependency between the absorption of VFA and Na. A decrease in the pH of solutions used to perfuse the colon of the pig, goat, and dog increased the rate of both VFA and Na absorption. Replacement of the VFA with Cl resulted in a marked decrease in the absorption of Na from the colon of the goat and the pig (Crump, Argenzio, and Whipp 1980), and replacement of Na with choline resulted in a reciprocal decrease in the absorption of VFA from the colon of the rat (Umesaki, Yajima, Yokokura, and Mutai 1979), sheep (Rübsamen and Engelhardt 1981), and pig (W. E. Roe, unpublished).

Results from the above studies of rumen epithelium and colonic mucosa could be explained by the model illustrated in Figure 6.3. Absorption of undissociated VFA would be aided in part by hydration of CO$_2$ in the lumen and partly by H secreted in exchange for Na. Bicarbonate, released in the lumen and transported into the lumen in exchange for Cl, would aid in the buffering of VFA as it is produced, and help limit its rate of absorption during periods

Daily absorption or appearance per kg body weight					
VFA (mmol)	Sodium (mmol)	Bicarbonate (mmol)	Water (ml)	BMR[a] (%)	Source
30	15	−15	40	23	Dobson (1959)
31	7	−12	18	33	Argenzio, Southworth, Lowe, and Stevens (1977)
55	26	−14	170	44	Argenzio and Whipp (1979)
40	20	−9	110	32	
9.8	3	−4.5	24	8.4	Ruppin et al. (1980)
31	29	NS[b]	190	26	Argenzio, Miller, and Engelhardt (1975)
7.5	8.3	NS	58	4	Herschel et al. (1981)

static system was used. The contribution of VFA absorption to BMR was calculated as if all VFA were absorbed as acetate at an energetic equivalent of 875 kJ (209 kcal)/mol. Absorption of an equimolar mixture of the three VFA would provide approximately twice the energy contribution of acetate alone.
[a]BMR, basal metabolic requirement.
[b]NS, not significant.
Source: Adapted from Stevens, Argenzio, and Clemens (1980).

of peak production. Intracellular metabolism of VFA would help provide H ions for the stimulation of Na absorption. Permeability of the blood-facing membrane to both the anion and undissociated forms of VFA also could account for the apparent "active" transport of organic acids and bases. Under normal conditions, this would help prevent accumulation of VFA within the cells.

Nitrogen recycling and protein synthesis

Stomach

Rumen microbes are capable of converting both protein and nonprotein nitrogenous compounds into microbial protein, which is digested in the abomasum and small intestine with subsequent absorption of the end products. Approximately two-thirds of the dietary protein of cattle on a normal ration can be digested by microbial enzymes, yielding peptides, amino acids, and ammonia. One major source of nonprotein nitrogen is endogenous urea, which enters the forestomach via both salivary secretion and direct diffusion

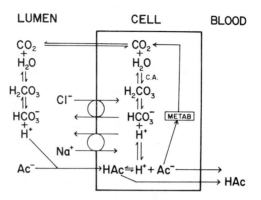

FIGURE 6.3 Hypothesis for the mechanisms of VFA, Na, Cl, and HCO_3 transport by rumen epithelium and colonic mucosa. High levels of CO_2 production from microbial fermentation in the lumen allow for its rapid hydration in the absence of carbonic anhydrase. This provides H ion for the nonionic diffusion of acetate or other VFA into the cell and releases HCO_3 into the lumen. Similar intracellular hydration of CO_2, derived from metabolism of VFA and other substrates, is catalyzed by carbonic anhydrase, providing HCO_3 and H that can be exchanged for the Cl and Na in the lumen. The relatively low levels of Cl normally present in the lumen could result in the more rapid secretion of H than HCO_3 into the lumen, which would aid in VFA absorption and favor release of cellular HCO_3 into the blood. Acetate is transported to the blood by diffusion of both the dissociated and undissociated forms of the fatty acid. Transport of Cl and Na to the blood (not depicted) is accomplished by diffusion of Cl down its electrochemical gradient and Na-K-ATPase transport of Na. (From Stevens, Argenzio, and Roberts, 1986.)

across its epithelium. This allows for both the synthesis of high-quality protein from low-quality diets and the conservation of nitrogen. The recycling of urea also allows for the conservation of water that would otherwise be required for its urinary excretion.

Utilization of nitrogen and fermentation of carbohydrate are interdependent. A shortage of dietary nitrogen can hamper the digestion of poor-quality roughage such as straw (Balch and Campling 1965). This may be due to the marked effect of nitrogen on bacterial numbers (Teather, Erfle, Boila, and Sauer 1980). The rate at which urea is recycled by rumen microorganisms is similarly dependent on an adequate supply of dietary nitrogen (Engelhardt 1978). Restriction of water increases the rate at which urea is recycled by temperate, tropical, and desert ruminants (Phillips 1961; Livingston, Payne, and Friend 1962; Utley, Bradley, and Boling 1970; Mousa, Ali, and Hume 1983).

The recycling of endogenous urea from blood to the digestive tract, with its utilization by gut microorganisms, has been shown to increase the ef-

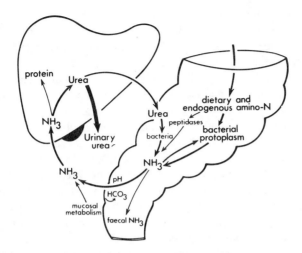

FIGURE 6.4 The origins and fates of intestinal ammonia. (From Wrong and Vince 1984.)

ficiency of nitrogen utilization by sheep (Cocimano and Leng 1967), camels (Schmidt-Nielsen, Schmidt-Nielsen, Houpt, and Jarnum 1957), and marsupial macropods such as the euro *Macropus robustus erubescens* (Brown 1969) and tammar wallaby (Lintern-Moore 1973; Kinnear and Main 1975; Kennedy and Hume 1978). A decrease in the concentration of dietary protein has been shown to increase the utilization of urea by the llama (Hinderer 1978) and wallaby (Kennedy and Hume 1978).

Hindgut

Microbes indigenous to the hindgut also synthesize protein. Significant amounts of amino acids were found to enter the large intestine of pigs on a soya bean meal diet (Holmes, Bayley, Leadbeater, and Horney 1974). Substantial quantities of endogenous substrate are provided in the form of digestive enzymes, desquamated mucosal cells, and diffusion of urea into the gut. Wrong and Vince (1984) reviewed information on the fate of nitrogenous compounds in the human large intestine. Most of the nitrogen that enters the large intestine from the ileum consists of urea and creatinine at levels equivalent to those of blood or protein derived from endogenous intestinal secretions, shed epithelial cells, or dietary residues (Fig. 6.4). The total quantity of endogenous amino acid nitrogen in the ileum of pigs was markedly increased by an increase in dietary fiber (Sauer, Stothers, and Parker 1977). Determinations of amino acid composition in the ileal contents of pigs suggested that these were derived predominantly from mucin and those amino acids that are least efficiently absorbed by the small intestine (Taverner, Hume, and Farrell 1981a,b).

There is substantial evidence that urea is secreted into the digestive tract of mammals that lack extensive foregut fermentation, as shown by high concentrations in the gut after the feeding of antibiotics to rats (Chao and Tarver 1953) or the establishment of germ-free conditions (Levenson, Crowley, Horowitz, and Malm 1959). The major site in animals that lack extensive foregut fermentation appears to be the hindgut. Urea is extensively recycled by the rabbit (Regoeczi, Irons, Koj, and McFarlane 1965), pony (Prior, Hintz, Lowe, and Visek 1974), and rock hyrax (Hume, Rübsamen, and Engelhardt 1980). Urea utilization by the hyrax increased with reduction of dietary protein or restriction of water. Wrong, Edmonds, and Chadwick (1981) point out that although there is evidence that the large intestine of the dog and the cecum of the rabbit are moderately permeable to urea, numerous studies indicate that relatively little urea passes into the experimentally cleansed large intestine of humans. Similar results were seen in studies of the perfused colon of sheep and goats (Engelhardt, Hinderer, Rechkemmer, and Becker 1984). However, this could be explained by urea hydrolysis within the large intestinal mucosa or at a juxtamucosal site, as suggested by Houpt and Houpt (1968) as an explanation for the apparent disappearance of urea as it passes across rumen epithelium. There appears to be little information on the recycling of nitrogen in the gut of nonmammalian vertebrates, other than the demonstration of microbial breakdown of uric acid in chickens (Bell and Bird 1966; Mead 1974; Mead and Adams 1975).

Urea is hydrolyzed to ammonia, which can be incorporated into microbial protein or absorbed and recycled to form nonessential amino acids. Although the levels of ammonia in feces are higher than those of the blood, endogenous urea is not the main precursor of human fecal ammonia, and most of the nitrogen in the digesta and feces is contained in bacterial protein (Wrong and Vince 1984; Wrong, Vince, and Waterlow 1985). Bonnafous and Raynaud (1968) provided evidence that microorganisms are digested in the proximal colon of the rabbit, and Wootton and Argenzio (1975) found that substantial amounts of microbial protein were digested within the large intestine of ponies.

Figure 6.5 shows changes in the volume and in the net appearance and disappearance of water, VFA, protein nitrogen, and urea plus ammonia in the four major compartments of the pony large intestine as a function of time after feeding. Equations derived from measurements of fluid marker transit between compartments allowed subtraction of the quantity of each substance added or removed by digesta transit between compartments. Each compartment underwent cyclic changes in volume according to elapsed time after feeding. This was especially marked in the ventral colon, which demonstrated a fivefold increase in volume during the first 8 hours after a meal. These cyclic changes in volume were associated with cyclic periods of net secretion and absorption of water that, in turn, correlated with net pro-

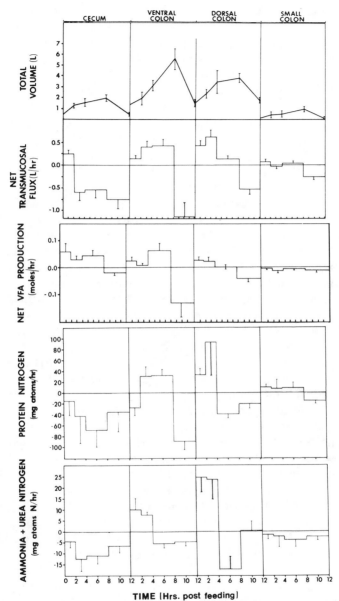

FIGURE 6.5 Volume, net transmucosal exchange of water, and net appearance and disappearance of VFA, protein, and urea + NH_3 nitrogen in the pony large intestine as a function of time after feeding. Animals were fed at 12-hour intervals. All values, other than volume, are corrected for exchange between segments resulting from digesta transit. (Modified from Argenzio and Southworth 1974; Argenzio et al. 1974; Wootton and Argenzio 1975.)

duction and absorption of VFA. With the exception of the cecum, VFA production was associated with an increase in protein nitrogen, presumably resulting from synthesis of microbial protein. The accompanying decrease in urea plus ammonia suggested that much of this was derived from urea diffusing across the mucosa. A net disappearance of both protein and nonprotein nitrogen during the last period of the cycle indicated in situ digestion of microbial protein and absorption of end products. These cyclic changes in the volume and quantities of VFA, protein, and nonprotein nitrogen provide another indication of the similarity between microbial digestive processes in the ruminant forestomach and large intestine.

The above results provide a good example of both the advantages and limitations of comparative information gained under any given set of circumstances. The use of ponies fed at 24-hour intervals allowed the demonstration of these cyclic activities by measurements of the appearance, disappearance, and transit of digesta contents. This would be difficult or impossible to measure in most other species, and the value of the pony large intestine to the general understanding of large-intestinal function results from differences rather than similarities in its construction. Furthermore, Clark and Argenzio (personal communication) found little evidence of this cyclic activity in ponies subjected to continuous feeding, the condition seen in the feral and, if left to their own preferences, domesticated states. Therefore, these pronounced cycles of microbial fermentation and the marked changes in fluid volume associated with these cycles do not appear to be the normal situation. This may explain a slower transit of digesta fluid in these studies as compared to those of other workers, as noted in Chapter 4. It also may explain the high incidence of large intestinal colic in horses that are fed concentrates only once or twice a day.

The concentration of nitrogen decreased as digesta progressed through the large intestine of the sheep (Hogan and Phillipson 1960; Goodall and Kay 1965), pony (Hintz et al. 1971), and rabbit (Rérat 1978). Active absorption of amino acids by large intestinal mucosa has been demonstrated in the newborn rat (Batt and Schachter 1969) and pig (James and Smith 1976; Jarvis, Morgan, Smith, and Wooding 1977), and passive absorption of amino acids has been shown in the rat (Binder 1970) and tortoise (Baillien and Schoffeniels 1961). Introduction of ^{15}N-labeled microbial protein into the cecum of the horse (Slade, Bishop, Morris, and Robinson 1971) and pig (Niiyama, Dequchi, Kagota, and Namioka 1979) resulted in the appearance of ^{15}N-labeled essential and nonessential amino acids in the cecal venous blood. However, infusion of ^{14}C-labeled alanine into the cecum of rats resulted in considerable breakdown and little absorption of alanine (Hoover and Heitmann 1975), and carbon skeletons of amino acids have been shown to be rapidly absorbed from the large intestine (Fordtran, Scroggie, and Polter 1964). Infusion of isotopically labeled leucine and isoleucine into the cecum of pigs failed to demonstrate absorption of either amino acid (Krawielitzki et al. 1983, 1984). Thus, it appears that there is no definite evidence for active absorption of amino acids by the large intestine of adult mammals.

Synthesis of vitamins

Rumen microorganisms are capable of synthesizing all of the B-complex vitamins required by sheep and cattle (Phillipson 1970). A B_{12} deficiency can be produced, but only if the diet is deficient in cobalt. Much of the thiamin, pantothenic acid, pyridoxine, and biotin are present in solution, and all of the B vitamins appear to be directly absorbed from the rumen (Dziuk 1984).

The presence of B vitamins in rat feces was recognized early (Steenbock, Seel, and Nelson 1923) and subsequent studies indicated that rats fed a diet deficient in biotin, riboflavin, pyridoxine, pantothenic acid, B_{12}, folic acid, or vitamin K did not show vitamin deficiencies if they were allowed to ingest their feces. Deficiencies of thiamin (Wostmann, Knight, and Reyniers 1958), pantothenic acid (Daft et al. 1963), folic acid (Luckey et al. 1955), biotin (Luckey et al. 1955), and vitamin K (Gustafsson 1948), and a susceptibility to riboflavin deficiency (Luckey et al. 1955) were demonstrated in germ-free rats. Most of the thiamin, riboflavin, and niacin in the rat cecum was present in bacteria (Mitchell and Isbel 1942) or, in the case of thiamin, as a pyrophosphate complexed with a large apoprotein (Wostmann and Knight 1961). A wide variety of enteric microorganisms are capable of taking up B_{12} from solutions (Giannella, Broitman, and Zamcheck 1971). However, absorption of thiamin from the cecum has been demonstrated in rats (Kasper 1962) and B_{12}, B_6, and pantothenic acid were found to be absorbed equally well after their oral or large-intestinal administration to human patients (Sorrell etal. 1971). The latter was not true for folates, biotin, ascorbate, or vitamins A and E.

Wrong et al. (1981) concluded that there was good evidence that thiamin, nicotinic acid (niacin, nicotinamide), riboflavin, pantothenic acid, biotin, pyridoxine, folic acid, vitamin B_{12}, and vitamin K are synthesized by microbes in the human large intestine and that all of these, except nicotinic acid, riboflavin, and pantothenic acid, are normally subject to some degree of direct absorption.

Coprophagy

The ingestion of feces is of nutritional importance in many animals. Robertson (1982) described widespread coprophagy among herbivorous and detrivorous fish on a Pacific coral reef. Autocoprophagy and intraspecies coprophagy were rare. Most of the interactions consisted of herbivores eating the feces of zooplanktivores and other carnivorous fish. In some species, this occurred to such a degree that fecal material constituted an important component of their diet. Hatchling green iguanas (*Iguana iguana*) regularly consume the feces of adult animals, a practice that presumably helps establish microbial colonization of their gut (Troyer 1984). Ingestion of maternal feces has been observed also in young golden hamsters (Dieterlein 1959), rats (Galef 1979), thoroughbred foals (Francis-Smith and Wood-Gush 1977), and koalas (Minchin 1973).

The reflex defecation elicited in the nursing young of many mammals by maternal licking of the anus is accompanied by ingestion of the feces. Marsupial females follow a similar practice by periodically cleaning their pouch. Ingestion of urine and feces from the offspring has been shown to conserve 50-80% of the water that otherwise would be lost by lactating female rodents and dingoes (Baverstock, Watts, and Spencer 1979).

Working sled dogs (Kronfeld 1973) and domestic cats (McCuistion 1966) on high-carbohydrate, cereal diets became hypoglycemic and practiced coprophagy after an overnight fast. It was suggested that coprophagy may provide amylase or other enzymes to these animals. It has been similarly suggested that the appearance of coprophagy in adult horses, within three months after they were placed on 6.2% crude protein diet, and its subsequent disappearance within a week after the protein level was raised to 10%, indicated an attempt to overcome a dietary protein deficiency (Schurg, Frei, Cheeke, and Holtan 1977). Coprophagy has been described also in shrews (Crowcroft 1952; Baxter and Meester 1982).

Because ingested feces displace food that would normally be taken into the stomach, the advantages of coprophagy depend on the relative nutrient content of the diet and the feces. As mentioned in Chapter 4, some animals periodically excrete and ingest feces that have a composition similar to that of their cecal contents. This has been termed "cecotrophy" (Harder 1950). Cecotrophy is seen in rabbits (Olsen and Madsen 1943; Hamilton 1955; Geis 1957), hares (Watson and Taylor 1955; Bookhout 1959; Pehrson 1983), and pika (Haga 1960). It has been described in numerous rodents, including the naked mole rat (Hill et al. 1957), beaver (Richard 1959), mountain beaver (Ingles 1961), ground squirrel (Turcek 1963), and chinchilla (Björnhag and Sjöblom 1977). It appears to be absent from other rodents such as the porcupine (McBee 1977) and capybara (Hörnicke and Björnhag 1979). Cecotrophy is practiced by the folivorous lemur *Lepilemur mustelinus leucopus* (Hladik et al. 1971) and the ringtail possum (Chilcott and Hume 1985). In some animals such as the koala (Minchin 1973) and rat (Leon 1974; Galef 1979), cecotrophy is seen only in the female during the weaning period.

The nutritional significance of coprophagy and cecotrophy was reviewed by Hörnicke and Björnhag (1979). Most of the detailed studies have been conducted on rats and rabbits. Although early studies claimed that cecotrophy was normally practiced by rats and mice, it now appears that there is relatively little difference in the composition of the two types of feces excreted by each of these animals. There is, however, a considerable difference between the two types of feces produced by rabbits (Eden 1940; Heisinger 1965; Henning and Hird 1972). In addition to dry, hard fecal pellets, rabbits produce soft feces with a strong mucous envelope, which are high in protein, vitamins, VFA, Na, K, and water (Table 6.4). The latter are produced during one or two periods totaling 8-10 hours per day (Jilge 1979). They are taken from the anus, swallowed without mastication, and stored in the stomach (within the membrane) for up to 6 hours after ingestion, where they maintain a relatively high pH due to $NaHPO_4$ and KH_2PO_4 buffers (Griffiths and Davies 1963). These pellets also contain bacterial amylase.

Nitrogen-rich cecotrophs also are produced by chinchilla during the daytime and by guinea pigs in short periods dispersed throughout the day and night (Holtenius and Björnhag, 1985). Other animals that practice cecotrophy show differences in the chemical composition of the two types of feces, but less difference in their size, shape, or consistency. Pika produce soft feces, at random during the day or night, that consist of an amorphous paste with no mucous envelope (Haga 1960). Some of these are directly ingested. The remainder are pasted to stones or other surfaces for later ingestion. The cecotroph is well formed in the naked mole rat, mountain beaver, and ground squirrel, but pulpy in the koala, beaver, and lepilemur. The stimulus for ingestion of cecotrophs is not known, but germ-free rabbits consume neither their hard nor their soft fecal pellets (Yoshida, Pleasants, Reddy, and Wostmann 1968). Volatile fatty acids, particularly butyric acid, have been implicated because of their strong odor and taste and their high concentration in soft feces (Henning and Hird 1972).

Cecotrophy can be extremely important to the nutrition of rabbits. It appears to provide B vitamins in excess of the animal's needs. Vitamin B_{12} was found to be synthesized at 100 times its daily requirement (Simnet and Spray 1961). Cecotrophs provided up to 30% of the total nitrogen intake of rabbits (Table 6.4), and the microbial protein was readily digestible with a high content of essential amino acids (Hörnicke 1981). Nitrogen balance was maintained in rabbits on an 8% protein diet by infusion of urea into the cecum (Salse, Crampes, and Raynaud 1977). Therefore, the practice of cecotrophy can greatly improve the biological value of protein in a diet. The VFA in the fecal pellets also would provide energy.

The function of coprophagy has been extensively studied in the rat. Osborne and Mendel (1911) made early mention of the fact that prevention of coprophagy decreased the growth rate of rats. However, coprophagy is difficult to prevent without the use of small quarters, large collars, or closely fit-

TABLE 6.4 Some components of rabbit feces and cecotrophs

Component	Feed	Feces Mean	Feces SD	Cecotrophs Mean	Cecotrophs SD	Ratio cecotrophs to feces	Reference
H_2O (%)	9.73	51.6 ±	3.4	70.1 ±	1.8	1.36	
Crude protein (%)	14.90	18.15 ±	0.57	28.15 ±	2.84	1.52	
True protein (%)	10.94	6.97 ±	0.52	18.55 ±	1.83	2.66	
NPN (%)	0.63	0.19 ±	0.04	1.60 ±	0.20	8.42	
Crude fiber (%)	19.7	29.6 ±	1.6	17.8 ±	2.4	0.60	Kandatsu,
SiO_2 (%)	2.92	4.99 ±	0.79	3.75 ±	0.65	0.75	Yoshihara, and
SO_3 (%)	0.68	0.50 ±	0.07	1.28 ±	0.21	2.56	Yoshida (1959)
MgO (%)	0.72	0.87 ±	0.06	1.28 ±	0.12	1.47	
CaO (%)	1.19	1.80 ±	0.37	1.35 ±	0.22	0.75	
Fe_2O_3 (%)	0.01	0.25 ±	0.06	0.26 ±	0.06	1.04	
Total P (%)	0.83	0.95 ±	0.17	1.54 ±	0.11	1.62	
Inorg. P (%)	0.16	0.60 ±	0.19	1.04 ±	0.26	1.73	
Organ. P (%)	0.65	0.35 ±	0.06	0.50 ±	0.15	1.43	
Cl (mmol/kg DM)		33 ±	4.5	55 ±	4	1.67	
Na (mmol/kg DM)		37.6 ±	8.2	104.7 ±	21.2	2.78	
K (mmol/kg DM)		84 ±	11.6	259.6 ±	29.5	3.09	
Total VFA (mmol/kg DM)		45		180		4.0	Bonnafous (1973)
Acetate (mmol/kg DM)		39.7 ±	8.8	123 ±	3	3.10	
Propionate (mmol/kg DM)		2.8 ±	1.7	10.7 ±	3	3.82	
Butyrate (mmol/kg DM)		2.7 ±	2	44.1 ±	5	16.33	
pH		7.91 ±	0.16	6.38 ±	0.10	33.88	
Cholesterol (%)		0.71 ±	0.13	0.40 ±	0.04	0.56	
Bacteria (10^{10}/g DM)		31 ±	1.5	142 ±	12	4.58	

Note: Values are given on dry matter (DM) basis; means with standard deviation.
Source: Hörnicke and Björnhag (1979).

ting harnesses or jackets. Barnes and co-workers (Barnes, Fiala, McGehee, and Brown 1957; Barnes and Fiala 1958; Barnes, Fiala, and Kwong 1963) showed that prevention of coprophagy was associated with a 15-25% depression in growth rate, even on rations adequate in dietary nutrients. Rats allowed to practice coprophagy maintained their growth rate only if allowed to ingest feces directly from the anus. Furthermore, prevention of coprophagy has been shown to result in changes in fecal microbes (Barnes 1962; Fitzgerald, Gustafsson, and McDaniel 1964). In spite of these problems, there is good evidence that microbial synthesis and coprophagy can provide rats with their requirements for vitamin K and at least part of their requirements for the B-complex vitamins.

Evolution of herbivores

The conditions most suitable for multiplication of microorganisms in the gastrointestinal tract of vertebrates appear to have been first met by the hindgut of adult amphibians. This provided a means for the conservation of endogenous nitrogen and carbohydrates and for the further digestion of food that escaped digestion or absorption by the upper gastrointestinal tract. However, the nutritional importance of this process is dependent upon the hindgut's capacity for retention of digesta and absorption of end products.

Parra (1978) examined the relationship between gut capacity and body weight in a large number of herbivorous mammals and found that the relative capacity of the gastrointestinal tract decreased with a decrease in body size. From this, and the fact that the rate of metabolism per unit body mass tends to increase with a decrease in body size, he concluded that microbial production of nutrients is limited in smaller species by the need for a sufficiently rapid rate of digesta transit. This would help explain the absence of herbivorous species among the small reptiles and birds. It also explains the marked advantages of selectively retaining fluid, microorganisms, and small plant particles in the cecum (or ceca) of species of relatively small body size. Cecotrophy provides an even greater advantage. A lessening of this restriction on gut capacity allowed larger herbivores to utilize the proximal colon or stomach as the principal site for microbial production of nutrients.

The evolution of ungulates was examined by Janis (1976), and the relationship between this and the evolution of plants was further examined by Van Soest (1982). The end of the Cretaceous and beginning of the Tertiary periods, approximately 70 million years ago, witnessed the extinction of dinosaurs, a beginning of the replacement of gymnosperms with angiosperms (flowering plants), and the appearance of the protoungulate ancestors of the perissodactyls and artiodactyls (Fig. 6.6). The widespread tropical forests of the time allowed broad selection of plant material. At first the perissodactyls were the predominant group of ungulates, but by the end of the Eocene Epoch of the Tertiary Period, the artiodactyls were in equal num-

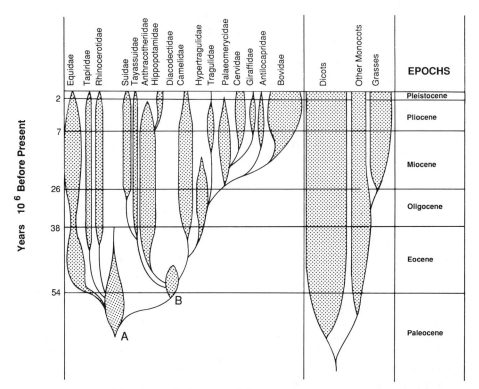

FIGURE 6.6 Schematic phylogeny of the ungulates and angiosperms. "A" represents the common ancestor of the artiodactyls and perissodactyls, and "B" represents the ancestral artiodactyl. The ungulate phylogeny is from Romer (1966) with adaptations from Janis (1976) and Hume and Warner (1980). The phylogeny of plants is from Van Soest (1982).

ber (Romer 1966). The cooler, drier climates of the Miocene Epoch favored the development of grasslands high in cellulose and low in lignin, and this was associated with the emergence of the artiodactyls as the predominant ungulates in both genera and species.

The earliest ungulates were probably selective feeders, such as the tapirs and small ruminants that are present today. Microbial digestion in the large intestine allows removal of readily available nutrients before their conversion to the less nutritive end products of microbial digestion. As Janis (1976) pointed out, this arrangement also allows for the more rapid transit of digesta on diets high in fiber content. However, prolonged retention of plant material in the stomach allows a greater extraction of energy, and the synthesis of proteins and vitamins from a high cellulose-low protein diet, prior to the presentation of these digesta to the small intestine. It also provides a

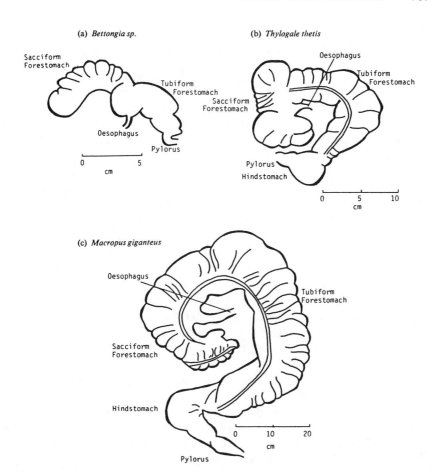

FIGURE 6.7 Stomachs of macropods showing variations in the relative length of the sacciform and tubiform segments of the forestomach. (From Hume and Warner 1980.)

means for destroying plant toxins that may be present in the diet of nonselective feeders. Retention would be aided by the development of a forestomach consisting of a cul-de-sac connected by a small orifice to the secretory compartment of the stomach, and further aided by development of the omasum. Rumination provides the additional advantage of reducing the time and energy necessary for feeding.

Van Soest (1982) estimated that the ruminant type of digestion would have an advantage over that of perissodactyls in animals with a body weight of 5-1800 kg. This fits quite closely to the range seen in present-day ruminants, with the exception of a few small selective feeders. Therefore, microbial digestion in the stomach appears to have set a limit on body size that was not experienced by perissodactyls such as the rhinoceros, which can range up to 2800 kg in body weight, or the elephants, which can reach weights of 5900 kg.

The limitations of body size on the evolution of microbial digestion in the forestomach of ruminants would apply to a lesser degree in animals such as the macropod marsupials, which retain particulate digesta within their forestomach for shorter periods of time. The same would be true for animals such as the macropods and sloths that have a lower rate of metabolism. However, the general rule appears to apply to all present-day mammals. Hume and Warner (1980) proposed that the earliest macropods may have had a forestomach similar to that of the small, more primitive rat-kangaroos (*Bettongia* sp.), which consists of a large sacciform cul-de-sac and a short tubiform region (Fig. 6.7). A relatively large sacciform segment is retained by many browsing macropods, but the grazing species have a short sacciform segment and a long segment of sacculated tubiform forestomach. The number of rodents also increased during the Miocene and Pliocene epochs (Simpson 1950). Although some rodents such as the voles have both an enlarged cecum and an enlarged, partially compartmentalized stomach, the hindgut remains the principal site for microbial digestion in these animals and even the largest rodents such as the capybara.

There appears to be little information available for speculation on evolution of the foregut in edentates, primates, sirenians, or whales. However, the general rule on the limitations of body size might suggest that the giant sloth (*Megatherium*), which was larger than the elephants of today (Vaughan 1986), had a digestive tract and diet quite different from those of the present species. It also suggests that the largest herbivorous dinosaurs such as the brontosaurs (*Apatosaurus* sp.), which are believed to have weighed up to 27,000 kg, utilized their hindgut for the microbial digestion of plants.

Summary

Microbes indigenous to the gastrointestinal tract can provide a major source of the nutrients required by herbivorous reptiles, birds, and mammals. Microbial fermentation of the structural carbohydrates of plants provides an additional source of energy. Microbial digestion and metabolism of nitrogenous compounds allows synthesis of additional essential amino acids and conservation of nitrogen and water by the recycling of endogenous urea through the gut. Microbial synthesis of vitamin K and B-complex vitamins can provide all of these substances required by some species. Although some adult fish and the larvae of some species of Amphibia can subsist on plant material, this ability may be due to extraction of nutrients from large quantities of ingested plant material, rather than microbial conversion of this material to nutrients.

The presence of VFA in the hindgut of carnivorous, omnivorous, and herbivorous reptiles, birds, and mammals suggests that microbial fermentation may be ubiquitous to the hindgut of all terrestrial vertebrates. This would be aided by the same mechanisms that retain digesta for the recovery of electrolytes and water, and would help conserve endogenous carbohydrates, including a major fraction of the mucus secreted by the gastrointestinal tract. However, the herbivorous reptiles show the first real evidence of an efficient system for the microbial production of nutrients by vertebrates. The largest dinosaurs were almost undoubtedly dependent on these processes at some site along their gastrointestinal tract. Among those reptiles that have survived extinction, herbivory appears to be limited to adults in a relatively small number of species. Its limitation to larger species suggests the need for a minimal gut capacity. The hindgut of herbivorous reptiles has a greater capacity, and in some species its proximal segment contains a cecal bulge and mucosal flaps that may aid in the retention of digesta. It also has been suggested that herbivory in reptiles may be limited by their inability to maintain a constant body temperature and triturate plant material efficiently. The latter two reasons would not explain the rarity of herbivorous birds. However, avian gut capacity may have been limited by requirements for flight. The herbivorous grouse and rheas have well-developed ceca, but grouse do not have an exceptional capacity for sustained flight and rheas do not fly at all.

Fossil evidence indicates that the earliest mammals were small carnivores. The evolution of an enlarged cecum and mechanisms for selective retention of fluid and small particles of digesta provides the basic requirements for microbial fermentation and synthesis of nutrients in the hindgut of smaller herbivorous mammals such as the koala, the lagomorphs, and many species of rodents. Efficient utilization of the end products would be increased by coprophagy and, especially by cecotrophy—that is, the selective ingestion of nutrient-rich feces. A larger body size allowed for a greater capacity of the large intestine and less dependence on the cecum in some herbivorous primates, the perissodactyls, and the proboscideans, in which digesta retention is aided by the presence of haustra and, in the latter two orders, permanent compartmentalization of the colon.

Use of the hindgut as the major site for microbial production of nutrients has the advantage of concentrating these efforts on a digesta that already has experienced the more efficient extraction of readily available nutrients by the upper gastrointestinal tract. However, use of the stomach for this purpose, in animals such as the kangaroos, sloths, colobus and langur monkeys, and in many species of artiodactyls, allows efficient utilization of protein and vitamins synthesized by these microorganisms, without recourse to coprophagy. It also provides an opportunity to digest plant toxins.

If one considers the possibility that whales may have also evolved from terrestrial herbivores, the herbivorous vertebrates achieved both the largest body size and the greatest structural complexities of the gastrointestinal tract. If these characteristics and general temperamental disposition were the only criteria, the herbivores could be considered the most highly evolved of all vertebrates.

7

Secretion and absorption of electrolytes and water

The digestive tract and its accessory glands secrete large quantities of Na, H, Cl, and HCO_3, and lesser amounts of K and PO_4 each day. As most of these secretions are isotonic to the animal's plasma, this represents the addition of large volumes of water that aid in both the digestion and transit of gut contents. Secretion of H provides the optimal pH for the activation and activity of pepsin within the stomach, and aids in the resorption of HCO_3 by some segments of the intestine. Bicarbonate secretion aids in the titration of H entering the small intestine and of the VFA produced by indigenous gut microorganisms. The digestive tract also absorbs large quantities of these ions and water. This allows dietary replacement of ions and water lost in secretions and excretions. However, a major function of these absorptive processes is to reabsorb and conserve the ions and water that are secreted or excreted into the digestive tract.

Electrolyte and water balance

The vertebrate body is composed largely of water. Although most of the body water is found in the cells, interstitial fluid, and plasma, a substantial percentage can be found in the gut contents of mammalian herbivores (Table 7.1). The electrolyte composition of seawater and the major body fluid compartments of humans is illustrated in Figure 7.1. Although the ionic composition of cell contents can differ with species and cell type, the major cation is K, with Mg and Na in much lower concentrations. The principal anions are organic phosphate and protein, with lower concentrations of SO_4 and HCO_3. The ionic composition of extracellular fluid (interstitial fluid and blood plasma) is quite different, with Na the principal cation and Cl and HCO_3 the major anions. Seawater contains much higher concentrations of Na and Cl. This fact is cited as evidence that animal life originated in the sea at a time before its NaCl concentration was increased to present levels by the leeching of this highly water-soluble salt from land. The latter also would explain the low levels of Na and high levels of the less soluble K salts presently found in soil and plants. The osmolality of seawater is approximately three times that of the cellular and extracellular body fluids of most vertebrates.

TABLE 7.1 Body fluid compartments

Compartment	Volume (L/100 kg body weight)	
	Man	Sheep
Intracellular	50	31.1
Extracellular	20	15.5
Plasma	(5)	(4.9)
Interstitial	(15)	(10.7)
Transcellular	2.8	22.2
Alimentary tract	(1.4)	(20)
Other[a]	(1.4)	(2.2)
Total	72.8	68.8

[a]Cerebrospinal fluid, synovial fluid, aqueous humor, and urine.

The earliest vertebrates are believed to have been freshwater fish whose major problem with respect to electrolyte and water balance is the elimination of excess water absorbed by the animal down an osmotic gradient. Freshwater fish do not drink water (Smith, Farinacci, and Breitweiser 1930). Excess water assimilated via the gills is excreted by the kidneys after glomerular filtration of plasma and resorption of ions and other required substances from the renal tubules.

Movement of fish into a marine environment resulted in the opposite problem—loss of body water down an osmotic gradient. This is resolved in hagfish, sharks, rays, and the coelacanth by the presence of body fluids that are either isotonic or hypertonic to seawater (Bentley 1982). The same is true for the crab-eating marine frog (*Rana cancrivora*). Salt glands that actively secrete NaCl are found in the rectum of some cartilaginous fish and the marine catfish (van Lennep and Lanzing 1967). However, the body fluids of marine teleosts have an osmolality and a Na and Cl content that are only slightly higher than those of freshwater species and, therefore, considerably lower than that of seawater.

Although marine teleosts drink large volumes of seawater, the osmolality and NaCl levels of their gastric contents are only slightly above the levels found in plasma (Smith et al. 1930). This is accomplished by esophageal absorption of Na and Cl. The esophagus of most vertebrates is lined with a mucus-secreting, stratified squamous epithelium that is impermeable to ions and water. However, when euryhaline fish such as eels and flounder, which are able to live in a wide range of salinity, are forced to adapt from freshwater to seawater, the stratified epithelium is replaced with a mitochondria-rich columnar epithelium (Yamamoto and Hirano 1978; Humbert, Kirsch, and Meister 1984) that actively absorbs Na in cotransport with Cl (Hirano and Mayer-Gostan 1976; Kirsch 1978; Parmelee and Renfro 1983). The salt-

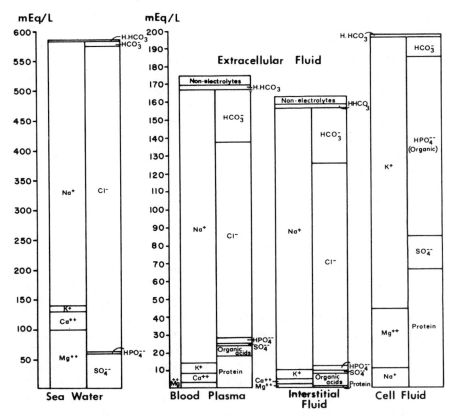

FIGURE 7.1 Electrolyte composition of seawater and human body fluid compartments. (From Gamble 1954.)

water-adapted epithelium of the flounder is permeable to water, which allows for an even more rapid development of osmotic equilibrium between blood and esophageal contents. Absorption of large quantities of Na and Cl requires mechanisms for their secretion with a minimal loss of water. Fish are unable to concentrate their urine. Therefore, glomerular filtration is a disadvantage to marine teleosts. The glomerulus is absent from the kidney of many of these species, and excess Na and Cl are actively secreted by the gills of these animals.

Salt glands have been described in many species of marine reptiles and birds. Their structure and function have been reviewed by van Lennep and Young (1979). Marine mammals that feed on other vertebrates can satisfy their water requirements from food sources plus that produced by lipid and protein metabolism (Ridgway 1972). Those that feed on marine invertebrates ingest considerable amounts of Na and Cl, and some captive marine mammals have been observed to drink saltwater. There appears to be no information on the drinking of seawater by those on a diet of marine inverte-brates. However, marine mammals are capable of concentrating their urine and, therefore, excreting excess Na and Cl via their kidney.

Terrestrial vertebrates must obtain and conserve water and Na from their diet. As mentioned in Chapter 4, the urine of amphibians, reptiles, and birds is excreted into the cloaca, and, in at least many species of reptiles and birds, refluxed the length of the hindgut. This allows the recovery of ions and water of both urinary and digestive system origin from the large intestine of these animals. The urinary and digestive systems exit separately from the body of most mammals. However, the development of the loop of Henle in the mam-malian nephron provides for the more efficient concentration of urine (Fig. 7.2) and an increase in the relative length of the colon in most mammals facilitates more efficient resorption of digestive tract secretions.

Electrolyte transport mechanisms

The earliest model of ion transport across an epithelial tissue was that de-scribed by Koefed-Johnson and Ussing (1958) for the frog skin. It assumed that Na passively diffused into the epithelial cell, via its outer membrane, down an electrochemical potential gradient maintained by the active transport of Na across the basal and lateral cell membranes. The Na pump in the latter membranes transported Na out of the cell in exchange for K. This model was established by the development of an in vitro system that allowed the study of ion transport under conditions in which the composition of the solutions bathing the tissue and the transepithelial electrical gradient could be modified. Experiments showed that frog skin bathed on both surfaces by an identical frog Ringer's solution and short-circuited to remove the electri-cal potential gradient normally present, actively transported Na from solu-tions bathing the outside to those bathing the inside of the skin. Further-more, the net transport of Na in the short-circuited tissue accounted for the entire ion current that would be required to explain the electrical potential (blood side positive) measured under open-circuited conditions. Therefore, the ability of frog skin to absorb Na and Cl from solutions containing low concentrations of NaCl was explained by active transport of Na and diffusion of Cl down the resulting electrochemical gradients. Certain aspects of this model have since been questioned (Ussing 1975). Transport of Na across the outer membrane does not appear to be entirely attributable to diffusion alone. There is evidence that Cl may be actively transported by the frog skin under certain conditions and that the Na-K exchange pump may be electro-genic—that is, it may demonstrate an exchange ratio greater than 1:1. However, this model and the techniques and procedures used to establish it have provided the basic approach for much of our present understanding of ion transport across epithelial tissue.

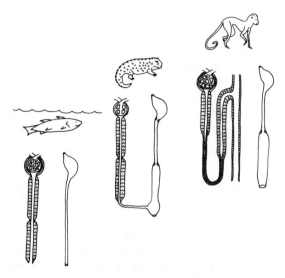

FIGURE 7.2 Development of the nephron and hindgut in relation to habitat. The nephrons of the fish, reptile, and bird kidneys are limited in their ability to concentrate urine. However, urinary electrolytes and water also are recovered by the large intestine of reptiles and birds. A few mammals have retained the cloaca, but the majority excrete their digesta and urine separately, and conservation of urinary water is largely accomplished by the loop of Henle and countercurrent multiplier system of the nephron. (Modified from Smith 1943 by Stevens 1977.)

Keynes (1969) reviewed information available on the mechanisms of salt and water transport across various types of epithelial tissue found in a wide range of invertebrate and vertebrate species. He concluded that there was good evidence for the presence of five or six different mechanisms for the active transport of univalent inorganic ions across epithelial cell membranes. The first was the electrogenic Na-K exchange pump described for the amphibian skin. This appeared to be ubiquitous to all animal cells and accounted for active transport of Na by the basolateral cell membranes of the urinary bladder, segments of kidney tubule, salt glands, salivary glands, intestinal mucosa, and stratified squamous rumen epithelium. The second consisted of paired pumps, capable of nonelectrogenic exchange of Na for either H or NH_4 and Cl for either HCO_3 or OH. This would account for transport of salt across the gills of freshwater Crustacea and fish, the anal papillae of insect larvae, and the epithelia of kidney tubules. The third, which provided electrogenic transport of Cl, was present in the gills of marine teleost fish, oxyntic gastric mucosa of vertebrates, acinar cells of salivary glands, and the intestine of some species. The fourth transported H, accounting for its secretion by the oxyntic gastric mucosa and possibly by the salivary glands of some mollusks. The fifth mechanism actively transported K alone (electrogenically), as exemplified by the silkworm midgut, the Malpighian tubules and labial gland of insects, and rumen epithelium. The sixth mechanism, first described by Diamond (1962), consisted of a pump that provided for nonelectrogenic coabsorption of Na and Cl from the gallbladder of the roach (*Rutilus rutilus*). Although more recent studies have provided additions, complications, and corrections to the list given by Keynes, it still appears that the transport of monovalent ions across epithelial cells of animals is performed by a relatively limited number of mechanisms.

Secretion and absorption

Salivary glands
Information on the electrolyte composition and volume of salivary secretions appears to be limited to relatively few mammalian species. Any discussion of species variations in the salivary glands is complicated by the fact that the different pairs of glands in a given species can differ in both structure and function. Saliva electrolyte composition also varies with the degree, type, and duration of the stimulus. Therefore, for comparative purposes, it is necessary to indicate the particular gland in question and the rate of secretion. Salivary secretion and its mechanisms have been the subject of a number of reviews (Burgen 1961, 1967; Ellison 1967; Leeson 1967; Phillipson 1970). Although these are confined largely to man, dog, cat, and rat, they cover salivary secretion in ruminants and a few other species.

There is considerable species variation in the size and location of the major salivary glands. For example, the submaxillary gland of some anteaters is extremely large and is provided with a storage bladder. The parotid gland is relatively large in ungulates (hoofed mammals), the beaver, manatee, and fruit-eating bats. Its size is approximately equal to that of the submaxillary gland in the rat, and smaller than the submaxillary gland in the insectivorous bat. Schneyer and Schneyer (1967) compared the rates of intensely stimulated flow per gram of tissue in man, rat, dog, cat, and sheep. The sublingual glands of these species secreted at a somewhat similar rate per unit gland weight, as did the parotid and submaxillary glands of man and the rat. However, the submaxillary glands of the dog, cat, and sheep secreted at rates that were approximately 7.5, 4.5, and 1.5 times greater than that of man, respectively. The parotids of the dog and sheep secreted at rates 7 and 4 times, respectively, that of man. Therefore, the marked species differences in volume flow of saliva are due to differences in secretory rate per gram of tissue, as well as the variations in glandular size.

Figure 7.3 depicts the salivary glands of the male adult rat. The acinus may contain secretory cells, yielding mucous, serous, or mixed secretions. Of the major salivary glands, the parotid is usually serous, whereas the submaxillary

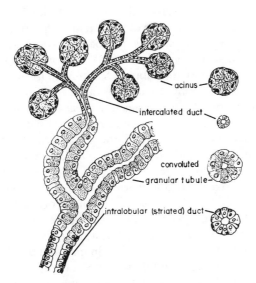

FIGURE 7.3 Organization of the submaxillary gland of the adult male rat. The granular tubule is interposed between the intercalated duct and the intralobular (striated) duct. (From Leeson 1967.)

(submandibular) and especially the sublingual are usually mucous. In some species, the secretory portion of the gland is in the form of acinar tubules, whereas in ruminants it is said to be tubular or "alveolar."

The secretory portion of the salivary glands connects to the intercalated duct, which is lined with small cuboidal cells containing few granules. The length or even the presence of this section of duct varies. This section is relatively long in ruminants. The intercalated duct is confluent with a duct that is lined with columnar cells that are striated in their basal portion and is therefore called the striated duct. This section of duct is rich in mitochondria and believed to be very active in the exchange of ions between blood and the lumen. Its development varies considerably among the glands in a given species and among species. In the monotremes, marsupials, and ruminants, it is extremely well developed. The last section of duct, called the extratubular or excretory duct, is lined with a double-layered or stratified squamous epithelium. The ducts of the salivary glands are richly supplied with blood. Burgen and Seeman (1958) concluded that the capillary flow runs countercurrent to the flow of duct saliva, providing an acinoductal portal system, which may act to recycle K from duct contents back to the acini.

The major inorganic ions in mammalian saliva are Na, K, Cl, and HCO_3, although PO_4 concentrations are high in ruminants. All of these seem to derive from the plasma, with the exception of HCO_3, which appears to be largely supplied by secretory cells. Parotid and submaxillary saliva of man, dog, and cat, as well as parotid saliva of the rat, are hypotonic at low rates of secretion. At higher rates of flow, they approach isotonicity, suggesting that an originally isotonic secretion is rendered hypotonic if sufficient time is allowed for ion resorption by the tubules. Saliva from the sheep parotid and cat sublingual glands is isotonic at all flow rates.

The ionic composition of saliva varies to a great extent with both species and the rate of salivary flow. Two general types of salivary secretion of inorganic ions that have been described in mammals are illustrated in Figure 7.4. Saliva from the parotid and submaxillary glands of man, dog, cat, and rat show a low Na:K ratio at low rates of flow (Fig. 7.4A). In man, and to some extent the dog, saliva excreted at low flow rates tends to contain less Cl and HCO_3 and slightly more PO_4 than plasma. The pH of parotid saliva in the dog (7.2) and man (6.0) is also lower than that of plasma during slower rates of secretion. However, at increasing rates of flow, there is a marked increase in the Na:K ratio (largely due to increased Na) in man and the dog, cat, and rat. In man and the dog, the increased flow rate is accompanied by a rise in Cl and HCO_3. Bicarbonate levels exceed that of plasma, and salivary pH may rise to levels of 7.8. Salivary PO_4 levels show a simultaneous decrease in man, but increase in the dog. These results do not hold for other salivary glands in the above species or the same gland in other species. For example, the sublingual gland of the dog and cat normally secretes Cl in concentrations higher than that of plasma, and mixed saliva of the rat has been recorded to have a pH of 8.9. Nevertheless, the general features described above for man and the dog appear to characterize one type of salivary secretion—that is, an initial secretion high in Na and HCO_3 and relatively low in K. The ducts then appear capable of replacing Na with K and removing HCO_3, the degree depending on the rate of salivary flow.

Secretion of saliva by the parotid gland of ruminants shows a number of characteristics quite different from those of the above species (Fig. 7.4B). Phillipson (1970) cites evidence that sheep can secrete from 8 to 16 liters and the cow from 98 to 190 liters of saliva per day. The parotid gland accounts for at least one-half of this in sheep and a major portion in cattle. The higher volumes are equivalent to, or exceed, the extracellular fluid volume of these animals.

The mechanisms of salivary secretion by the sheep parotid gland were extensively studied and reviewed by Blair-West and co-workers (1965, 1970). Saliva is spontaneously secreted in the absence of stimulation or innervation. This appears to have been otherwise noted only in the cat sublingual and rabbit submaxillary glands. The parotid saliva of sheep was isotonic, with a relatively high Na:K ratio, PO_4 concentration (40 mEq), HCO_3 concentration (80 mEq), and pH (8.2) even at low rates of flow. The Cl concentration was relatively low (20 mEq). When the flow rate was increased by stimulation,

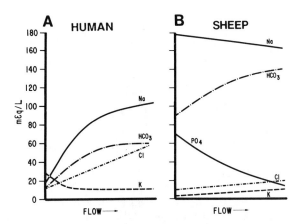

FIGURE 7.4 Concentration of major electrolytes in saliva of humans (from Thaysen, Thorn, and Schwartz 1954) and sheep (from Argenzio 1984) as the salivary flow rate is increased.

the Na:K ratio increased still further. The concentration of HCO_3 increased to levels of 120-140 mEq, and Cl levels rose slightly. Yet neither the osmolality nor the pH appeared to change. The diet of feral herbivores tends to have a very low Na:K ratio, and the large amounts of Na in the saliva are normally recovered by efficient absorption from the gut. However, in Na-deficient sheep, part of the Na in salivary secretions was replaced by K (Blair-West et al. 1970). This response was initiated by aldosterone, and the degree of response correlated with the degree of Na deficit. The P_{CO_2} levels dropped between the arterial and venous blood supply of the sheep parotid, indicating that the glandular cells utilized both arterial CO_2 and that supplied by cellular metabolism. Information on the structure and function of camelid salivary glands suggests few differences from the sheep (Engelhardt and Höller 1982).

Stomach
The four different regions of epithelium that may be found in the stomach of vertebrates were described briefly in Chapter 1. Proper gastric glandular mucosa, which has been the most intensively studied, contains glands that secrete HCl and pepsinogen. One cell appears to be responsible for secretion of both the acid and the enzyme by the stomach of fish (Materazzi and Menghi 1975; Bondi, Menghi, Palatroni, and Materazzi 1982). The same was said to be true for adult amphibians, reptiles, and birds, although other studies suggest that HCl and pepsinogen can be secreted by separate areas of the amphibian esophagus and stomach. The proper gastric glandular mucosa of mammals contains compound tubular glands with oxyntic or parietal cells that secrete HCl and neck chief cells, which secrete pepsinogen. Secretion of

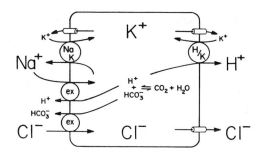

FIGURE 7.5 A model proposed for secretion of HCl by gastric parietal cells. The apical membrane contains an Na-H exchange pump and conductive pathways for passive transport of K and Cl. The basolateral membrane contains an Na-K pump, a K conductance, and separate Na-H and Cl-HCO$_3$ exchange mechanisms. (From Reenstra, Bettencourt, and Forte 1987.)

HCl by the oxyntic cell is explained by the presence of a system that provides for the separate active transport of H and passive diffusion of Cl into the lumen of the stomach (Fig. 7.5). The transport of Cl is electrogenic, but the transport of H consists of its nonelectrogenic exchange for K absorbed from the lumen. Intracellular hydration of CO$_2$, catalyzed by carbonic anhydrase, is believed to result in equal quantities of H for secretion into the lumen and HCO$_3$ for release into the blood in exchange for Cl.

Although HCl secretion has been correlated with the presence and location of carbonic anhydrase in the oxyntic cells of mammals, the dual-purpose oxynticopeptic cells of amphibians and birds contained little or no carbonic anhydrase, whereas the surface mucosal cells of frogs, birds, and mammals contained relatively high levels of this enzyme (Gay, Schraer, and Shanabrook 1981). This suggested that the surface cells might be the source of the acid secreted by lower vertebrates. However, surface epithelial cells in the proper gastric mucosa of the frog, guinea pig, and dog also have been shown to secrete HCO$_3$ (Flemström 1977; Garner and Flemström 1978; Kauffman, Reeve, and Grossman 1980), partly in exchange for Cl absorption (Flemström, Heylings, and Garner 1982).

Hogben, Kent, Woodward, and Sill (1974) examined the relative percentages of parietal, chief, and mucosal cells in the proper gastric mucosa of man and the dog, cat, guinea pig, and frog, and related this to the maximal rate of H secretion per unit weight of parietal cell. Values were calculated both per unit body weight and per unit surface area to allow interspecies comparison. There were significant species differences in the relative percentages of parietal versus chief cells and the maximal secretory rate per gram weight of parietal cell. However, after reviewing the possible error inherent in the various experimental procedures used to obtain these data, they concluded that there may be no significant difference in the secretory capacities of the parietal cells among the mammalian species they examined. Merritt and Brooks (1970) examined variations in basal and histamine-induced secretion of gastric acid by the pig, dog, cat, rat, and human stomach.

As mentioned in Chapter 1, a region of cardiac glandular mucosa is found near the gastroesophageal junction of reptiles, some adult amphibians, and most mammals. Its secretions are difficult to analyze in most species because of the small area occupied by this tissue and often a lack of clear demarcation between proper gastric and cardiac glandular regions. For example, the proper gastric glands of the human stomach simply decrease in number in a gradual blend with the cardiac glandular region. However, a few species have extensive areas of cardiac glandular mucosa. Studies on the secretions of isolated pouches prepared from the cardiac gland region of the swine stomach, which occupies approximately one-third of the total gastric mucosal surface, showed secretion of a slightly alkaline fluid that was low in mucus and protein, but high in HCO_3 (Höller 1970a,b). In pigs weighing 30-40 kg, at least 500-600 ml were secreted over a 24-hour period. This secretion was inhibited by gastrin and was therefore minimal during the cephalic or early gastric phases of gastric digestion (see Chapter 8). This could account for the finding of Argenzio and Southworth (1974), which showed a much higher pH in the cranial one-half of the pig stomach 2 and 4 hours after a meal (see Fig. 7.11). Secretion of a buffer could serve a protective function, preventing damage to gastric mucosa when HCl and pepsin are not needed for digestion, but evidence that a substantial degree of microbial fermentation occurs in the cranial one-half of the pig stomach during these periods of higher digesta pH (see Chapter 6) suggests an additional need for protection from organic acids.

When the cardiac pouch of the pig was directly infused with a solution containing HCl or NaCl, an increased secretion of HCO_3 was associated with both a rise in the pH and a loss in Cl from the solution. A similar secretion of HCO_3 and absorption of Cl was noted by the cardiac glandular sac mucosa of the llama forestomach (Eckerlin and Stevens 1973). However, Rübsamen and Engelhardt (1978) found that although Na and Cl were rapidly absorbed from these glandular pouches, there was little net appearance of HCO_3. In vivo studies of the segment of the third compartment of the llama forestomach, which contains a similar cardiac glandular type of mucosa, showed that it absorbed Na and appeared to secrete Cl, but HCO_3 levels were not measured (Engelhardt, Ali, and Wipper 1979). Therefore, cardiac glandular mucosa may serve a critical buffering capacity in at least some mammalian species. This may help to provide the higher pH required for survival of the symbiotic gastric microbes and help prevent damage to the gastric epithelium by either HCl or organic acids. The characteristic gastric ulcer of swine, a separation between the cardiac and nonglandular stratified squamous epithelium (Curtin, Goetsch, and Hollandbeck 1976), may be due to a malfunction of this mechanism. Surface cells of the proper gastric region also have been shown to secrete HCO_3, which is held in the mucous layer, providing a means of preventing the back-diffusion of H.

As noted in Chapter 3, nonglandular, stratified squamous epithelium occupies at least some portion of the stomach of species belonging to a wide range of mammalian orders. Secretion and absorption of electrolytes by the stratified squamous epithelium of the ruminant stomach can be studied in the absence of the glandular secretions found in tissue lining the remainder of the gastrointestinal tract. It has been estimated that sheep can secrete 1.2-1.5 moles of Na in their saliva each day (Kay 1960), and that one-half of this can be reabsorbed by the rumen (Dobson 1959). Chloride was reabsorbed from concentrations as low as 6-7 mM (Sperber and Hydén 1952; Hydén 1961). After it was demonstrated that Na was absorbed against an electrochemical gradient from the rumen of anesthetized sheep (Dobson 1959), in vitro studies of short-circuited rumen epithelium from the cow, goat, and sheep showed that this tissue actively absorbed both Na and Cl (Stevens 1964; Ferreira, Harrison, and Keynes 1964; Harrison, Keynes, and Zurich 1968). Active secretion of much smaller amounts of K also was indicated in studies of the sheep rumen epithelium (Ferreira et al. 1964). Subsequent studies of short-circuited rumen epithelium from sheep and cattle, in which the Na of the bathing solutions was replaced with choline or the Cl was replaced with acetate or SO_4, led to the conclusion that 1) approximately one-third of the net Na transport was due to its electrogenic transport; 2) a significant portion of the Cl was actively absorbed in nonelectrogenic exchange for HCO_3; and 3) two-thirds of the Na was absorbed actively, but nonelectrogenically, either in cotransport with Cl or in exchange with H (Fig. 7.6). In vivo studies of sheep and goats showed that the omasum also absorbed Na and water (Engelhardt and Hauffe 1975). However, it appeared to secrete Cl in association with a net disappearance of HCO_3.

Pancreatic and biliary secretion

Exocrine cells of the pancreas are arranged in acini in a manner comparable to the salivary gland, apart from containing zymogen cells that secrete the pancreatic enzymes in addition to the centroacinar cells that secrete a fluid with a Na, K, Cl, and HCO_3 composition similar to that of plasma (Davenport 1982). As with saliva, the ˍˍrolyte composition of the pancreatic juice varies with species and, in most species, with the rate of flow. Intermittent feeders such as humans, dogs, and cats secrete chiefly during the digestive phase after a meal, and the aqueous component of their pancreatic juice is derived largely from the centroacinar cells. Continuous feeders such as the sheep, rabbit, rat, and horse continuously secrete a pancreatic fluid that includes a larger aqueous fraction derived from the zymogen acinar cells.

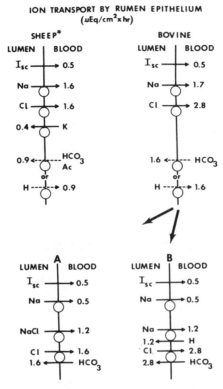

ION TRANSPORT BY RUMEN EPITHELIUM
$(\mu Eq/cm^2 \times hr)$

FIGURE 7.6 Diagrammatic representation of net ion transport and short-circuit current (SCC) observed in rumen epithelium. All values are expressed in $\mu Eq/cm^2 \cdot hr$. The tissue is represented as a single membrane bathed on both surfaces with identical solutions of HCO_3-Ringer's. The first two diagrams (top) summarize results obtained from studies of sheep (Ferreira, Harrison, Keynes, and Zurich 1972) and bovine (Chien and Stevens 1972) tissue. The dashed lines indicate alternative explanations given for the discrepancy between net ion transport and SCC. The possibility of fatty acid transport is indicated by the Ac symbol. Diagrams A and B (bottom) represent two alternatives that could account for the effects of ion substitution, observed with bovine rumen epithelium. (From Stevens 1973.)

In most species that have been studied, the concentration of HCO_3 is high (120-150 mM) at the most rapid rates of pancreatic flow, but decreases at slow rates of flow. This appears to be due to secretion of HCO_3 and a reciprocal absorption of Cl by centroacinar and proximal tubular duct cells and a reversal of this pattern (evident at the slower rates of flow) by cells of the distal extratubular ducts. Pancreatic flow rates have been reported as high as 2 L/day in man (Davenport 1982) and 0.5-1.0 L/day per 100 kg body weight in sheep, and 3-5 L/day per 100 kg body weight in cattle (Argenzio 1984).

However, the horse showed a continuous, relatively rapid rate of pancreatic fluid secretion even without stimulation, and daily secretions of 10-12 L/100 kg body weight (Alexander and Hickson 1970). Furthermore, the levels of Cl and HCO_3 in equine pancreatic fluid remained relatively constant at all rates of flow.

Secretion of HCO_3 by the pancreas is associated with its disappearance from the fluid perfusing the organ. This may be due to secretion of H into the perfusate in exchange for transport of Na into the cell—in some respects, a reversal of the process of H secretion by the oxyntic cells of the stomach.

Biliary secretions consist of two major components: the bile salt-dependent secretory flow that arises from the liver parenchymal cells, and the bile salt-independent secretions that appear to be provided mostly by the enterohepatic canaliculi and bile ducts (Davenport 1982; Argenzio 1984). The electrolyte secretion has a Na:K ratio similar to that of plasma, and a HCO_3:Cl ratio that increases with an increase in secretory rate in a manner similar to that noted with salivary and pancreatic secretions. Bile is stored in the gallbladder of many species for release during the digestive phase following a meal. The gallbladder of some species can concentrate bile 20-30 fold during interdigestive periods by absorption of Na and Cl. However, little absorptive ability is apparent in other species such as the pig and ruminants. Bile is continuously secreted into the duodenum of the horse and other species lacking a gallbladder.

Intestine

Much of the following information on electrolyte and water transport is drawn from Powell's (1987) more recent comparative review of transport mechanisms in the small and large intestine of vertebrates. The intestine is a major site for the secretion and the principal site for the absorption of electrolytes and water. The small intestine of the rat, rabbit, and human is quite permeable to passive diffusion of monovalent ions and water in its most proximal segment, but demonstrates a decreasing gradient of permeability between the duodenum and terminal ileum. The proximal colon is less permeable than the ileum, and the distal colon is the least permeable of all. Most of the intestine's permeability to ions resides at the epithelial cell attachments or junctions near their mucosal surface. This permeability can be visualized as due to the presence of pores, some of which are selectively permeable to cations. For example, the small intestine shows a selective permeability order of $P_K > P_{Na} > P_{Cl}$ and the colon appears to show a similar selective permeability to these three ions.

The intestine appears to be able to transport water against an osmotic gradient. Curran (1968) proposed a three-compartment model that would account for this. A slightly different model was later proposed by Diamond (1977), and an entirely different mechanism, consisting of a countercurrent multiplier system of vessels within the subepithelial tissue of the villus, was suggested for the rat and dog (Bond, Levitt, and Levitt 1977; Lundgren 1984). Regardless of the exact mechanism, each of these models assumes that the absorption of water against a transepithelial osmotic gradient is the result of an osmotic gradient produced by the absorption of ions into an intervening compartment rather than active transport of water molecules.

The study of intestinal secretion and absorption of ions and water is complicated by the fact that these two functions occur simultaneously in different cells. Sodium, Cl, HCO_3, and water are secreted by cells of the crypts of Lieberkühn (Field 1978; Welsh, Smith, Fromm, and Frizzell 1982). Secretions by the crypts of the small intestine appear to result from the active transport of Cl and passive transport of Na into the lumen. Colonic secretion appears to result in a net movement of K, Cl, and HCO_3 into the lumen. Ions secreted by the epithelial cells of the villi in exchange for the absorption of other ions are discussed below.

Absorption of Na is the major force for the absorption of water from the digestive tract. Sodium is absorbed from the lumen into the intestinal cell by a number of different processes that involve its passive or active transport across the apical cell membrane and down a concentration gradient. The low concentration of Na within the cell is maintained by its active transport out of the cell by means of the Na pump located in the basal and lateral membranes of the enterocyte. The pump consists of membrane Na-K-ATPase, which produces a net transport of charge (Rose, Nahrwold, and Koch 1977). Its metabolic fuels are lipid, glucose, and glutamine. Considering the large quantities of ketone bodies produced from the metabolism of VFA by rumen and large intestinal epithelium, it is interesting to note that the RQ values for rabbit ileal and dog jejunal mucosa suggested that ketone bodies were the major fuels for interdigestive periods (Frizzell, Markscheid-Kaspi, and Schultz 1974; Lester and Grim 1975). The proximal colon of rabbits and humans utilizes glucose, glutamine/glutamate, and VFA, whereas the distal colon uses mostly butyrate.

Sodium can be absorbed from the lumen along with glucose or amino acids. This has been demonstrated in the small intestine of mammals, birds, and fish, and in the midgut of a number of invertebrates. Studies of man and a variety of laboratory animals indicate that Na absorption is stimulated by increasing the glucose concentration of lumen contents up to levels of 50 mM. Addition of glucose resulted in a fourfold increase in the absorption of Na from the jejunum. It appears that Na and glucose are transported by a mechanism that provides for a 2:1 coupling. Although galactose also provides a stimulus to Na absorption, it had a lesser effect. The addition of amino acids to the lumen of the small intestine provides a stimulatory effect that is both similar and additive to that of glucose. Sodium and amino acids are absorbed by the separate, noncompetitive mechanisms of amino acid transport. Amino acid absorption has been shown to be stimulated by the presence of Cl and Na in the lumen of the fish intestine (Bogé, Rigal, and Pérès 1983). Amino acids also have been shown to stimulate Na absorption by the large intestine of birds (Skadhauge 1980), and neonatal dogs (Robinson 1976) and pigs (James and Smith 1976). Although earlier studies suggested that dipeptides may be absorbed by Na-dependent mechanisms, this has not been confirmed by studies using brush border vesicles. Sodium-coupled cotransport of monosaccharides and amino acids into the enterocyte could be driven by Na transport down its electrochemical gradient. However, there is considerable evidence that the cotransport of Na may result from passive flow of monosaccharides or amino acids and water (solvent drag).

A relatively large percentage of the Na that is absorbed by the rabbit jejunum appears to enter the cells by non-carrier-mediated passive diffusion (Gunther and Wright 1983). This does not appear to be a major mechanism for Na entrance into the cells of the ileum, cecum, or proximal colon, and there appears to be little or no passive diffusion of Na into the cells of the distal colon or rectum (Powell 1978).

Another mechanism of Na absorption by the intestinal cell is the one that is linked to the absorption of Cl, in a manner similar to that described for rumen epithelium. The presence of such a system was demonstrated in studies of the rabbit ileum (Nellans, Frizzell, and Schultz 1973; Frizzell, Nellans, Rose, Markscheid-Kaspi, and Schultz 1973), and the colon of rats (Binder and Rawlins 1973), rabbits (Sellin and De Soignie 1984), monkeys (Powell et al. 1982), and humans (Hawker, Mashiter, and Turnberg 1978). Table 7.2 compares results from in vitro studies of Na and Cl transport across short-circuited rumen epithelia and across mucosa from the large intestine of the human, rat, pony, rabbit, and pig. The cecum and ventral (proximal) colon of the pony showed an active secretion of Cl, and this, along with active absorption of Na, could account for the short-circuit current (I_{sc}). However, both Na and Cl were actively absorbed by each of the remaining tissues, and, with the exception of the distal colon of the rabbit, the active transport of Na was substantially greater than that accountable to I_{sc}. Binder and Rawlins (1973) found that replacing Na with choline in the solutions bathing colonic tissue of the rat reduced the rate of Cl absorption, whereas replacing Cl with isothionate reduced the rate of Na absorption, in a manner similar to that noted for rumen epithelium.

The nonelectrogenic transport of Na and Cl was initially assumed to represent their cotransport as NaCl. However, studies by Turnberg, Bieberdorf, Morawski, and Fordtran (1970) showed that this could be due to simultaneous exchange of Na for H and Cl for HCO_3, coupled in varying degrees to the hydration of intracellular CO_2 (Fig. 7.7). This interpretation has been supported by studies of transport across brush border membrane vesicles (Murer, Hopfer, and Kinne 1976; Liedtke and Hopfer 1977, 1982; Powell and

TABLE 7.2 Net Na and Cl fluxes and short-circuit current across rumen and colon epithelium

Species	Na zJ_{net} L → B	Cl zJ_{net} L → B	I_{sc} L → B	References
Ox rumen	1.7	−2.8	0.5	Chien and Stevens (1972)
Sheep rumen	1.6	−1.6	0.5	Ferreira et al. (1972)
Human colon	10.1	−2.4	7.0	Archampong, Harris, and Clark (1972)
	4.6	−1.4	2.9	Rask-Madsen and Hjelt (1977)
Rat colon	8.8	−9.1	1.0	Binder and Rawlins (1973)
	3.6	−3.6	0.7	Stoebel and Goldner (1975)
Pony cecum	0.9	0.7	1.5	Giddings, Argenzio, and Stevens (1974)
ventral colon	0.5	0.4	0.8	Argenzio, Southworth, Lowe, and Stevens (1977)
Rabbit colon	3.0	−1.6	2.8	Frizzel, Koch, and Schultz (1976)
Pig colon	6.0	−4.5	1.5	Argenzio and Whipp (1983)

The header for the ion fluxes columns reads: Ion fluxes and short-circuit current[a] ($\mu Eq/cm^2/hr$)

[a]Net fluxes of ions (J_{net}) from lumen (L) to blood (B) are multiplied by the + or − signs (z) of the ion to allow comparison of their algebraic sum with the short-circuit current (I_{sc}), which is expressed in similar flux units.

Source: Stevens, Argenzio, and Roberts (1986).

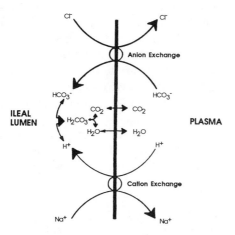

FIGURE 7.7 A double-exchange Cl-HCO$_3$, Na-H mechanism that could explain the relationship of ion movements across the mammalian ileum. (From Turnberg et al. 1970.)

Fan 1984; Knickelbein et al. 1985). The interrelationship between Na and VFA transport by colonic epithelium also can be explained partly by an associated exchange of Na for H (see Fig. 6.3).

Marine teleosts also demonstrate cotransport of Na and Cl into their intestinal cells, but this process includes the cotransport of K in a ratio of Na:K:2Cl (Palfrey and Rao 1983). In the absence of monosaccharide- or amino acid-stimulated Na transport, this resulted in a net transport of Cl in excess of Na and, therefore, a transepithelial electrical potential with mucosa positive to serosa, which is opposite to that seen in the intestine of amphibians, reptiles, birds, and mammals. A mechanism accounting for the cotransport of these three ions is depicted in Figure 7.8. However, the finding that HCO_3 was needed for optimal transport (Rao et al. 1981) and the demonstration of Na-H exchange (Huang and Chen 1971) suggest that the interdependency of these ions could be as readily explained by a combination of Na-H, K-H, and Cl-HCO_3 exchanges.

The interrelationship between absorption of Na and VFA and the net appearance of HCO_3 in the colon (discussed in Chapter 6) could result from the uptake of H (secreted into the lumen in exchange for Na and resulting from intraluminal production of H_2CO_3) via absorption of undissociated organic acid (see Fig. 6.3). Appearance of HCO_3 in the rabbit ileum also has been attributed partly to absorption of H or secretion of OH (Hubel 1974). An interdependence between Na absorption and the *disappearance* of HCO_3 has been demonstrated in the jejunum of a number of mammals, and there is strong evidence that it may be attributed to secretion of H, in exchange for Na, and conversion of the HCO_3 of lumen contents into CO_2 and water.

Although much of the disappearance of HCO_3 from the jejunum can be attributed to its titration by H, not all of the HCO_3 disappearance can be related to Na absorption. White and co-workers (Gunter-Smith and White 1979; White and Imon 1981, 1983; White 1982; Imon and White 1984; White 1985) demonstrated an apparent electrogenic absorption of HCO_3 by the jejunum of the amphibian congo eel (*Amphiuma*) that was independent of Na, but dependent on K in the lumen. Inhibition of this mechanism by omeprazole, a specific inhibitor of K-H-ATPase, suggested that the disappearance of HCO_3 may be due to K-H rather than Na-H exchange. There is evidence for a similar process in the rat jejunum (Lucas 1976), rabbit ileum (Smith, Cascairo, and Sullivan 1985), and the human and rabbit colon (Davis, Morawski, Santa Ana, and Fordtran 1983; Garcia, Campos, and Lopez 1984).

A mechanism for the nonelectrogenic absorption of Cl in exchange for HCO_3, like that described for cardiac mucosa and rumen epithelium, has been demonstrated in the ileum and colon of many mammals. However, studies of the rabbit ileum and pig colon showed that the appearance of HCO_3 in the lumen was accompanied by a decrease in P_{CO_2} (Hubel 1974;

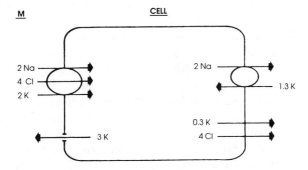

FIGURE 7.8 A model that would account for Na, K, and Cl transport by flounder intestine. The Na-K-2Cl cotransport mechanism. (From Frizzell et al. 1984.)

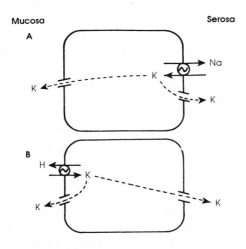

FIGURE 7.9 Models for transcellular K transport by the intestinal cell. A) K secretion in which the Na-K-ATPase creates an electrochemical gradient for K across the cell membrane with subsequent K diffusion across conductance channels. The direction of net transport in this model could be governed by relative K conductances at the two cell borders. B) K absorption in which a K-H-ATPase mediates entry across the apical membrane to achieve intracellular concentrations above electrochemical equilibrium. The basolateral and apical membrane K conductances will determine the direction of net K movement across the epithelium. (From Smith and McCabe 1984.)

Argenzio and Whipp 1981, 1983), suggesting that HCO_3 appearance was initiated by H absorption or OH secretion.

Although active absorption of K has been demonstrated in the amphibian small intestine (Imon and White 1984), potassium transport in the small intestine of mammals can be attributed primarily to passive diffusion (Phillips and Code 1966; Turnberg 1971). Studies of the rat indicated that under normal conditions, K was secreted by the proximal colon and absorbed by the distal colon (Foster, Hayslett, and Binder 1984), but during Na depletion, K was secreted by both segments of the colon. Many results obtained from studies of K secretion and absorption could be explained by passive diffusion of K through a paracellular shunt path down its electrochemical gradient. However, Smith and McCabe (1984) concluded that these results could be equally explained by active uptake of K into the intestinal cell by a K-Na- or K-H-ATPase, and a difference in the relative permeability of the apical and basolateral membranes to the passive efflux of K (Fig. 7.9).

The pH of gastrointestinal contents

Many functions of the digestive system are highly dependent upon the pH of gastrointestinal contents. Pepsinogen is activated to pepsin by the low pH of gastric contents, and the activities of this and other enzymes are pH dependent. Microbial digestion is equally dependent on the maintenance of a suitable pH. In addition, the intestinal mucosa is susceptible to ulceration if the HCl secreted into the stomach or the VFA produced by microbial digestion is not neutalized. Digesta pH is controlled principally by the secretion, neutralization, or absorption of H; by the secretion of HCO_3 or PO_4; and, in some segments of the tract, by the production and absorption of VFA and NH_3. Each of these factors must be considered in the interpretation of digesta pH measurements.

Figures 7.10, 7.11, and 7.12 show the pH of digesta in serial segments of the gastrointestinal tracts of the dog, pig, and pony 2, 4, 8, and 12 hours after feeding. These results are from the same experiments described in Chapter 6 (Fig. 6.2) and include the organic acid concentrations measured at these same times. It can be seen that the pH of gastric contents in all three species showed marked variations with time after feeding. In the dog and pony, the pH was highest just prior to the meal and decreased to its lowest values 8 hours after feeding. The periods of lowest pH did not correlate with the periods of highest organic acid production. Therefore, it appeared that the low pH was due primarily to HCl secretion rather than organic acid production. There was no significant difference between values obtained from the oral versus aboral halves of the dog or pony stomach.

Results from measurements of digesta pH in the pig stomach were quite

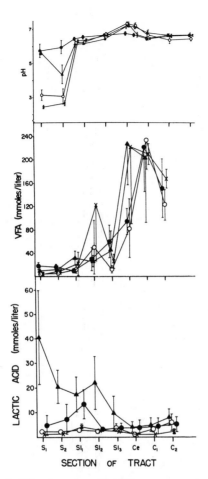

FIGURE 7.10 Mean (±SE) values for pH, VFA concentration, and lactic acid concentration in various segments of the gastrointestinal tract of dogs fed a meat diet. Symbols correspond to four time periods after feeding: 0 or 12 hr (●); 2 hr (▲); 4 hr (○); 8 hr (×). Segments of tract are as follows: cranial (S_1) and caudal (S_2) halves of stomach; proximal (SI_1), middle (SI_2), and distal (SI_3) thirds of small intestine; cecum (Ce); and proximal (C_1) and distal (C_2) colon. (From Banta et al. 1979.)

different. The pH of digesta in the oral and aboral halves of the stomach were low prior to feeding, but pH levels in the oral half rose to significantly higher values 2 and 4 hours after the meal. Volatile fatty acids and lactic acid (LA) levels were highest in the oral half of the pig stomach, and, although VFA concentrations showed little variation with time, LA levels were highest during the period of high pH. Therefore, as in the dog and pony, it appeared that organic acid production did not contribute to the lowering of pH.

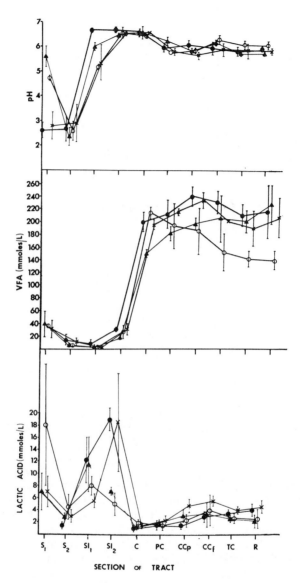

FIGURE 7.11 Mean (±SE) values for pH, VFA concentration, and lactic acid concentration in various segments of the gastrointestinal tract of pigs fed a conventional, high-concentrate diet. Symbols correspond to those in Figure 7.10. Segments of tract are as follows: cranial (S_1) and caudal (S_2) halves of stomach; proximal (SI_1) and distal (SI_2) small intestine; cecum (C); proximal (PC), centripetal (CCp), centrifugal (CCf), and terminal (TC) colon; and rectum (R). (From Argenzio and Southworth 1974.)

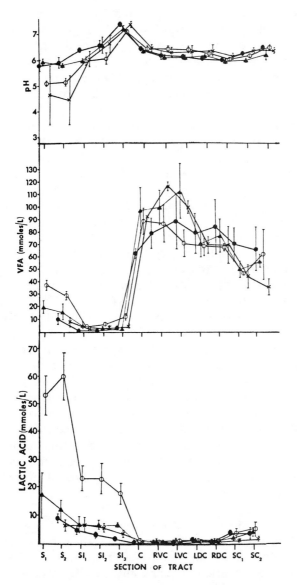

FIGURE 7.12 Mean (±SE) values for pH, VFA concentration, and lactic acid concentration in various segments of the gastrointestinal tract of ponies fed a conventional, pelleted horse diet. Symbols correspond to those in Figure 7.10. Segments of tract are as follows: cranial (S_1) and caudal (S_2) halves of stomach; proximal (SI_1), middle (SI_2), and distal (SI_3) thirds of small intestine; cecum (C); right ventral colon (RVC) and left ventral colon (LVC); left dorsal colon (LDC) and right dorsal colon (RDC); and proximal small colon (SC_1) and distal small colon (SC_2). (From Argenzio et al. 1974.)

Alexander and Davies (1963) had noted a similar inverse relationship between the concentrations of LA and HCl in the stomach contents of pigs, and Cranwell, Noakes, and Hill (1968) cited evidence that LA may inhibit the secretion of HCl. Infusion of acetic acid into the stomach of the dog (Babkin, Hebb, and Kreuger 1941) and cat (Flemström and Frenning 1968) resulted in a marked inhibition of gastric secretion; however, high concentrations (167-170 mM) were used, and this effect was associated with changes in the permeability of the gastric mucosa. Stimulation of HCO_3 secretion by the cardiac mucosa offers another explanation for the high pH values recorded in the oral half of the pig stomach. Bicarbonate was secreted at its slowest rate during the early hours after a meal, suggesting that gastrin may inhibit its secretion (Höller 1970b). However, these measurements were made in isolated gastric pouches or sham-fed animals. In the studies illustrated in Figure 7.11, in which food was allowed to enter the stomach, the highest pH was noted 2 hours after the meal at a time when gastrin levels would presumably still be quite high. This suggests that the rate of HCO_3 secretion by cardiac mucosa is subject to factors other than gastrin. A third possibility, in the intact animal, could be that a rise in pH and net appearance of HCO_3 resulted from the absorption of H with VFA.

Borch-Madsen (1946) produced experimental achlorhydria (absence of HCl) in 2- to 3-month-old pigs by surgically removing the proper gastric glandular area of the stomach. Surprisingly, the stomach of these animals still maintained a relatively low pH. Gastric contents of achlorhydric pigs contained significantly higher concentrations of both LA and VFA in comparison to control animals. Lactic acid concentrations were higher than those of the VFA—more than twice as high during some periods following feeding. The animals with experimental achlorhydria remained healthy, showing no signs of pernicious anemia. However, he pointed out that human patients may show no signs of pernicious anemia until many years after the development of achlorhydria. It would be interesting to know if hypochlorhydric or achlorhydric conditions in man result in a similar increase in organic acid concentrations.

In the first and middle segment of the dog small intestine and all three segments of the pig and pony small intestines, the pH was highest just prior to feeding. The pH of the duodenum was lower subsequent to feeding, presumably due to the addition of gastric chyme. The ileal contents of the dog and pony demonstrated a pH significantly higher than that of any other segment of gastrointestinal tract—presumably due to secretion of HCO_3 (or absorption of H) by this segment.

The pH of large intestinal contents varied with time after feeding, within a range of approximately 6.0-6.5 in the dog, 5.5-6.5 in the pig, and 5.8-6.5 in the pony. These variations showed no definite relationship to the concentrations of VFA present. This could be due to the rapid titration of VFA, by ileal effluent and large intestinal secretions, and the rapid absorption of these organic acids from intestinal contents.

Enterocirculation

The ions and water that are absorbed from the intestine obviously represent a mixture of those presented in the diet and the secretions of the digestive system. Figure 7.13 gives the mean values for the osmolality and the concentrations of major cations and anions found along the equine gastrointestinal tract during a 24-hour period between meals. The concentrations of a given ion in gut contents is dependent upon the quantity in the diet; the quantity added by secretions of the digestive system; and the quantity absorbed relative to the volumes of water ingested, secreted, or absorbed. Gastric chyme had a higher osmotic activity than plasma as a result of dissolution and digestion of food in the stomach. High concentrations of Na were derived largely from salivary secretions, and most of the Cl was provided by secretion of HCl. The osmotic activity was reduced to levels equivalent to that of plasma in the proximal intestine, as a result of secretion of water down osmotic gradients, as well as absorption of nutrients and the addition of isotonic glandular secretions. The concentration of Na remained high throughout most of the intestinal tract. This was due to the addition of Na in the pancreatic and biliary secretions and the absorption of Na and water at a relatively constant ratio. The more distal segments of the colon showed a decrease in Na concentration with a reciprocal increase in K levels due to a more efficient absorption of Na, microbial release of K from the digesta, and secretion of K into the lumen.

The extremely high concentrations of Cl in stomach contents were reduced by Cl absorption along the small intestine with a reciprocal increase in HCO_3 levels—especially in the ileum. Bicarbonate is the major anion in pancreatic and biliary secretions and is also released by Cl-HCO_3 exchange in the ileum. Bicarbonate is the predominant anion found in the ileum of the horse and pig (Alexander 1962) and human (Fordtran 1973). However, PO_4 appears to be the principal anion in the ileum of the dog, cat, rabbit, and guinea pig (Alexander 1965). The markedly reduced concentration of HCO_3 in the equine cecum and colon was associated with a reciprocal increase in the production of VFA and release of PO_4 by microbial digestion. Volatile fatty acids are rapidly absorbed, but because these organic acids are the principal anions responsible for osmotic absorption of water their concentrations remain relatively high throughout the large intestine.

It can be seen that although the digestive tract must continuously assimilate inorganic ions and water of dietary origin, in order to replace their continuous loss, a major amount of the total energy utilized in the absorption of

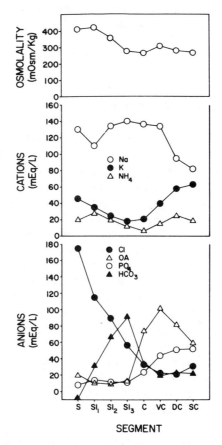

FIGURE 7.13 Digesta osmolality and ion concentrations along the gastrointestinal tract of the pony. Data represent mean values obtained from four measurements taken over a 12-hour period between meals in ponies fed a conventional, pelleted diet. Segments of the gastrointestinal tract are similar to those in the previous figure. The concentrations of PO_4 were calculated under the assumption that the pK_a of NaH_2PO_4 was 6.8 and utilizing the mean pH of digesta in each segment. At the pH of large intestinal contents, NH_3 and the organic acids (OA) would exist primarily in their ionized form. Concentrations of HCO_3 were calculated as the difference in concentration of measured cations and anions. (Modified from Argenzio and Stevens 1975.)

electrolytes and water is directed toward simple reabsorption of secretions. This was clearly demonstrated by Soergel and Hofmann (1972). According to their calculations, some 5.9 liters of water are secreted each day into the gut of humans subjected to a 24-hour fasting period. This is equivalent to approximately 40% of the individual's extracellular fluid volume (ECFV) or

12% of the total body water (TBW). The tolerance of man to an uncompensated fluid loss is indicated by the appearance of clinical signs of dehydration with a 15% loss of ECFV or a 20% loss in TBW. Death usually results from a 30% and 35% loss of fluid from these two body fluid compartments, respectively.

The estimated quantity of ions in these daily secretions consists of approximately 600 mEq Na, 130 mEq H, 475 mEq Cl, and 200 mEq HCO_3. These values would represent approximately 30% of the Na and Cl, and 45% of the HCO_3 normally present in the extracellular fluid. Soergel and Hofmann (1972) estimated that 98% of the ions and water secreted into the digestive tract were reabsorbed, principally by the jejunum, ileum, and colon. However, even these estimates represent minimal values for the quantity of ions and water secreted into the digestive tract of man, since they were obtained under fasting conditions and assumed no secretion of electrolytes or water into the large intestine.

The significance of the above findings to the maintenance of a normal fluid and acid-base balance is obvious. Excessive sequestration of these secretions in any segment of the gut, as a result of obstruction, or excessive loss due to emesis or diarrhea can be life threatening. Human patients suffering from cholera may lose 1 liter of fecal fluid per hour. Under these conditions, fluid therapy is the most critical, immediate consideration. The patient may not survive long enough to benefit from other treatment.

As indicated in Table 7.1, the relative volume of gastrointestinal contents can be much greater in herbivores. Estimates for the secretion of salivary (Kay 1960), abomasal (Hill 1965), pancreatic (Magee 1961; Taylor 1962), and biliary (Harrison 1962) secretions into the digestive tract of sheep give a total volume of 12-22 L/day. These would be equivalent to approximately two to three times the ECFV of the animal. Estimates of the daily volume of water contained in the ileal effluent and fecal excretion (Kay and Pfeffer 1970) indicate that the sheep large intestine had an absorptive load equivalent to approximately one-half the ECFV—without consideration of any additional secretions by the cecum and colon. Alexander and Hickson (1970) estimated that the daily parotid, biliary, and pancreatic secretions of a 100-kg pony contributed 30 liters of fluid to the digestive tract. Argenzio et al. (1974a) concluded from their studies that 12 liters of ileal effluent entered the large intestine of a pony of this size each day, and that this was supplemented by the secretion of an additional 7.5 liters of fluid, with excretion of only 0.9 liters of water in the feces. Assuming a distribution of body water similar to that of sheep, these secretions would be equivalent to approximately 2.5 times the ECFV of the pony, and over one-half of this would be reabsorbed by the large intestine.

The above use of data obtained from a variety of procedures and techniques should be viewed with considerable caution. Results for the human digestive tract were obtained from individuals previously starved for 24 hours. This fact and the absence of information on secretions of the human and sheep large intestine would result in an underestimate of fluid absorption in both species. The data on secretions into the upper digestive tract of sheep and ponies were obtained from many sources and under a variety of conditions. The values for large-intestinal secretion in the pony are calculated for animals fed at 24-hour intervals, and, as noted in Chapter 6, the volume of these secretions appears to be much less in animals subjected to a more normal continuous feeding regime. Nevertheless, these estimates do indicate the marked differences between herbivorous and nonherbivorous species.

Summary

Secretion of electrolytes and water provides the fluid and pH necessary for digestion. The fluid initially secreted into the ducts of the parotid and submaxillary salivary glands, the pancreas, and the biliary tract of mammals has an electrolyte composition similar to that of plasma, except for higher concentrations of HCO_3 and, in the case of salivary glands of some species, higher levels of PO_4. The composition then may be adjusted as these secretions pass through the salivary, pancreatic, and biliary ducts, or the bile is stored in the gallbladder. At rapid rates of salivary and pancreatic flow, and in animals lacking a gallbladder, this results in the addition of large volumes of buffered solutions to the digestive tract. Slower rates of salivary and pancreatic secretion, and the intermittent release of concentrated bile from the gallbladder, act to conserve Na, HCO_3, and water until they are needed for digestive processes.

Gastric secretion of HCl appears to be an innovation of vertebrates that serves principally for the initial digestion of protein. This requires protection of gastric mucosa by the impermeability of apical membranes to H and the secretion of HCO_3, and by the protection of upper intestinal mucosa by titration of H with HCO_3. Excess HCO_3 is then removed by more distal segments of the small intestine and replaced by Cl.

Production of large quantities of VFA in the stomach of some mammalian herbivores and in the large intestine of most terrestrial vertebrates presents an even greater potential problem than HCl, with regard to the damage of gut epithelial cells. These organic acids are not only capable of reducing the pH of digesta to levels that would be damaging to these cells; they are also rapidly absorbed at a rate that increases with a lowering of digesta pH. Damage is prevented under normal conditions by the addition of highly buffered saliva to the stomach, highly buffered ileal effluent to the large intestine, and secretion of HCO_3 into both segments of the digestive tract. The rapid absorption of VFA, at the levels of pH normally maintained in the gastrointestinal tract, and their metabolism by gut epithelium provide additional safety measures.

Absorption of electrolytes from the digestive tract involves a limited number of mechanisms that appear to have evolved prior to the appearance of vertebrates. The major differences among vertebrates seem to be in the segments of gut involved and the degree of absorption that takes place. The esophagus of marine fish adjusts the osmolality and the NaCl content of ingested sea water by absorption of Na and Cl and secretion of water. The hindgut of reptiles and birds absorbs electrolytes and water of both digestive secretions and urinary excretions. Furthermore, the hindgut of terrestrial vertebrates must conserve electrolytes and water to a greater extent than that of aquatic species, and the voluminous secretions of herbivores require additional compensation. The ultimate need for efficiency of this process is seen in desert herbivores.

Comparative studies of secretion and absorption of electrolytes and water have contributed a great deal to our present understanding of the underlying mechanisms. More extensive studies of these mechanisms should provide a rich source of future information.

8 Neurohumoral control

Many of the motor, secretory, and absorptive activities of the digestive system are under neuroendocrine control. Functions that involve the most cranial and caudal ends of the digestive tract such as mastication, salivary secretion, deglutition, and defecation are primarily or entirely under the control of the nervous system. Functions involving the remainder of the digestive system are coordinated by the nervous system, hormones, or a combination of both. Nerve stimulation tends to produce rapid and transient effects, whereas release of hormones tends to result in a slower, but more prolonged, response. The latter is especially true for endocrine secretions, which act on distant target organs, as opposed to paracrine secretions, which have only a local effect. However, the distinction between whether a given substance serves as a neurotransmitter, a neuromodulator, or a hormone has become difficult with the discovery that many of these agents are found in both neurons and endocrine cells, and that their function can vary with both site in the digestive tract and animal species.

The understanding of neurohumoral control mechanisms has rapidly advanced, as a result of the chemical characterization of many of the agents involved and the use of immunoassay techniques to locate their site of production. However, many of these agents are present in multiple molecular forms, and Van Noorden and Polak (1979) point out that economy of nature often adapts one molecule to many uses that depend on both its structure and target organ. Furthermore, the site on a molecule that determines its antigenic response may have evolved separately from the amino acid sequence that determines its biological activity. Therefore, immunoreactivity does not necessarily indicate the presence of a given neurotransmitter or hormone, nor does the absence of immunoreactivity necessarily indicate its absence.

Nervous control

The innervation of the mammalian digestive system has been reviewed by Gabella (1981), Wood (1981), and Davenport (1982). It consists of neurons intrinsic to the salivary glands, pancreas, gallbladder, and digestive tract, and the extrinsic neurons that connect these to the central nervous system. Cell

220

bodies of neurons intrinsic to the digestive tract are located in the ganglia of the submucosal plexus, and also the myenteric plexus, which is situated between the circular and longitudinal muscle layers of the gut. Receptors of afferent neurons can sense changes in the pH, osmolality, the chemical composition of digesta, or the degree of gut wall distension. This information can be transferred to efferent neurons to provide a local reflex response, or conducted via the extrinsic afferent nerve supply to the central nervous system, where it is joined by information gathered in other parts of the body. The extrinsic nerve trunks also contain visceral efferent fibers, which provide for central control or modulation of the digestive system.

The visceral efferent or autonomic innervation of the digestive system has been classically divided into the parasympathetic and sympathetic nervous systems (Fig. 8.1). The parasympathetic division is often referred to as the craniosacral system, because of the location of its preganglionic nerve cell bodies in the brain and sacral spinal cord. Preganglionic parasympathetic nerve fibers of man travel via the glossopharyngeal and facial nerves to the salivary glands; the vagal and recurrent laryngeal nerves, to the smooth muscle segment of the esophagus; and the vagal nerves, to the stomach, pancreas, gallbladder, small intestine, and proximal segment of the large intestine. These fibers synapse with neurons located in ganglia present in the glands or walls of the gallbladder and digestive tract. The postganglionic neurons innervate muscle or epithelial cells directly or by way of internuncial neurons. The sacral segment of the spinal cord provides similar parasympathetic innervation to the more distal segments of the large intestine via the pelvic nerves.

Preganglionic neurons of the sympathetic division originate in the thoracolumbar region of the spinal cord. Most of these nerve fibers synapse with neurons whose cell bodies are located in paravertebral ganglia or the prevertebral cervical, celiac, and mesenteric ganglia. Postganglionic fibers travel by sympathetic nerves to the digestive system, with the exception of postganglionic fibers from the celiac ganglia of some mammals such as the dog and rat, which join the vagus to form a vagosympathetic trunk. Postganglionic sympathetic fibers terminate on blood vessels, muscles, and secretory cells, but the vast majority of those supplying the digestive tract terminate in the myenteric plexus.

It once was believed that the terminals of all preganglionic parasympathetic and sympathetic neurons, and the terminals of all postganglionic parasympathetic neurons that supply the digestive system, were cholinergic, releasing acetylcholine, which was excitatory to postsynaptic neurons or effector cells. All postganglionic sympathetic neurons were believed to be adrenergic, releasing norepinephrine as an antagonistic inhibitory agent. Although this appears to be true for the autonomic control of pancreatic secretion, innervation of the salivary glands and digestive tract has been shown

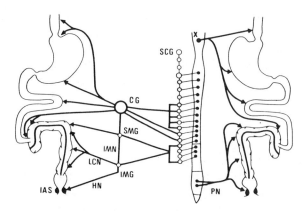

FIGURE 8.1 Schema of the extrinsic efferent innervation of the mammalian gut. The sympathetic innervation is represented to the left, the parasympathetic innervation to the right. This representation is a synthesis of various data and may present variations according to different species. Superior cervical ganglion (SCG); celiac ganglion (CG); superior mesenteric ganglion (SMG); inferior mesenteric ganglion (IMG); intermesenteric nerve (IMN); lumbar colonic nerves (LCN); hypogastric nerves (HN); vagus dorsal motor nucleus and vagus nerve (X); pelvic nerves (PN); internal anal sphincter (IAS). (From Roman and Gonella 1981.)

to consist of a much more complex arrangement. Adrenergic neurons provide excitatory stimuli to the salivary glands and muscle cells of sphincters and the muscularis mucosa. Furthermore, postganglionic adrenergic neurons appear to produce their inhibitory effect on gut muscle by preventing the release of acetylcholine from cholinergic neurons, rather than by a direct effect on postsynaptic membranes. The cranial parasympathetic nerve supply to stomach muscle was found to contain also postganglionic, nonadrenergic inhibitory fibers, which are responsible for receptive relaxation during feeding. This led to the discovery that transmission and modulation of information by the enteric nervous system involve a variety of neurotransmitting and modulating agents.

Descriptions of the autonomic nervous system of fish (Campbell 1970), reptiles (Berger and Burnstock 1979), and birds (Bennett 1974) are not as complete as those for mammals. However, Burnstock (1969) concluded that vagal innervation of the digestive tract of fish did not extend beyond the stomach and that the postganglionic neurons innervated by the vagus were entirely nonadrenergic, noncholinergic, and inhibitory, whereas the spinal autonomic innervation of the fish gastrointestinal tract consisted of both cholinergic excitatory and adrenergic inhibitory neurons (Figs. 8.2 and 8.3). Autonomic innervation of the gastrointestinal tract of adult amphibians appeared to be similar to that of fish, except that some sympathetic neurons joined the vagus to form a vagosympathetic trunk, similar to that noted in the

FISH

AMPHIBIANS

REPTILES

MAMMALS

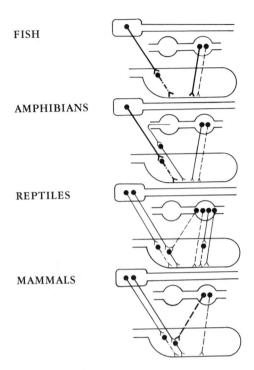

FIGURE 8.2 Diagrammatic representation of the autonomic cholinergic excitatory (—), adrenergic (----), and nonadrenergic inhibitory (-· · · -) nerves to the stomach of vertebrates. Note that the vagal parasympathetic outflow is purely inhibitory to the fish and amphibian stomach and is opposed by excitatory cholinergic sympathetic fibers. In reptiles and mammals, the cholinergic excitatory nerves have been switched to the parasympathetic outflow, with sympathetic fibers becoming adrenergic and inhibitory. Adrenergic modulation of intramural ganglion cell activity is rudimentary in reptiles and strongly developed in mammals. The diagram depicting the innervation of the reptile stomach is largely conjectural. Intramural neurons that are independent of the extrinsic nerve supply are not included in the diagrams. (From Burnstock 1969.)

dog and rat, and there was evidence of a sacral parasympathetic nerve supply. The autonomic nervous system of reptiles appeared to be similar to that of mammals, with the complete exchange of cholinergic excitatory function from the sympathetic to the parasympathetic outflow, retention of some vagal nonadrenergic, noncholinergic inhibitory innervation, and with the definite presence of sacral parasympathetic innervation to the hindgut.

The digestive tract of birds is provided with a vagal, thoracolumbar, and sacral innervation similar to that of mammals, but the extent of vagal innervation and the characteristics of the sacral innervation are unclear (Bennett

FISH

AMPHIBIANS

REPTILES

MAMMALS

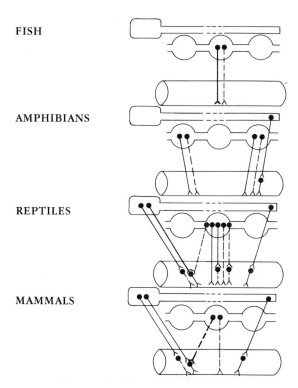

FIGURE 8.3 Diagrammatic representation of the autonomic cholinergic excitatory (—), adrenergic (----), and nonadrenergic inhibitory (-·-·-) nerves to the intestine of vertebrates. The diagram illustrating the reptile condition is partly conjectural, being based largely on the results of fluorescent histochemical studies. Note that the vagus nerve does not extend far enough down the gut in lower vertebrates to influence most of the intestine, whereas in mammals it extends at least as far as the ileocolonic junction. Adrenergic modulation of ganglion activity appears first in the reptiles and is strongly developed in mammals. It is not known whether adrenergic nerve terminals envelop postganglionic sacral parasympathetic neuron cell bodies. A separate sacral parasympathetic outflow in amphibians is depicted, although this is debatable. Intramural neurons that are independent of the extrinsic nerve supply are not included, although it should be noted that in the large intestine there is evidence that intramural nonadrenergic inhibitory neurons, as well as cholinergic neurons, are involved in local reflex pathways. (From Burnstock 1969.)

1974). The ganglionic nerve of Remack, which extends the length of the avian small and large intestine, receives fibers from the vagus, as well as the thoracolumbar and sacral spinal cord. There appears to be disagreement over the extent of intestinal innervation by the vagus (Tindall 1979), but there is evidence that cholinergic neurons project orally from the sacral spinal cord (Hodgkiss 1984).

Salivary, gastric, pancreatic, and intestinal secretion are under varying degrees of autonomic control. The salivary glands of mammals receive their cholinergic stimulation via the glossopharyngeal and facial nerves, and their adrenergic stimulation of blood vessels and secretory cells from sympathetic innervation via the superior cervical ganglia. The salivary glands of birds (Bennett 1974) and reptiles (Berger and Burnstock 1979) appear to be provided with similar innervation. The cephalic phase of gastric secretion is stimulated by the sight and taste of food. Reflex stimulation of vagal nerves results in the production of HCl and pepsinogen, and release of the hormone gastrin from pyloric mucosa. Gastrin then stimulates release of HCl, which appears to provide the stimulus for pepsinogen secretion. Distension of the stomach with food results in the further release of gastrin via local nerve reflexes. Geese (Karpov 1919), ducks (Walter 1939), and barn owls (Smith and Richmond 1972) are reported to have a cephalic phase of gastric secretion, but there appears to be disagreement over its presence in chickens (Burhol 1982; Duke and Bedbury 1985). Vagal stimulation resulted in the secretion of small volumes of gastric juice, high in HCl and pepsinogen, by the tortoise (*Testudo graeca*) and lizard (*Trachydosaurus rugosus*) (Wright, Florey, and Sanders 1957), but Barrington (1957) found no evidence of a cephalic phase for gastric secretion in fish. The gastric phase of gastric secretion appears to be common to all vertebrates studied.

Preganglionic vagal nerve fibers innervate postganglionic cholinergic neurons in the pancreas of mammals, which provides a cephalic phase of pancreatic secretion. These postganglionic neurons also are connected to neurons in the ganglia of the intestinal tract. Cholinergic innervation of pancreatic cells has been observed in the snake *Elaphe quadrivirgata* (Watari 1968), but there appears to be no evidence of vagal stimulation of the fish pancreas (Barrington 1957). Evidence for autonomic control of electrolyte transport by the intestinal mucosa has been reviewed by Powell and Tapper (1979). They concluded that the cholinergic response of the small and large intestine was secretion of Cl, and the adrenergic response was absorption of Na and Cl, with perhaps an associated inhibition of HCO_3 secretion.

In addition to the nonadrenergic inhibition of gastric muscle, there now is evidence for nonadrenergic inhibition of muscle in the small intestine of the guinea pig, mouse, rabbit, and cat, and for noncholinergic excitatory innervation of muscle in the small intestine of the guinea pig, rabbit, dog, cat, bird, reptile, amphibian, and cyclostome (see Burnstock 1972). Nonadrenergic inhibitory responses also have been demonstrated in the large intestine of the guinea pig, rat, mouse, Mongolian gerbil, chimpanzee, and man (see Burnstock 1986), as well as the hindgut of the chicken (Mishra and Raviprakash 1981; Komori and Ohashi 1982; Meldrum and Burnstock 1985).

The discovery of nonadrenergic, noncholinergic inhibition of gastric muscle led to the proposal that a purine nucleotide, ATP, served as the neurotransmitter (Burnstock, Campbell, Satchell, and Smythe 1970). This was followed by evidence of aminergic neurons, which released 5-hydroxytryptamine (5-HT), dopamine, and γ-aminobutyric acid (GABA); and peptidergic neurons that released enkephalins, vasoactive intestinal polypeptide (VIP), substance P, bombesin/gastrin-releasing peptide, neurotensin, cholecystokinin (CCK), and neuropeptide Y/pancreatic polypeptide. Table 8.1 lists nonadrenergic, noncholinergic neurotransmitters that have been found in the gut and their possible physiological roles.

Substance P was the first peptide identified in both the gut and the central nervous system (von Euler and Gaddum 1931). It is a potent vasodilator and a strong stimulus for salivary secretion (see Van Noorden and Polak 1979). It also has been shown to stimulate contraction of intestinal muscle and stimulate the secretion or inhibit the absorption of electrolytes by the intestine of mammals (Powell and Tapper 1979). It appears to function as a neurotransmitter or a modulator of synaptic transmission in both the central and autonomic nervous system. Substance P-like immunoreactivity has been identified in the gut of birds, reptiles, amphibians, and teleosts (Buchan, Polak, and Pearse 1980; Rawdon and Andrew 1981; Brodin et al. 1981; Holmgren, Vaillant, and Dimaline 1982; Buchan, Lance, and Polak 1983; Rombout and Reinecke 1984).

Vasoactive intestinal polypeptide is found in the cell body and terminals of nerves throughout the digestive tract of mammals (Keast, Furness, and Costa 1985). Infusion of VIP into the blood usually results in relaxation of gastrointestinal smooth muscle, inhibition of gastric acid secretion, and stimulation of biliary and intestinal secretion. However, its concentration in the blood does not rise after a meal, and its presence in neurons of the gut and spinal cord (Gibson et al. 1984) suggests that it serves only as a neurotransmitter in mammals.

Vasoactive intestinal polypeptide-like immunoreactivity has been reported in the gut of birds, reptiles, amphibians, teleosts, and chondrichthyeans (Langer, Van Noorden, Polak, and Pearse 1979; Buchan et al. 1980; Falkmer et al. 1980; Fouchereau-Peron et al. 1980; Vaillant, Dimaline, and Dockray 1980; Buchan et al. 1981; Fontaine-Perus, Chanconie, Polak, and Le Douarin 1981; Rawdon and Andrew 1981; Reinecke, Schlüter, Yanaihara, and Forssmann 1981; Holmgren et al. 1982; Buchan et al. 1983; Holmgren and Nilsson 1983; el Salhy 1984; Rombout and Reinecke 1984; Rawdon 1984). Vasoactive intestinal polypeptide-like activity was found in the nerve cells of all vertebrates except cyclostomes, but it also was present in the gut epithelial cells of vertebrates other than mammals (Reinecke et al. 1981; Rawdon 1984); therefore, it is considered to serve both as a neurotransmitter and a primitive hormone (Van Noorden and Polak 1979).

Somatostatin is another neurotransmitter/hormone that has been iden-

TABLE 8.1 Possible roles of nonadrenergic, noncholinergic neurotransmitters in gut

Putative transmitters	Nerves to smooth muscle					Nerves to mucosal secretory cells	Sensory nerves	Nerves modulating release of	
	Interneurons	Nonsphincteric	Sphincteric	Musc. mucosa	Blood vessels			ACh	NE
Adenosine triphosphate	–	I	I	–	I	–	–	→	→
5-Hydroxytryptamine	E	E	–	–	I	I	–	←	–
Dopamine	I	I	I	–	I	–	–	→	→
γ-Aminobutyric acid	I/E	–	–	–	–	–	–	→	→
Peptides									
Enkephalin/Endorphin	I	–	–	–	–	I	–	→	–
Somatostatin	I	–	–	–	–	I	–	→	–
Vasoactive intestinal polypeptide/ Peptide HI	I/E	I	I	I	I	E	–	←	–
Substance P	E	E	–	E	I	E	–	←	–
Gastrin-releasing peptide/Bombesin	E	E	–	–	–	–	–	–	–
Neurotensin	E	–	–	–	–	–	–	←	–
Cholecystokinin/Gastrin	E	–	–	–	–	I	–	←	–
Neuropeptide Y/Pancreatic polypeptide	–	–	–	–	E	–	–	–	→

Note: Excitatory (E) or inhibitory (I) effects and modulating effects on release of acetylcholine or norepinephrine. Dash (–) indicates no effect.
Source: Burnstock (1986).

tified in the gastrointestinal tract and pancreas of a wide range of vertebrates (Van Noorden and Polak 1979). It appears to serve a strictly paracrine function as a hormone. Its reported effects include the inhibition of 1) gastric and pancreatic secretion, 2) myoelectric activity in the stomach and intestine, and 3) glucose and amino acid absorption from the intestine. Somatostatin appears to act, at least in part, by inhibiting release of other hormones such as gastrin and motilin. Somatostatin-like immunoreactivity has been identified in the gut of mammals, birds, reptiles, amphibians, teleosts, chondrichthyeans, and cyclostomes (Falkmer and Östberg 1977; Alumets, Sundler, and Håkanson 1977; Langer, Van Noorden, Polak, and Pearse 1979; Seino, Porte, and Smith 1979a; Buchan et al. 1980; Rawdon and Andrew 1981; Holmgren et al. 1982; Holmgren and Nilsson 1983).

The possible roles of the other neurotransmitters listed in Table 8.1 have been reviewed by Burnstock (1986). Enkephalin (Epstein, Lindberg, and Dahl 1980; Saffrey, Polak, and Burnstock 1982), neurotensin (Saffrey et al. 1982), and γ-aminobutyric acid (Saffrey, Marcus, Jessen, and Burnstock 1983) have been reported in the gut of chickens. Neurotensin-like immunoreactivity has been demonstrated in the gut of birds, reptiles, amphibians, and freshwater, stomachless teleosts (Sundler et al. 1977a,b; Reinecke et al. 1980). Cholecystokinin, bombesin, and pancreatic polypeptide (PP) will be discussed later under their role as hormones. The subject of neurotransmitting and neuromodulating agents is further complicated by the fact that some neurons produce more than one of these agents, and one of these may modulate release of the other. This includes neurons that release ATP and norepinephrine, ATP and substance P, and acetylcholine and VIP. Furthermore, the neurotransmitter produced by an individual neuron can change during embryonic development (Black and Patterson 1980).

Endocrine control

Endocrine control of the mammalian digestive system has been reviewed by Solcia et al. (1981), Walsh (1981), and Davenport (1982). Bentley (1982) has reviewed the comparative endocrinology of vertebrates. The role of gastrointestinal peptide hormones in ruminants has been recently reviewed by Titchen (1986). As with the nervous control, information on the endocrinology of the digestive system has expanded rapidly since the early 1970s, with some major changes in our understanding of the action of those hormones that were previously described, and the addition of many other putative and candidate hormones. The activities of the digestive system are controlled by a large number of peptides produced by endocrine, paracrine, and nerve cells. This led to the theory that the hormones of the digestive system evolved originally from neuroectodermal tissue (Pearse 1969). Although this does not appear to hold true for many of these hormones, some families of hormones appear to have evolved from similar ancestral proteins, and some hormones may have served originally as neurotransmitting or neuromodulating agents.

Hormones

Gastrin family

Gastrin is synthesized by and stored in the G cells of the pyloric glandular and duodenal mucosa of mammals. The structural characteristics of gastrin have been determined for a number of species. As many as six different gastrinlike molecules have been isolated from a single species, and many amino acid substitutions are found in the gastrins of different animals. However, the C-terminal pentapeptide is common to all forms of gastrin. Gastrin can be released by vagal stimulation, distention of the stomach, or the presence of peptides, amino acids, and Ca salts in gastric contents. The stimulus for gastrin release by the duodenal mucosa is not clear.

In addition to its effect on gastric secretion, gastrin has been shown to stimulate gastric blood flow, contraction of the circular muscle of the stomach and lower esophageal sphincter, and growth of gastric mucosa, intestinal mucosa, and pancreatic tissue. There is evidence that gastrin stimulates secretion or inhibits the absorption of electrolytes by the intestine (Powell and Tapper 1979). Systemic administration of pentagastrin, a synthetic peptide with gastrin activity, inhibited the motility of the reticulorumen of sheep (McLeay and Titchen 1970, 1975) and cattle (Ruckebusch 1971). Subsequent studies in sheep showed this to be a direct effect on the brain (Grovum and Chapman 1982; Nicholson 1982). Intracerebroventricular administration of pentagastrin, tetragastrin, or gastrin-17 also had an inhibitory effect on rumination in sheep (Honde and Bueno 1984).

Gastrin appears to be absent from the gut of lampreys (Holmquist, Dockray, Rosenquist, and Walsh 1979) and dogfish (Vigna 1979). However, gastrinlike activity has been found in the stomach of the coho salmon (*Oncorhynchus kisutch*) (Vigna 1979) and rainbow trout (*Salmo gairdneri*) (Holmgren et al. 1982), the stomach and intestine of the perch (*Perca fluviatilis*) and catfish (*Ameiurus nebulosus*), the intestine of the stomachless carp (*Cyprinus carpio*) (Noaillac-Depeyre and Hollande 1981), and the rectum of the dogfish (Holmgren and Nilsson 1983). Gastrinlike immunoreactivity has been demonstrated in the stomach of the frog (*Rana berlandieri*), salamander (*Salamandra salamandra*), crocodile (*Crocodylus niloticus*), and alligator (*Alligator mississippiensis*) (Dimaline et al. 1982; Buchan et al. 1983) and at the junction of the gizzard and duodenum of the chicken (Larsson et al. 1974; Vigna 1984), duck (Larsson and Rehfeld 1977), and turkey (Dockray 1979a). Pentagastrin was a stronger stimulus for pepsin secretion than the secretion of HCl by the cells that produce both of these in the chicken (Burhol 1982).

Cholecystokinin was first described as a hormone produced in the duodenal and jejunal mucosa of mammals, released by the presence of fat in the duodenum and responsible for contraction of the gallbladder and relaxation of the choledochal sphincter. This was followed by the discovery that the presence of HCl and the digestive products of fat and protein in the duodenum released a hormone called pancreozymin (PZ), which stimulated the secretion of pancreatic enzymes. Pancreozymin proved to be identical to CCK with the result that it was subsequently labeled CCK-PZ and then simply, CCK. Cholecystokinin also has been shown to augment the action of secretin on pancreatic HCO_3 secretion, delay gastric emptying, and exert a trophic (growth) effect on pancreatic tissue. There is evidence that CCK additionally may serve as a short-term satiety factor (Denbow 1982; Savory and Hodgkiss 1984).

The structure of CCK can vary with species. However, the C-terminal octapeptide possesses all of the hormone activity, and the C-terminal pentapeptide is identical to that of gastrin, causing confusion in the immunoreactive assays. Although infusion of CCK stimulated gastric acid secretion and gastrin infusions stimulated the release of pancreatic enzymes in birds and mammals (Burhol 1982), these may have been pharmacological, rather than physiological effects. Cholecystokinin appears to be a major peptidergic neurotransmitter in both the central nervous system and the gut (Vanderhaeghen, Signeau, and Gept 1975; Rehfeld 1980; Burnstock 1986).

Cholecystokinin-like immunoreactivity has been demonstrated in the intestine of birds, reptiles, amphibians, teleosts, chondrichthyeans, and cyclostomes (Barrington and Dockray 1970, 1972; Nilsson 1970, 1973; Östberg, Van Noorden, Pearse, and Thomas 1976; Dockray 1979b; Vigna 1979; Dimaline et al. 1982; Holmgren et al. 1982; Buchan et al. 1983; Dimaline 1983; Holmgren and Nilsson 1983; Vigna, Fischer, Morgan, and Rosenquist 1985), and in the stomach of the crocodile (Dimaline et al. 1982) and chicken (Vigna 1984). Caerulein is a decapeptide initially found in frog skin, with a structure identical to that of the C-terminal decapeptide of CCK, except for a single amino acid substitution. Caerulein-like activity has been identified in the stomach of teleosts and amphibians (Larsson and Rehfeld 1977, 1978; Dimaline 1983).

Secretin family

Secretin was the first hormone to be discovered, when Bayless and Starling (1902) showed that acid placed in a denervated loop of the upper intestine of the anesthetized dog resulted in secretion of pancreatic fluid. It was demonstrated later that secretin stimulates the release of HCO_3-containing fluid from the pancreas and liver. Although secretin has only a weakly stimulatory effect on pancreatic enzyme secretion and although CCK is a weak stimulus for pancreatic and biliary secretion of fluids, the two hormones, when released together, have a synergistic effect on both types of secretion. Secretin has been reported to inhibit gastric secretion of HCl in mammals, but stimulates the secretion of HCl and pepsin in birds (Burhol 1982).

Secretin is a peptide containing 27 amino acid residues, many of which are common to VIP, gastric inhibitory polypeptide (GIP), and glucagon. Secretin-like activity has been identified in the gut of birds, reptiles, teleosts, and cyclostomes (Barrington and Dockray 1970; Dockray 1974, 1975, 1978; Nilsson, Carlquist, Jörnvall, and Mutt 1980; Rawdon and Andrew 1981; Buchan et al. 1983). No similar activity was found in the gut of the salamander (Buchan et al. 1980).

Vasoactive intestinal polypeptide shares nine of its 28 amino acids with secretin. The biological activity and probable role as a hormone and neurotransmitter have been discussed.

Gastric inhibitory polypeptide

Gastric inhibitory polypeptide is released from the intestinal mucosa of mammals by the presence of glucose, fat digestion products, HCl, and amino acids in the lumen. This hormone inhibits gastric acid secretion and, in pharmacological doses, slows gastric emptying, causes mesenteric vasodilation, and stimulates the release of glucagon and insulin. The rise in plasma GIP concentration after a meal supports its role as a true hormone. The peptide isolated from pig intestinal mucosa contains 43 amino acids, and ten of the first 27 are identical to those of porcine secretin. There appear to be significant differences between the GIP of human and porcine origin (Bacarese-Hamilton, Adrian, and Bloom 1984). Gastric inhibitory polypeptide has been considered as a prime candidate for the role of enterogastrone, a hormone proposed to be responsible for the inhibitory effect of fat ingestion on gastric acid secretion. Neurohumoral controls of gastric and pancreatic secretion and the release of bile are illustrated in Figure 8.4.

Gastric inhibitory polypeptide-like immunoreactivity has been reported in the pancreas and intestinal mucosa of the dogfish (el-Salhy 1984), but was not demonstrated in the gut of the salamander (Buchan et al. 1980) or reptiles (Buchan, Ingman-Baker, Levy, and Brown 1982; Buchan et al. 1983). However, porcine GIP was shown to stimulate HCO_3 secretion by the duodenum of the bullfrog (*Rana catesbeiana*) (Flemström and Garner 1980).

Motilin

Motilin, which has been isolated from the intestinal mucosa of mammals, appears to stimulate the interdigestive migrating myoelectric complexes in the stomach and small intestine of the dog (Lee, Park, Chang, and Chey 1983; Nakaya et al. 1983; Usellini et al. 1984; Borody, Byrnes, and Titchen 1984),

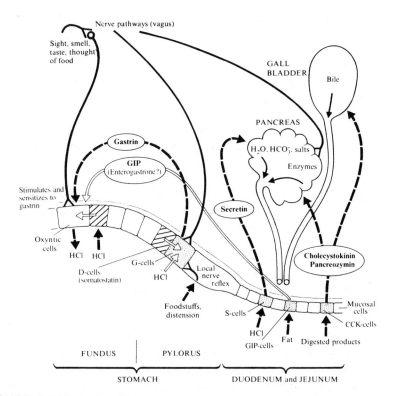

FIGURE 8.4 The role of hormones in controlling gastric acid secretion, pancreatic secretion of salts and enzymes, and the contraction of the gallbladder. Gastrin, from the pylorus, initiates secretion of hydrochloric acid by the oxyntic cells in the fundus. The duodenal-jejunal hormones, secretin and cholecystokinin-pancreozymin (CCK), initiate the secretion of pancreatic juice and enzymes, respectively. Gastric-inhibitory peptide (GIP), from the duodenum-jejunum, inhibits gastric acid secretion. The open arrows indicate an inhibitory effect; the dashed closed arrows, a stimulation. (From Bentley 1982.)

but not those of the pig (Borody et al. 1984). Motilin-like immunoreactivity has been identified in the intestine of the chicken (Rawdon and Andrew 1981) and Japanese quail (*Coturnix c. japonica*), but not in the lizard (*Anolis carolinensis*) or the Pacific hagfish (*Eptatretus stouti*) (Seino, Porte, Yanaihara, and Smith 1979b).

Pancreatic polypeptide
Pancreatic polypeptide (PP) was first extracted from the pancreas of chickens by Kimmel, Hayden, and Pollock (1975) and shown to be a potent

stimulus of acid and pepsin secretion (Kimmel, Pollock, and Hayden 1978). It was subsequently isolated from the pancreas of several mammals (Floyd, Fajans, Pek, and Chance 1977). Although cattle, sheep, pigs, dogs, and humans appear to have a PP identical in molecular structure, 20 of the 36 amino acids in mammalian PP differ from those of the chicken hormone (Walsh 1981). Furthermore, mammalian PP has little or no effect on gastric secretion of either the chicken or mammals. It has been shown, however, to increase gastrointestinal motility in the dog and to inhibit pancreatic enzyme and HCO_3 secretion in both the dog and the human.

Pancreatic polypeptide-like immunoreactivity was demonstrated in the pancreas of ten additional species of bird, as well as reptiles and amphibians (Langslow, Kimmel, and Pollock 1973). It has been identified in the intestine of the chicken (Alumets, Håkanson, and Sundler 1978; Rawdon and Andrew 1981), the gut and pancreas of teleosts, and the intestine of cyclostomes (Van Noorden and Polak 1979). However, it was not found in the gut of the salamander (Buchan et al. 1980) or alligator (Buchan et al. 1983).

Bombesin/gastrin-releasing peptide

Bombesin is another peptide, like caerulein, that was initially isolated from amphibian skin. A bombesin-like peptide that showed a potent gastrin-releasing activity was isolated from intestinal tissue of the pig and labeled gastrin-releasing peptide (GRP) (McDonald et al. 1979). Bombesin-like immunoreactivity has been identified in the gut and brain of many mammals and the gut of birds, reptiles, amphibians, teleosts, and chondrichthyeans (Timson et al. 1979; Vaillant, Dockray, and Walsh 1979; Buchan et al. 1980; Rawdon and Andrew 1981; Holmgren et al. 1982; Buchan et al. 1983; Holmgren and Nilsson 1983). Administration of bombesin to the Atlantic codfish, *Gadhus morhua*, stimulated gastric secretion of acid (Holstein and Humphrey 1980).

Prostaglandins

The prostaglandins, a series of oxygenated, unsaturated, cyclic fatty acids, also have a variety of hormonelike actions on motility and the transport of electrolytes across gastrointestinal mucosa. Evidence that prostaglandins may play a role in the motor activity associated with the production of soft feces by the rabbit large intestine (Pairet et al. 1986) was mentioned earlier. They also have been proposed to be the stimulus for inhibition of gastric secretion during the time that larvae are housed in the stomach of the gastric brooding frog (Shearman et al., 1984). Recent findings suggest that prostaglandins produced by intestinal epithelium can play a very important role in the control of electrolyte secretion and absorption by intestinal mucosa (Cooke 1987).

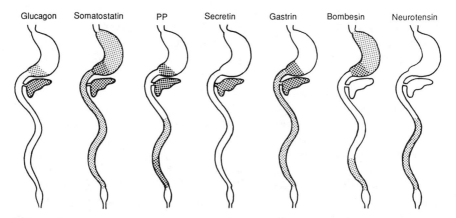

FIGURE 8.5 Diagram showing the population of endocrine cells containing different peptide immunoreactivities in the gastroentericpancreatic system of *Rana catesbeiana*. (From Fujita et al. 1981.)

Candidate hormones

Other peptides, including neurotensin and enkephalin, also have been proposed as candidates for mammalian gut hormones (Van Noorden and Polak 1979; Walsh 1981). The role of somatostatin as a paracrine hormone was discussed with the nervous system. Many of the hormones postulated for mammals appear to be present in the digestive tract of birds (Table 8.2) and lower vertebrates (Fig. 8.5).

Evolution of gut peptides

The evolution of peptides that serve as neurotransmitters, neuromodulators, or hormones in the gut of vertebrates is discussed by Barrington (1982) and Krieger (1983). Barrington stressed the need for caution in interpreting fragmentary evidence collected from few species with no fossil support, and pointed out that many of the adaptations are determined by evolution of receptors and modulation of the programing of receptor cells, rather than changes in the molecular structure of the peptides. However, substances similar to the peptides found in the vertebrate nervous and endocrine systems appear to provide intercellular communication in higher plants and protozoa (Fig. 8.6). Peptides that serve as neurotransmitters or neurohormones are present also in other invertebrates such as the coelenterates and annelids. Therefore, some of the substances secreted by the nervous and endocrine systems appear to have been present before these systems evolved.

The gastrin family of peptides is believed to have been derived from a single ancestral molecule, rather than through parallel evolution. Chole-

TABLE 8.2 Distribution of cells showing immunoreactivities for gut peptides and serotonin in the gastrointestinal tract of the chicken

	SRIF	APP	PYY[a]	GLUC	SEC	VIP	GAS	CCK	NT	BN	SP	ENK[b]	MOT	5HT[c]
Proventriculus	••	•		••					•	•				•
Gizzard										•			•	
Pylorus	••• ••			•			••• •••		•• ••				•	•
Duodenum	•	•	•	•	••	•• •	•	•[d]	•• •		•• •	•	•	•• ••
Upper ileum			•	•		•• •		•	• •		•• •	•		•• •
Lower ileum	•	•		•		•• •		•	•			•		•• •
Cecum						•					•	•		••
Rectum						•			••			•		•• ••

Abbreviations: SRIF, somatostatin; APP, avian pancreatic polypeptide; PYY, polypeptide YY; GLUC, glucagon; SEC, secretin; VIP, vasoactive intestinal peptide; GAS, gastrin; CCK, cholecystokinin; NT, neurotensin; BN, bombesin; SP, substance P; ENK, leu-enkephalin; MOT, motilin; 5HT, serotonin.

[a]El-Salhy et al. (1982), recently hatched chicks.

[b]Alumets et al. (1978), chickens.

[c]Unpublished observations, chicks at hatching.

[d]Larsson and Rehfeld (1977), chickens. All other data from Rawdon and Andrew (1981), chicks at hatching.

Source: Rawdon (1984).

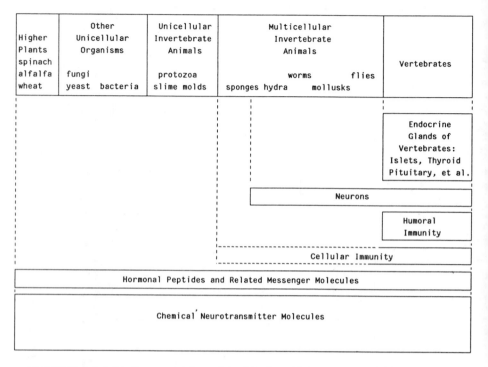

FIGURE 8.6 Evolutionary origins of the biochemical elements of the endocrine, the immunolgic, and the nervous system. (From Roth et al. 1985.)

cystokinin-like peptides have been demonstrated in the gut of species belonging to all classes of vertebrates, and CCK is the only member of this hormone family found in cyclostomes. Although extracts from the intestine of hagfish stimulate contraction of the guinea pig gallbladder (Vigna 1979; Vigna and Gorbman 1979), contraction of the hagfish gallbladder is not stimulated by either these extracts or mammalian CCK, suggesting an absence of the appropriate receptors. Radioimmunoassay and immunostaining studies indicate that a separate gastrinlike peptide appeared at the divergence of amphibians and reptiles, but bioassay procedures capable of discriminating between CCK and gastrin indicate an earlier divergence between elasmobranch and teleost fish. Nevertheless, the gallbladder of the coho salmon responded to both CCK and gastrin (Vigna and Gorbman 1977), suggesting that its receptors may not be able to distinguish between the two.

Among the peptides of the secretin family, VIP has been considered to be the ancestral form, because of its presence in nervous tissue of prochordates and its wide range of activity in mammals. However, secretin-like activity has been reported in prochordates and molluscs.

Pancreatic polypeptide-like immunoreactivity has been identified in the intestinal epithelium of cyclostomes, and in the gut and pancreas of teleosts and birds. A similar substance has been found in the gut of prochordates and the nervous system of an earthworm, mollusc, and insect. Substance P has been identified in the nervous system of a coelenterate (*Hydra*), prochordates, cyclostomes, teleosts, amphibians, and birds. The intestinal mucosa of prochordates and species from all of the major classes of vertebrates also demonstrates immunoreactivity to somatostatin.

Summary

Neurohumoral control of the digestive system is effected by a wide range of substances secreted by nerves and endocrine cells. These substances, which include amino acids, amines, purines, and peptides, act on the effector cells or interact on the release of one another. Many of these agents, or substances similar in structure, appear to have been present before the evolution of vertebrates. Others such as gastrin seem to have evolved in vertebrates from molecules that may have served a different purpose. Although the nervous system appeared earlier than the classical glandular endocrine system, vertebrate-like hormones are found very early in the phylogeny of invertebrates.

As with previous topics, comparative data on neurohumoral control are limited to only a few species in each of the major classes of vertebrates. Interpretation is often further hampered by a lack of information on the molecular structure of the substance produced in a given species and its physiological effects in the same animal. However, the procedures and techniques that now are available for these studies should yield much additional information on the function and evolution of these systems and the general evolution of vertebrates.

9 Conclusions

The vertebrate digestive system has a number of general characteristics common to most species. Food material is procured and prepared for ingestion by the headgut (mouthparts and pharynx), passed through the esophagus, and stored in the stomach, where it undergoes physical digestion and the initial process of protein digestion by pepsin and HCl. Gastric digesta are released into the midgut, which is the principal site for digestion by endogenous enzymes and absorption of nutrients. Digesta then pass into the hindgut, which aids in the reabsorption of electrolytes and water. The sequential and episodic events that are associated with the ingestion and digestion of food, the transit of digesta, and the absorption or excretion of end products are controlled and integrated by the autonomic nervous system and/or by hormones secreted by endocrine or paracrine cells. However, there are many variations in these processes that can be related to the classification, diet, environment, or other physiological characteristics of an animal.

The gills of many fish require adaptations, the gill rakers, which prevent the escape of food particles. In the cyclostomes and in some species within the more advanced classes of fish, the stomach is absent, and the pancreas varies in its form and distribution along the gut. The hindgut of fish is relatively short, and can be difficult to distinguish from the midgut in many species. These characteristics, with the exception of the diffuse pancreas, are shared by amphibian larvae. Salivary glands, which secrete mucus and often a variety of other substances, first appear in the adult amphibians and reach their major complexity in some mammals. A stomach is present in adult amphibians and the more advanced classes of vertebrates, although its functions are distributed among the crop, proventriculus, and ventriculus in birds. The midgut's capacity for digestion and absorption is expanded by the presence of a spiral valve or pyloric ceca in some fish, ridges and folds in reptiles, and villi in birds and mammals. Digestive enzymes can vary among the classes of vertebrates. The digestive tract of amphibians, reptiles, and birds (and a few species of fish and mammals) terminates in a cloaca, which allows the hindgut to share in the conservation of urinary electrolytes and water. A cecum is present in the hindgut of a few reptiles and many mammals, and paired ceca are present in most birds. These variations in the digestive system of the different classes of vertebrates are associated with changes in the mechanisms responsible for its neurohumoral control.

The mammalian digestive system shows a few general characteristics that are absent from lower vertebrates. The masticatory apparatus is better organized for the trituration of food than that of most other vertebrates. In species belonging to ten of the mammalian orders, the stomach is at least partially lined with stratified squamous epithelium. Eight orders include species with a stomach that is expanded into a voluminous organ, which is compartmentalized by sacculations or permanent structural arrangements. The hindgut of mammals tends to be longer than that of other vertebrates in relation to body length. Small herbivores that feed principally on plant fiber have a large and usually sacculated cecum. With the exception of those with a voluminous stomach, large mammalian herbivores on this type of diet have a voluminous colon, which is sacculated and, in some species, compartmentalized.

Species within each class show a wide range of adaptations to their diet or dietary regimen. These are most evident in the headgut. For example, the mouthparts and/or pharynx demonstrate elaborate arrangements for microphagous filter-feeding in some species of fish, amphibian larvae, birds, and mammals. The mouthparts of anteaters belonging to several mammalian orders have evolved similar arrangements of the teeth, jaws, and tongue. Teeth may be constructed for grasping, puncturing, or tearing, or for the grinding of plant material. The stomach and hindgut of mammals show numerous structural adaptations that aid in the retention of food or digesta, and the complement and quantities of digestive enzymes can vary with the diet of a particular species.

Other adaptations appear to be related to the environment. Absorption of NaCl by the esophagus adjusts the osmolality of seawater to that of the body fluid in many marine fish. The various adaptations of the hindgut for the conservation of electrolytes and water by terrestrial vertebrates provide other examples. Adaptations of the digestive system to other characteristics of an animal include the use of the stomach as "uterus" in the gastric-brooding frog, and the distribution of weight along the digestive tract of birds as an aid to flight.

An understanding of species variations in the digestive system is essential for the care and maintenance of domesticated species that serve as companion animals (pets) or for the production of food or fiber. It is equally necessary for the care and maintenance of captive wild animals and the preservation of wild species. Many diseases of domesticated and captive herbivores can be traced to the intermittent feeding of high-concentrate diets to animals with a digestive system designed for practically continuous feeding on bulk diets high in plant fiber. Similar problems can arise from the over-feeding of rapidly digestible diets to carnivores whose ancestors quickly ingested their prey, often with long intervals between meals.

More information on feeding habits and food chains is needed. Some large species of wild African herbivores are endangered because of restrictions in their migratory grazing patterns, due to expanding farmlands or attempts to facilitate their observation by tourists. These animals require extensive rangelands to obtain enough food and to take advantage of interdependencies between those that feed on mature plants and those that feed on the earlier, more succulent growth. Migration also allows for the dispersal of feces containing the eggs of gut parasites. The survival of many species depends on the balance among herbivores, plants, and predators. More information is needed on the role of wax esters in the food chain of marine animals. Plankton, insects, and rodents may merit more serious consideration as a direct source of food for humans and domesticated animals.

One of the major contributions of comparative physiology is the information it provides for the better understanding of human physiology. Because of the cost and inherent dangers of conducting studies on human volunteers, much of this information must be obtained from the examination of other animals. The majority has been derived from use of common laboratory animals, principally the dog, rabbit, and four or five species of rodent. Yet, with respect to the digestive system, the dog is a carnivore with a relatively simple hindgut, and the remaining species are omnivores or herbivores with a hindgut that includes a large cecum and is designed for coprophagy and cecotrophy. Therefore, some characteristics of the human digestive system might be better examined in animals with a diet and digestive tract more similar to those of humans such as the domestic pig and certain species of nonhuman primates. However, the concentration on similarities as a major criterion for the choice of "animal models" misses the point that much of the understanding of basic mechanisms has come from the study of differences rather than similarities in structure and function.

Good animal models, like beauty, are often in the eye of the beholder. For example, our present knowledge of how the human kidney works is partly based on studies of the aglomerular kidney of marine fish and the separate blood supply to the glomerulus and renal tubule of birds. In a similar fashion, much of our understanding of the human digestive system is derived from studies of ion transport across frog skin and the gallbladder of a fish (the roach), microbial digestion and metabolism in the ruminant forestomach, the motor functions of the cat and rabbit colon, and the digestive enzymes in a wide range of vertebrates. The same is true for studies of animals that lack a stomach, a gallbladder, or the ability to produce lactase, and those that practice coprophagy. Important clues on the function and malfunction of neurohumoral agents can be gained from knowledge of their evolution. Yet, if one considers the number of species of fish (21,700), amphibians (4000), reptiles (6250), birds (8600), and mammals (4150) available for study, it is apparent that most species and many entire groups of vertebrates have not been subjected to rigorous study or have not been studied at all. Furthermore, the researcher with a broad interest in comparative zoology is facing extinction because of increasing specialization and a lack of funding. Thus, more effective communication and cooperation between those investigators whose primary interest is the animals themselves, and those whose principal concern is the basic physiology mechanisms will be needed to accomplish these goals.

References

Adam, H. 1963. Structure and histochemistry of the alimentary canal. In *The Biology of Myxine,* ed. A. Brodal and R. Fänge, pp. 256-288. Oslo: Universitetsforlaget.

Akester, A. R., Anderson, R. S., Hill, K. J., and Osboldiston, G. W. 1967. A radiographic study of urine flow in the domestic fowl. *Br. Poult. Sci.* 8: 209-212.

Alexander, F. 1962. The concentration of certain electrolytes in the digestive tract of the horse and pig. *Res. Vet. Sci.* 3: 78-84.

Alexander, F. 1965. The concentration of electrolytes in the alimentary tract of the rabbit, guinea pig, dog and cat. *Res. Vet. Sci.* 6: 238-244.

Alexander, F., and Davies, M. E. 1963. Production and fermentation of lactate by bacteria in the alimentary canal of the horse and pig. *J. Comp. Pathol. Ther.* 73: 1-8.

Alexander, F., and Hickson, J. C. D. 1970. The salivary and pancreatic secretions of the horse. In *Physiology of Digestion and Metabolism in the Ruminant,* ed. A. T. Phillipson, pp. 375-389. Newcastle upon Tyne: Oriel Press.

Al-Hussaini, A. H. 1946. The anatomy and histology of the alimentary tract of the bottom-feeder, *Mulloides auriflamma* (Forsk.). *J. Morphol.* 78: 121-153.

Al-Hussaini, A. H. 1947. The anatomy and histology of the alimentary tract of the plankton-feeder, *Atherina forskali* Rüpp. *J. Morphol.* 80: 251-286.

Al-Hussaini, A. H. 1949a. On the functional morphology of the alimentary tract of some fish in relation to differences in their feeding habits: anatomy and histology. *Q. J. Microsc. Sci.* 90: 109-139.

Al-Hussaini, A. H. 1949b. On the functional morphology of the alimentary tract of some fish in relation to differences in their feeding habits: cytology and physiology. *Q. J. Microsc. Sci.* 90: 323-354.

Alliot, E. 1967. Absorption intestinale de l'*N*-acétylglucosamine chez la petite roussette: *Scylliorhinus canicula. C.R. Soc. Biol. (Paris)* 161: 2544-2546.

Allison, M. J. 1984. Microbiology of the rumen and small and large intestines. In *Dukes' Physiology of Domestic Animals,* 10th ed., ed. M. J. Swenson, pp.340-350. Ithaca: Cornell University Press.

Alpers, D. H., and Solin, M. 1970. The characterization of rat intestinal amylase. *Gastroenterology* 58: 833-842.

Alumets, J., Håkanson, R., and Sundler, F. 1978. Distribution ontogeny and ultrastructure of pancreatic polypeptide (PP) cells in the pancreas and gut of the chicken. *Cell Tissue Res.* 194: 377-386.

Alumets, J., Sundler, F., and Håkanson, R. 1977. Distribution, ontogeny and ultrastructure of somatostatin immunoreactive cells in the pancreas and gut. *Cell Tissue Res.* 185: 465-480.

Andrew, W. 1959. Going to great lengths—the alimentary tract of vertebrates. In *Textbook of Comparative Histology,* pp. 227-295. New York: Oxford University Press.

Andrew, W. 1963. Mucus secretion in cell nests and surface epithelium. *Ann. N.Y. Acad. Sci.* 106: 502-517.

Annison, E. F., Hill, K. J., and Kenworthy, K. J. 1968. Volatile fatty acids in the digestive tract of the fowl. *Br. J. Nutr.* 22: 207-216.

Antony, M. 1920. Über die sprecheldrüsen der Vögel. Zool. *Jahrb. Agt. Anat. Onlog. Tiere* 41: 547-660.

Anuras, S., and Christensen, J. 1975. Electrical slow waves of the colon do not extend into the caecum. *Rendic. Gastroent.* 7: 56-59.

Archampong, E. Q., Harris, J., and Clark, C. G. 1972. The absorption and secretion of water and electrolytes across the healthy and diseased human colonic mucosa measured in vitro. *Gut* 13: 880-886.

Argenzio, R. A. 1984. Secretory functions of the gastrointestinal tract. In *Dukes' Physiology of Domestic Animals,* 10th ed., ed. M. J. Swenson, pp. 290-300. Ithaca: Cornell University Press.

Argenzio, R. A., Lowe, J. E., Pickard, D. W., and Stevens, C. E. 1974a. Digesta passage and water exchange in the equine large intestine. *Am. J. Physiol.* 226: 1035-1042.

Argenzio, R. A., Miller, N., and Engelhardt, W. v. 1975. Effect of volatile fatty acids on water and ion absorption from the goat colon. *Am. J. Physiol.* 229: 997-1002.

Argenzio, R. A., and Southworth, M. 1974. Sites of organic acid production and absorption in gastrointestinal tract of the pig. *Am. J. Physiol.* 228: 454-460.

Argenzio, R. A., Southworth, M., Lowe, J. E., and Stevens, C. E. 1977. Interrelationship of Na, HCO_3 and volatile fatty acid transport by equine large intestine. *Am. J. Physiol.* 233: E469-E478.

Argenzio, R. A., Southworth, M., and Stevens, C. E. 1974. Sites of organic acid production and absorption in the equine gastrointestinal tract. *Am. J. Physiol.* 226: 1043-1050.

Argenzio, R. A., and Stevens, C. E. 1975. Cyclic changes in ionic composition of digesta in the equine intestinal tract. *Am. J. Physiol.* 228: 1224-1230.

Argenzio, R. A., and Stevens, C. E. 1984. The large bowel—a supplementary rumen? *Proc. Nutr. Soc.* 43: 13-23.

Argenzio, R. A., and Whipp, S. 1979. Interrelationship of sodium, chloride, bicarbonate and acetate transport by the colon of the pig. *J. Physiol. (Lond.)* 295: 265-381.

Argenzio, R. A., and Whipp, S. C. 1981. Effect of *Escherichia coli* heat-stable enterotoxin, cholera toxin and theophylline on ion transport in porcine colon. *J. Physiol. (Lond.)* 320: 469-487.

Argenzio, R. A., and Whipp, S. C. 1983. Effect of theophylline and heat-stable enterotoxin of *Escherichia coli* on transcellular and paracellular ion movement across isolated porcine colon. *Can. J. Physiol. Pharmacol.* 61: 1138-1148.

Ash, P. W., and Dobson, A. 1963. The effect of absorption on the acidity of rumen contents. *J. Physiol. (Lond.)* 169: 39-61.

Babkin, B. P., Hebb, C. O., and Kreuger, L. 1941. Changes in the secretory activity of the gastric glands resulting from the application of acetic acid solutions to the gastric mucosa. *Q. J. Exp. Physiol.* 31: 63-78.

Bacarese-Hamilton, A. J., Adrian, T. E., and Bloom, S. R. 1984. Human and porcine immunoreactive gastric inhibitory polypeptides (IR-GIP) are not identical. *FEBS Lett.* 168: 125-128.

Baillien, M., and Schoffeniels, E. 1961. Origine des potentiels bioélectriques de l'épithélium intestinal de la tortue grecque. *Biochim. Biophys. Acta* 53: 537-548.

Baker, F., Nasr, H., Morrice, F., and Bruce, J. 1950. Bacterial breakdown of structural starches and starch products in the digestive tract of ruminant and non-ruminant mammals. *J. Pathol. Bacteriol.* 62: 617-638.

Balch, C. C., and Campling, R. C. 1965. Rate of passage of digesta through the ruminant digestive tract. In *Physiology of Digestion in the Ruminant,* ed. R. W. Dougherty, pp.108-130. Washington, D.C.: Butterworths.

Banta, C. A., Clemens, E. T., Krinsky, M. M., and Sheffy, B. E. 1979. Sites of organic acid production and pattern of digesta movement in the gastrointestinal tract of dogs. *J. Nutr.* 109: 1592-1600.

Barcroft, T., McAnally, R. A., and Phillipson, A. T. 1944. Absorption of volatile acid from the alimentary tract of the sheep and other animals. *J. Exp. Biol.* 20: 120.

Bar-Eli, A., White, H. B., and Van Vunakis, H. 1966. Proteolytic enzymes from *Mustelus canis* fundic mucosae. *Fed. Proc.* 25: 745.

Barker, J. M. 1961. The metabolism of carbohydrate and volatile fatty acids in the marsupial, *Setonix brachyurus. Q. J. Exp. Physiol.* 46: 54-68.

Barnard, E. A. 1969a. Biological function of pancreatic ribonuclease. *Nature* 221: 340-344.

Barnard, E. A. 1969b. Ribonucleases. *Annu. Rev. Biochem.* 38: 677-732.

Barnard, E. A., and Prosser, C. L. 1973. Comparative biochemistry and physiology of the digestive system. In *Comparative Animal Physiology,* 3rd ed., ed. C. L. Prosser, pp. 133-164. Philadelphia: W. B. Saunders.

Barnes, R. D. 1974. *Invertebrate Zoology,* 3rd ed. Philadelphia: W. B. Saunders.

Barnes, R. H. 1962. Nutritional implications of coprophagy. *Nutr. Rev.* 10: 289-291.

Barnes, R. H., and Fiala, G. 1958. Effects of the prevention of coprophagy in the rat. I. Growth studies. *J. Nutr.* 64: 533-540.

Barnes, R. H., Fiala, G., and Kwong, E. 1963. Decreased growth rate resulting from prevention of coprophagy. *Fed. Proc.* 22: 125-133.

Barnes, R. H., Fiala, G., McGehee, B., and Brown, A. 1957. Prevention of coprophagy in the rat. *J. Nutr.* 63: 489-498.

Barrington, E. J. W. 1957. The alimentary canal and digestion. In *The Physiology of Fishes,* Vol. 1: *Metabolism,* ed. M. E. Brown, pp. 109-161. New York: Academic Press.

Barrington, E. J. W. 1962. Digestive enzymes. *Adv. Comp. Physiol. Biochem.* 1: 1-65.

Barrington, E. J. W. 1982. Evolutionary and comparative aspects of gut and brain peptides. *Med. Bull.* 38: 227-232.

Barrington, E. J. W., and Dockray, G. J. 1970. The effect of intestinal extracts of lampreys (*Lampetra fluviatilis* and *Petromyzon marinus*) on pancreatic secretion in the rat. *Gen. Comp. Endocrinol.* 14: 170-177.

Barrington, E. J. W., and Dockray, G. J. 1972. Cholecystokinin-pancreozymin-like activity in the eel *Anguilla anguilla. Gen. Comp. Endocrinol.* 19: 80-87.

Batt, E. R., and Schachter, D. 1969. Developmental pattern of some intestinal transport mechanisms in newborn rats and mice. *Am. J. Physiol.* 216: 1064-1068.

Bauchop, T., and Martucci, R. W. 1968. Ruminant-like digestion of the langur monkey. *Science* 161: 698-700.

Baverstock, P. R., Watts, C. H. S., and Spencer, L. 1979. Water-balance of small lactating rodents. V. The total water-balance picture of the mother-young unit. *Comp. Biochem. Physiol.* 63A: 247-252.

Baxter, R. M., and Meester, J. 1982. The captive behaviour of the red musk shrew, *Crocidura f. flavescens* (I. Geoffroy, 1827)(Soricidae: Crocidurinae). *Mammalia* 46: 11-27.

Bayless, W. M., and Starling, E. H. 1902. The mechanism of pancreatic secretion. *J. Physiol. (Lond.)* 28: 325-353.

Beerten-Joly, B., Piavaux, A., and Goffart, M. 1974. Quelques enzymes digestifs chez un prosimian, *Perodicticus potto. C. R. Soc. Biol. (Paris)* 168: 140-143.

Behmann, H. 1973. Vergleichend- und funktionell-anatomische Untersuchungen am Caecum und Colon myomorpher Nagetiere. *Z. wiss. Zool.* 186: 173-294.

Bell, D. J., and Bird, T. P. 1966. Urea and volatile base in the caeca and colon of the domestic fowl: the problem of their origin. *Comp. Biochem. Physiol.* 18: 735-744.

Bell, F. R. 1958. The mechanism of regurgitation during the process of rumination in the goat. *J. Physiol. (Lond.)* 142: 503-515.

Bell, G. H., Emslie-Smith, D., and Paterson, C. R. 1980. *Textbook of Physiology,* 10th ed. New York: Churchill Livingstone.

Bennett, T. 1974. Peripheral and autonomic nervous systems. In *Avian Biology,* Vol. 4, ed. D. S. Farner, J. R. King, and K. C. Parkes, pp. 1-77. New York: Academic Press.

Bensadoun, A., and Rothfeld, A. 1972. The form of absorption of lipids in the chicken, *Gallus domesticus. Proc. Soc. Exp. Biol. Med.* 141: 814-817.

Bensley, R. R. 1902-03. The cardiac glands of mammals. *Am. J. Anat.* 2: 105-156.

Benson, A. A., and Lee, R. F. 1975. The role of wax in oceanic food chains. *Sci. Am.* 232: 77-86.

Benson, A. A., Lee, R. F., and Nevenzel, J. C. 1972. Wax esters: major marine metabolic energy sources. In *Current Trends in the Biochemistry of Lipids,* ed. J. Ganguly and R. M. S. Smellie, pp. 175-187. New York: Academic Press.

Bentley, P. J. 1982. *Comparative Vertebrate Endocrinology.* Cambridge: Cambridge University Press.

Berger, P. J., and Burnstock, G. 1979. Autonomic nervous system. In *Biology of the Reptilia,* Vol. 10, ed. C. Gans, R. G. Northcutt, and P. Ulinski, pp. 1-57. New York: Academic Press.

Bertin, L. 1958. Appareil digestif. In *Traité de Zoologie,* Vol.13, ed. P.-P. Grassé, pp. 1249-1302. Paris: Masson.

Bidder, A. M. 1976. New names for old: the cephalopod "mid-gut gland." *J. Zool. Lond.* 180: 441-443.

Binder, H. J. 1970. Amino acid absorption in the mammalian colon. *Biochim. Biophys. Acta* 219: 503-506.

Binder, H., and Rawlins, C. 1973. Electrolyte transport across isolated large intestinal mucosa. *Am. J. Physiol.* 225: 1232-1239.

Björnhag, G. 1972. Separation and delay of contents in the rabbit colon. *Swed. J. Agric. Res.* 2: 125-136.

Björnhag, G. 1981. The retrograde transport of fluid in the proximal colon of rabbits. *Swed. J. Agric. Res.* 11: 63-69.

Björnhag, G. 1987. Comparative aspects of digestion in the hindgut of mammals. The colonic separation mechanism (CSM) (a review). *Dtsch. Tierärztl. Wochenschr.* 94: 33-36.

Björnhag, G., and Sjöblom, L. 1977. Demonstration of coprophagy in some rodents. *Swed. J. Agric. Res.* 7: 105-114.

Björnhag, G., and Sperber, I. 1977. Transport of various food components through the digestive tract of turkeys, geese and guinea fowl. *Swed. J. Agric. Res.* 7: 57-66.

Björnhag, G., Sperber, I., and Holtenius, K. 1984. A separative mechanism in the large intestines. *Can. J. Anim. Sci.* 64: 89-90.

Black, I. B., and Patterson, P. H. 1980. Developmental regulation of neurotransmitter phenotype. *Curr. Top. Dev. Biol.* 15: 27-40.

Black, J. L., and Sharkey, M. J. 1970. Reticular groove (*Sulcus reticuli*)—an obligatory adaptation ruminant-like herbivores. *Mammalia* 34: 294-302.

Bläckberg, L., Hernell, O., and Olivecrona, T. 1981. Hydrolysis of human milk fat globules by pancreatic lipase. *J. Clin. Invest.* 67: 1748-1752.

Blain, A. W., and Campbell, K. N. 1942. A study of digestive phenomena in snakes with the aid of the roentgen ray. *Am. J. Roentgenol. Radiat. Ther.* 48: 229-239.

Blair-West, J. R., Bott, E., Boyd, G. W., Coghlan, J. P., Denton, D. A., Goding, J. R., Weller, S., Wintour, M., and Wright, R. D. 1965. General biological aspects of salivary secretion in ruminants. In *Physiology of Digestion in the Ruminant,* ed. R. W. Dougherty, pp. 198-220. Washington, D.C.: Butterworths.

Blair-West, J. R., Coghlan, J. P., Denton, D. A., and Wright, R. D. 1970. Factors in sodium and potassium metabolism. In *Physiology of Digestion in the Ruminant,* ed. A. T. Phillipson, pp. 350-361. Newcastle upon Tyne: Oriel Press.

Bogé, G., Rigal, A., and Pérès, G. 1983. Analysis of two chloride requirements for sodium-dependent amino acid and glucose transport by intestinal brush-border membrane vesicles of fish. *Biochim. Biophys. Acta* 729: 209-218.

Bolton, W. 1962. Digestion in the crop of the fowl. *Proc. Nutr. Soc.* 21: XXIV.

Bolton, W. 1965. Digestion in the crop of the fowl. *Br. Poult. Sci.* 6: 97-102.

Bond, J., Levitt, D., and Levitt, M. 1977. Quantitation of countercurrent exchange during passive absorption from the dog small intestine. *J. Clin. Invest.* 59: 308-318.

Bondi, A. M., Menghi, G., Palatroni, P., and Materazzi, G. 1982. Ultrastructural observations on the morphology of the stomach of *Salmo irideus* and on the localization of carbonic anhydrase during development. *Anat. Anz.* 151: 473-481.

Bonnafous, R. 1973. Quelques aspects de la physiologie colique en relation avec la dualité de l'excrétion fécale chez le lapin. Ph.D. thesis, Toulouse.

Bonnafous, R., and Raynaud, P. 1968. Mise en evidence d'une activité lysante du colon proximal sur les microorganismes du tube digestif du lapin. *Arch. Sci. Physiol.* 22: 57-64.

Bookhout, T. A. 1959. Reingestion by the snowshoe hare. *J. Mammal.* 40: 250.

Borch-Madsen, P. 1946. *Resorptionens Storrelse ved experimentelt fremkaldt Achylia gastrica.* Copenhagen: Ejvind Christensens.

Borody, T. J., Byrnes, D. J., and Titchen, D. A. 1984. Motilin and migrating myoelectric complexes in the pig and the dog. *Q. J. Exp. Physiol.* 69: 875-890.

Botha, G. S. 1958. A note on the comparative anatomy of the gastroesophageal junction. *Acta Anat.* 34: 52-84.

Botha, G. S. 1962. *The Gastro-Oesophageal Junction.* Boston: Little, Brown.

Bottino, N. R. 1975. Lipid composition of two species of Antarctic krill: *Euphausia superba* and *E. crystallorophias. Comp. Biochem. Physiol.* 50B: 479-484.

Brambell, F. W. R. 1970. *The Transmission of Passive Immunity from Mother to Young.* New York: American Elsevier.

Brambell, M. R. 1972. Mammals: their nutrition and habitat. In *Biology of Nutrition,* Vol. 18, ed. R. N. T-W-Fiennes, pp. 613-652. Elmsford, N.Y.: Pergamon Press.

Brandt, C. S., and Thacker, E. J. 1958. A concept of rate of food passage through the gastrointestinal tract. *J. Anim. Sci.* 17: 218-223.

Britton, S. W. 1941. Form and function in the sloth. *Q. Rev. Biol.* 16: 190-207.

Brockerhoff, H., and Hoyle, R. J. 1965. Hydrolysis of triglycerides by the pancreatic lipase of a skate. *Biochim. Biophys. Acta* 98: 435-436.

Brockerhoff, H., and Jensen, R. G. 1974. *Lipolytic Enzymes.* New York: Academic Press.

Brodin, E., Alumets, J., Hakanson, R., Leander, S., and Sundler, F. 1981. Immunoreactive substance P in the chicken gut: distribution, development and possible functional significance. *Cell Tissue Res.* 216: 455-469.

Brown, G. D. 1969. Studies on marsupial nutrition. VI. The utilization of dietary urea by the euro or hill kangaroo, *Macropus robustus* (Gould). *Aust. J. Zool.* 17: 187-194.

Brown, K. M., and Moog, F. 1967. Invertase activity in the intestine of the developing chick. *Biochim. Biophys. Acta* 132: 185-187.

Brown, R. H. 1962. The anatomy of the alimentary tract of three genera of bats. Master's thesis, University of Arizona, Tucson.

Brown, W. R. 1968. Rumination in the adult. A study of two cases. *Gastroenterology* 54: 933-939.

Browne, T. G. 1922. Some observations on the digestive system of the fowl. *J. Comp. Pathol. Ther.* 35: 12-32.

Bryant, M. P. 1977. Microbiology of the rumen. In *Dukes' Physiology of Domestic Animals,* 9th ed., ed. M. J. Swenson, pp. 287-304. Ithaca: Cornell University Press.

Buchan, A. M. J., Ingman-Baker, J., Levy, J., and Brown, J. C. 1982. A comparison of the ability of serum and monoclonal antibodies to gastric inhibitory polypeptide to detect immunoreactive cells in the gastroenteropancreatic system of mammals and reptiles. *Histochemistry* 76: 341-349.

Buchan, A. M. J., Lance, V., and Polak, J. M. 1983. Regulatory peptides in the gastrointestinal tract of *Alligator mississipiensis*. An immunocytochemical study. *Cell Tissue Res.* 231: 439-449.

Buchan, A. M. J., Polak, J. M., Bryant, M. G., Bloom, S. R., and Pearse, A. G. E. 1981. Vasoactive intestinal polypeptide (VIP)-like immunoreactivity in anuran intestine. *Cell Tissue Res.* 216: 413-422.

Buchan, A. M. J., Polak, J. M., and Pearse, A. G. E. 1980. Gut hormones in *Salamandra salamandra*. An immunocytochemical and electron microscopic investigation. *Cell Tissue Res.* 211: 311-343.

Buchner, P. 1965. *Endosymbiosis of Animals with Plant Microorganisms*. New York: Interscience.

Buddington, R. K., and Diamond, J. M. 1987. Pyloric ceca of fish: a "new" absorptive organ. *Am. J. Physiol.* 252: G65-G76.

Bueno, L., and Ruckebusch, Y. 1974. The cyclic motility of the omasum and its control in sheep. *J. Physiol. (Lond.)* 238: 295-312.

Burgen, A. S. V. 1961. The comparative physiology of the salivary glands. In *Physiology of the Salivary Glands,* ed. A. S. V. Burgen and N. G. Emmelin, pp. 267-272. Baltimore: Williams & Wilkins.

Burgen, A. S. V. 1967. Secretory processes in salivary glands. In *Handbook of Physiology,* Sec. 6: *Alimentary Canal,* Vol. 2: *Secretion,* ed. C. F. Code and W. Heidel, pp. 561-579. Washington, D.C.: American Physiological Society.

Burgen, A. S. V., and Seeman, P. 1958. The role of the salivary duct system in the formation of saliva. *Can. J. Biochem. Physiol.* 36: 119-143.

Burhol, P. G. 1982. Regulation of gastric secretion in the chicken. *Scand. J. Gastroenterol.* 17: 321-323.

Burns, W. A., Flores, P. A., Moshyedi, A., and Albacete, R. A. 1970. Clinical conditions associated with columnar lined esophagus. *Am. J. Dig. Dis.* 15: 607-615.

Burnstock, G. 1969. Evolution of the autonomic innervation of visceral and cardiovascular systems in vertebrates. *Pharmacol. Rev.* 21: 247-324.

Burnstock, G. 1972. Purinergic nerves. *Pharmacol. Rev.* 24: 509-581.

Burnstock, G. 1986. The changing face of autonomic neurotransmission. *Acta Physiol. Scand.* 126: 67-91.

Burnstock, G., Campbell, G., Satchell, D. G., and Smythe, A. 1970. Evidence that adenosine triphosphate or a related nucleotide is the transmitter substance released by non-adrenergic nerves in the gut. *Br. J. Pharmacol.* 40: 668-688.

Calaby, J. H. 1958. Studies on marsupial nutrition. II. The rate of passage of food residues and digestibility of crude fibre and protein by the quokka, *Setonix brachyurus* (Quoy and Gaimard). *Aust. J. Biol. Sci.* 11: 571-580.

Calligan, J. J., Costa, M., and Furness, J. B. 1985. Gastrointestinal myoelectric activity in conscious guinea pigs. *Am. J. Physiol.* 249: G92-G98.

Calloway, D. H. 1968. Gas in the alimentary canal. In *Handbook of Physiology,* Sec. 6: *Alimentary Canal,* Vol. 5: *Bile; Digestion; Ruminal Physiology,* ed. C. F. Code and W. Heidel, pp. 2839-2859. Washington, D.C.: American Physiological Society.

Campbell, G. 1970. Autonomic nervous systems. In *Fish Physiology,* Vol. 4, ed. W. S. Hoar and D. J. Randall, pp. 109-132. New York: Academic Press.

Campbell, G., and Burnstock, G. 1968. Comparative physiology of gastrointestinal motility. In *Handbook of Physiology,* Sec. 6, *Alimentary Canal,* Vol. 4: *Motility,* ed. C. F. Code and W. Heidel, pp. 2213-2266. Washington, D.C.: American Physiological Society.

Cannon, W. B. 1902. The movements of the intestines studied by means of roentgen rays. *Am. J. Physiol.* 6: 251-277.

Carr, D. H., McLeay, L. M., and Titchen, D. A. 1970. Factors affecting reflex responses of the ruminant stomach. In *Physiology of Digestion and Metabolism in the Ruminant,* ed. A. T. Phillipson, pp. 35-41. Newcastle upon Tyne: Oriel Press.

Castle, E. J. 1956a. The rate of passage of foodstuffs through the alimentary tract of the goat. 1. Studies on adult animals fed on hay and concentrates. *Br. J. Nutr.* 10: 15-23.

Castle, E. J. 1956b. The rate of passage of foodstuffs through the alimentary tract of the goat. 2. Studies on growing kids. *Br. J. Nutr.* 10: 115-125.

Chao, F. C., and Tarver, H. 1953. Breakdown of urea in the rat. *Proc. Soc. Exp. Biol. Med.* 84: 406-409.

Chien, W.-J., and Stevens, C. E. 1972. Coupled active transport of Na and Cl across forestomach epithelium. *Am. J. Physiol.* 223: 997-1003.

Chilcott, M. J., and Hume, I. D. 1985. Coprophagy and selective retention of fluid digesta: their role in the nutrition of the common ringtail possum, *Pseudocheirus peregrinus. Aust. J. Zool.* 33: 1-15.

Chivers, D. J., and Hladik, C. M. 1980. Morphology of the gastrointestinal tract of primates: comparisons with other mammals in relation to diet. *J. Morphol.* 166: 337-386.

Christensen, J. 1971. The controls of gastrointestinal movements: some old and new views. *N. Engl. J. Med.* 285: 85-98.

Christensen, J. 1981. Motility of the colon. In *Physiology of the Gastrointestinal Tract,* Vol. 1, ed. L. R. Johnson, J. Christensen, M. I. Grossman, E. D. Jacobson, and S. G. Schultz, pp. 445-471. New York: Raven Press.

Christensen, J., Anuras, S., and Hauser, R. L. 1974. Migrating spike bursts and electrical slow waves in the cat colon: effect of sectioning. *Gastroenterology* 66: 240-247.

Christensen, J., Caprilli, R., and Lund, G. F. 1969. Electric slow waves in circular muscle of cat colon. *Am. J. Physiol.* 217: 771-776.

Clarke, R. T. J. 1977. Protozoa in the rumen ecosystem. In *Microbial Ecology of the Gut,* ed. R. T. J. Clarke and T. Bauchop, pp. 251-275. New York: Academic Press.

Claypole, E. J. 1897. The comparative histology of the digestive tract. *Trans. Am. Microsc. Soc.* 83-91.

Clemens, E. T. 1977. Sites of organic acid production and patterns of digesta movement in the gastrointestinal tract of the rock hyrax. *J. Nutr.* 107: 1954-1961.

Clemens, E. T. 1980. The digestive tract: insectivore, prosimian, and advanced primate. In *Comparative Physiology: Primitive Mammals,* ed. K. Schmidt-Nielson, L. Bolis, and C. R. Taylor, pp. 90-99. Cambridge: Cambridge University Press.

Clemens, E. T. 1982. Comparison of polyethylene glycol and dye markers in nutrition research. *Nutr. Res.* 2: 323-334.

Clemens, E. T., and Maloiy, G. M. O. 1982. The digestive physiology of three East African herbivores: the elephant, rhinoceros and hippopotamus. *J. Zool.* 198: 141-156.

Clemens, E. T, and Phillips, B. 1980. Organic acid production and digesta movement in the gastrointestinal tract of the baboon and Sykes monkey. *Comp. Biochem. Physiol.* 66A: 529-532.

Clemens, E. T., and Stevens, C. E. 1979. Sites of organic acid production and patterns of digesta movement in the gastrointestinal tract of the raccoon. *J. Nutr.* 109: 1110-1116.

Clemens, E. T., Stevens, C. E., and Southworth, M. 1975a. Sites of organic acid production and patterns of digesta movement in the gastrointestinal tract of swine. *J Nutr.* 105: 759-768.

Clemens, E. T., Stevens, C. E., and Southworth, M. 1975b. Sites of organic acid production and pattern of digesta movement in the gastrointestinal tract of geese. *J. Nutr.* 105: 1341-1350.

Cloudsley-Thompson, J. L. 1972. The classification and study of animals by feeding habits. In *Biology of Nutrition,* Vol. 18, ed. R. N. T-W-Fiennes, pp. 439-470. Elmsford, N.Y.: Pergamon Press.

Cocimano, M. R., and Leng, R. A. 1967. Metabolism of urea in sheep. *Br. J. Nutr.* 21: 353-371.

Code, C. F., and Schlegel, J. F. 1968. Motor action of the esophagus and its sphincters. In *Handbook of Physiology,* Sec. 6: *Alimentary Canal,* Vol.4: *Motility,* ed. C. F. Code and W. Heidel, pp. 1821-1839. Washington, D.C.: American Physiological Society.

Cohen, T., Gertler, A., and Birk, Y. 1981a. Pancreatic proteolytic enzymes from carp (*Cyprinus carpio*). I. Purification and physical properties of trypsin, chymotrypsin, elastase and carboxypeptidase B. *Comp. Biochem. Physiol.* 69B: 639-646.

Cohen, T., Gertler, A., and Birk, Y. 1981b. Pancreatic proteolytic enzymes from carp (*Cyprinus carpio*). II. Kinetic properties and inhibition studies of trypsin, chymotrypsin and elastase. *Comp. Biochem. Physiol.* 69B: 647-653.

Coleman, G. S. 1980. Rumen ciliate protozoa. *Adv. Parasitol.* 18: 121-173.

Colin, D. A. 1972. Relations entre la nature de l'alimentation et l'importance de l'activité chitinolytique du tube digestif de quelques téleostéens marins. *C. R. Soc. Biol. (Paris)* 166: 95-98.

Cooke, H. J. 1987. Neural and humoral regulation of small intestinal electrolyte transport. In *Physiology of the Gastrointestinal Tract,* 2nd ed., Vol. 2, ed. L. R. Johnson, J. Christensen, M. Jackson, E. D. Jacobson, and J. Walsh, pp. 1307-1350. New York: Raven Press.

Corbet, P. S. 1960. The food of a sample of crocodiles (*Crocodilus niloticus* L.) from Lake Victoria. *Proc. Zool. Soc. Lond.* 133: 561-572.

Cork, S. J., and Hume, I. D. 1978. Volatile fatty acid production rates in the caecum of the greater glider. *Bull. Aust. Mammal Soc.* 5: 24-25.

Cork, S. J., and Hume, I. D. 1983. Microbial digestion in the koala (*Phascolarctos cinereus,* Marsupialia), an aboreal folivore. *J. Comp. Physiol. Psychol.* 152B: 131-135.

Cork, S. J., and Warner, A. C. I. 1983. The passage of digesta markers through the gut of a folivorous marsupial, the koala *Phascolarctos cinereus. J. Comp. Physiol. Psychol.* 152B: 43-51.

Cork, S. J., Warner, A. C. I., and Harrop, C. J. F. 1977. Preliminary study of the rate of passage of digesta through the gut of the koala. *Bull. Aust. Mammal Soc.* 4: 24.

Cornelius, C., Dandrifosse, G., and Jeuniaux, C. 1975. Biosynthesis of chitinases by mammals of the order Carnivora. *Biochem. Syst. Ecol.* 3: 121-122.

Cranwell, P. D., Noakes, D. E., and Hill, K. J. 1968. Observations on the stomach content of the suckling pig. *Proc. Nutr. Soc.* 27: 26A.

Crompton, A. W. 1980. Biology of the earliest mammals. In *Comparative Physiology: Primitive Mammals,* ed. K. Schmidt-Nielsen, L. Bolis, and C. R. Taylor, pp. 1-12. Cambridge: Cambridge University Press.

Crompton, A. W., and Parker, P. 1978. Evolution of the mammalian masticatory apparatus. *Am. Sci.* 66: 192-201.

Crompton, A. W., Taylor, C. R., and Jagger, J. A. 1978. Evolution of homeothermy in mammals. *Nature* 272: 333-336.

Cross, D. G. 1969. Aquatic weed control using grass carp. *J. Fish Biol.* 1: 27-30.

Crowcroft, P. 1952. Refection in the common shrew. *Nature* 170: 627.

Crump, M. H., Argenzio, R. A., and Whipp, S. C. 1980. Effects of acetate on absorption of sodium and water from the pig colon. *Am. J. Vet. Res.* 41: 1563-1568.

Cummings, J. F., Munnell, J. F., and Vallenas, A. 1972. The mucigenous glandular mucosa in the complex stomach of two New World camelids, the llama and guanaco. *J. Morphol.* 137: 71-109.

Curran, P. 1968. Twelfth Bowditch Lecture: Coupling between transport processes in intestine. *Physiologist* 11: 3-23.

Curtin, T. M., Goetsch, G. D., and Hollandbeck, R. 1963. Clinical and pathologic characterization of esophagogastric ulcers in swine. *J. Am. Vet. Med. Assoc.* 143: 854-860.

Daft, F. S., McDaniel, E. G., Hermann, L. G., Pomine, M. K., and Hegner, J. R. 1963. Role of coprophagy in utilization of B vitamins synthesised by intestinal bacteria. *Fed. Proc.* 22: 129-133.

Dahlqvist, A., and Thomson, D. L. 1963. Separation and characterization of two rat-intestinal amylases. *Biochem. J.* 89: 272-277.

Dandrifosse, G. 1963. La sécrétion de chitinase par la muqueuse gastrique isolée. *Ann. Soc. R. Zool. Belg.* 92: 199-201.

Dandrifosse, G. 1974. Digestion in reptiles. *In Chemical Zoology,* Vol. 9, ed. M. Florkin and B. T. Sheer, pp. 249-275. New York: Academic Press.

Danielli, J. R., Hitchcock, M. W. S., Marshall, R. A., and Phillipson, A. T. 1945. The mechanism of absorption from the rumen as exemplified by the behavior of acetic, propionic and butyric acids. *J. Exp. Biol.* 22: 75-84.

Davenport, H. W. 1982. *Physiology of the Digestive Tract,* 5th ed. Chicago: Year Book Medical Publishers.

Davis, D. D. 1961. Origin of the mammalian feeding mechanism. *Am. Zoologist* 1: 229-234.

Davis, G., Morawski, S., Santa Ana, C., and Fordtran, J. 1983. Evaluation of chloride/bicarbonate exchange in the human colon in vivo. *J. Clin. Invest.* 71: 201-207.

de Lahunta, A., and Habel, R. E. 1986. Intestines. In *Applied Veterinary Anatomy,* pp. 246-256. Philadelphia: W. B. Saunders.

Dellow, D. W. 1979. Physiology of digestion in the macropodine marsupials. Ph.D. thesis, University of New England, Armidale, N.S.W., Australia.

Dellow, D. W., and Hume, I. D. 1982. Studies on the nutrition of macropodine marsupials. IV. Digestion in the stomach and the intestine of *Macropus giganteus, Thylogale thetis* and *Macropus eugenii. Aust. J. Zool.* 30: 767-777.

Dellow, D. W., Nolan, J. V., and Hume, I. D. 1983. Studies on the nutrition of macropodine marsupials. V. Microbial fermentation in the forestomach of *Thylogale thetis* and *Macropus eugenii. Aust. J. Zool.* 31: 433-443.

Denbow, D. M. 1982. Eating, drinking and temperature responses to intracerebroventricular cholecystokinin in the chick. *Peptides* 3: 739-743.

Denis, C., Jeuniaux, C., Gerebtzoff, M. A., and Goffart, M. 1967. La digestion stomacale chez un paresseux: l'unau *Choloepus hoffmanni* (Peters). *Ann. Soc. R. Zool. Belg.* 97: 9-29.

Diamond, J. M. 1962. The mechanism of solute transport by the gall-bladder. *J. Physiol. (Lond.)* 161: 474-504.

Diamond, J. M. 1977. Twenty-First Bowditch Lecture: The epithelial junction: bridge, gate, and fence. *Physiologist* 20: 10-18.

Dieterlein, F. 1959. Das Verhalten des syrischen Goldhamsters (*Mesocricetus auratus* Waterhouse). *Z. Tierpsychol.* 16: 47-103.

Dimaline, R. 1983. Is caerulein amphibian CCK? *Peptides* 4: 457-462.

Dimaline, R., Rawdon, B. B., Brandes, S., Andrew, A., and Loveridge, J. P. 1982. Biologically active gastrin/CCK-related peptides in the stomach of a reptile, *Crocodylus niloticus;* identified and characterized by immunochemical methods. *Peptides* 3: 977-984.

Dirksen, G. 1970. Acidosis. In *Physiology of Digestion and Metabolism in the Ruminant,* ed. A. T. Phillipson, pp. 612-625. Newcastle upon Tyne: Oriel Press.

Dobson, A. 1959. Active transport through the epithelium of the reticulo-rumen sac. *J. Physiol. (Lond.)* 146: 235-251.

Dockray, G. J. 1974. Extraction of a secretin-like factor from the intestines of Pike (*Esox lucius*). *Gen. Comp. Endocrinol.* 23: 340-347.

Dockray, G. J. 1975. Comparative studies on secretin. *Gen. Comp. Endocrinol.* 5: 203-210.

Dockray, G. J. 1978. Evolution of secretin-like hormones. In *Gut Hormones,* ed. S. R. Bloom, pp. 64-67. New York: Churchill Livingstone.

Dockray, G. J. 1979a. Comparative biochemistry and physiology of gut hormones. *Annu. Rev. Physiol.* 41: 83-95.

Dockray, G. J. 1979b. Cholecystokinin-like peptides in avian brain and gut. *Experientia* 35: 628-630.

Donta, S. T., and Van Vunakis, H. 1970a. Chicken pepsinogens and pepsins. Their isolation and properties. *Biochemistry* 9: 2791-2797.

Donta, S. T., and Van Vunakis, H. 1970b. Immunochemical relationships of chicken pepsinogens and pepsins. *Biochemistry* 9: 2798-2802.

Doty, R. W. 1968. Neural organizations of deglutition. In *Handbook of Physiology*, Sec. 6: *Alimentary Canal*, Vol. 4: *Motility*, ed. C. F. Code and W. Heidel, pp. 1861-1902. Washington, D.C.: American Physiological Society.

Dougherty, R. W. 1968. Physiology of eructation in ruminants. In *Handbook of Physiology*, Sec. 6: *Alimentary Canal*, Vol. 5: *Bile; Digestion; Ruminal Physiology*, ed. C. F. Code and W. Heidel, pp. 2695-2698. Washington, D.C.: American Physiological Society.

Dougherty, R. W. 1977. Physiopathology of the ruminant digestive tract. In *Dukes' Physiology of Domestic Animals*, 9th ed., ed. M. J. Swenson, pp. 305-312. Ithaca: Cornell University Press.

Drawert, F., Kuhn, H.-J., and Rapp, A. 1962. Reaktions-Gaschromatographie. III. Gaschromatographische Bestimmung der niederflüchtigen Fettsaüren im Magen von Schlankaffen (*Colobinae*). *Hoppe-Seylers Z. Physiol. Chem.* 329: 84-89.

Dubos, R. 1966. The microbiota of the gastrointestinal tract. *Gastroenterology* 51: 868-874.

Duellman, U. E., and Trueb, L. 1986. *Biology of Amphibians*. New York: McGraw-Hill.

Duke, G. E., and Bedbury, H. P. 1985. Lack of a cephalic phase of gastric secretion in chickens. *Poult. Sci.* 64: 575-578.

Duke, G. E., Dziuk, H. E., and Hawkins, L. 1969. Gastrointestinal transit times in normal and bluecomb diseased turkeys. *Poult. Sci.* 48: 835-842.

Duke, G. E., Jegers, A. A., Loff, G., and Evanson, O. A. 1975. Gastric digestion in some raptors. *Comp. Biochem. Physiol.* 50A: 649-656.

Duncan, D. L. 1953. The effects of vagotomy and splanchnicotomy on gastric motility in the sheep. *J. Physiol. (Lond.)* 119: 157-169.

Dziuk, H. E. 1971. Reverse flow of gastrointestinal contents in turkeys. *Fed. Proc.* 30: 610.

Dziuk, H. E. 1984. Digestion in the ruminant stomach. In *Dukes' Physiology of Domestic Animals*, 10th ed., ed. M. J. Swenson, pp. 320-339. Ithaca: Cornell University Press.

Eadie, J. M., and Mann, S. O. 1970. Development of the rumen microbial population: high starch diets and instability. In *Physiology of Digestion and Metabolism in the Ruminant*, ed. A. T. Phillipson, pp. 335-347. Newcastle upon Tyne: Oriel Press.

Eckerlin, R. H., and Stevens, C. E. 1973. Bicarbonate secretion by the glandular saccules of the llama stomach. *Cornell Vet.* 63: 436-445.

Eden, A. 1940. Coprophagy in the rabbit: origin of 'night' faeces. *Nature* 145: 628-629.

Edmund, A. G. 1969. Dentition. In *Biology of the Reptilia*, Vol. 1, ed. C. Gans, A. d'A. Bellairs, and T. S. Parsons, pp. 117-200. New York: Academic Press.

Eggermont, E. 1969. The hydrolysis of the naturally occurring α-glucosides by the human intestinal mucosa. *Eur. J. Biochem.* 9: 483-487.

Ehle, F. R., Robertson, J. B., and Van Soest, P. J. 1982. Influence of dietary fibers on fermentation in the human large intestine. *J. Nutr.* 112: 158-166.

Ehrlein, H. J. 1970. Untersuchungen über die Motorik des Labmagens der Ziege unter besonderer Berücksichtigung des Pylorus. *Zentralbl. Veterinaermed. Reihe A* 17: 481-497.

Ehrlein, H. J., and Engelhardt, W. v. 1971. Untersuchungen über die Magenmotorik beim Lama. *Zentralbl. Veterinaermed. Reihe A* 18: 181-191.

Ehrlein, H. J., and Hill, H. 1969. Motorik und Nahrungstransport des Psalters (Omasum) der Ziege. *Zentralbl. Veterinaermed. Reihe A* 16: 573-596.

Ehrlein, H. J., Reich, H., and Schwinger, M. 1983. Colonic motility and transit of digesta during hard and soft feces formation in rabbits. *J. Physiol. (Lond.)* 338: 75-86.

Einhorn, A. H. 1977. Rumination syndrome. In *Pediatrics,* 16th ed., ed. A. M. Rudolph, H. L. Barnett, and A. H. Einhorn, pp. 987-988. New York: Appleton-Century-Crofts.

Elliot, W. B. 1978. Chemistry and immunology of reptilian venoms. In *Biology of the Reptilia,* Vol. 8, ed. C. Gans, pp. 163-436. New York: Academic Press.

Elliott, T. R., and Barclay-Smith, E. 1904. Antiperistalsis and other muscular activities of the colon. *J. Physiol. (Lond.)* 31: 272-304.

Ellison, S. A. 1967. Proteins and glycoproteins of saliva. In *Handbook of Physiology,* Sec. 6: *Alimentary Canal,* Vol. 2: *Secretion,* ed. C. F. Code and W. Heidel, pp. 531-559. Washington, D.C.: American Physiological Society.

el-Salhy, M. 1984. Immunocytochemical investigation of the gastro-entero-pancreatic (GEP) neurohormonal peptides in the pancreas and gastrointestinal tract of the dogfish *Squalus acanthias. Histochemistry* 80: 193-205.

el-Salhy, M. E., Wilander, L., Grimelius, L., Terenius, L., Lundberg, J. M., and Tatemoto, K. 1982. The distribution of polypeptide YY (PYY)-immunoreactive and pancreatic polypeptide (PP)-immunoreactive cells in the domestic fowl. *Histochemistry* 75: 25-30.

Elsden, S. R., Hitchcock, M. W. S., Marshall, R. A., and Phillipson, A. T. 1946. Volatile acid in the digesta of ruminants and other animals. *J. Exp. Biol.* 22: 191-202.

el-Shazly, K. 1952a. Degradation of protein in the rumen of the sheep. 1. Some volatile fatty acids, including branched-chained isomers, found in vivo. *Biochem. J.* 51: 640-647.

el-Shazly, K. 1952b. Degradation of protein in the rumen of the sheep. 2. The action of rumen micro-organisms on amino acids. *Biochem. J.* 51: 647-653.

Engelhardt, W. v. 1978. Adaptation to low protein diets in some mammals. *Proceedings of the Zodiac Symposium on Adaptation.* Wageningem, Netherlands: Center Agricultural Publications.

Engelhardt, W. v., Ali, K. E., and Wipper, E. 1979. Absorption and secretion in the tubiform forestomach (compartment 3) of the llama. *J. Comp. Physiol. Psychol.* 132: 337-341.

Engelhardt, W. v., and Hauffe, R. 1975. Role of the omasum in absorption and secretion of water and electrolytes in sheep and goats. In *Digestion and Metabolism in the Ruminant,* ed. I. W. McDonald and A. C. I. Warner, pp. 216-230. Armidale N.S.W.: University of New England Publishing Unit.

Engelhardt, W. v., Hinderer, S., Rechkemmer, G., and Becker, G. 1984. Urea secretion into the colon of sheep and goat. *Q. J. Exp. Physiol.* 69: 469-475.

Engelhardt, W. v., and Holler, H. 1982. Salivary and gastric physiology of camelids. *Verh. Dtsch. Zool. Ges.* 1982: 195-204.

Engelhardt, W. v., Wolter, S., Lawrenz, H., and Hemsley, J. A. 1978. Production of methane in two non-ruminant herbivores. *Comp. Biochem. Physiol.* 60A: 309-311.

Epstein, M. L., Lindberg, I., and Dahl, J. L. 1980. Development of enkephalin in chick brain, gut, adrenal gland and Remak's ganglion. *Soc. Neurosci. Abstr.* 6: 618.

Erdman, S., and Cundall, D. 1984. The feeding apparatus of the salamander *Amphiuma tridactylum:* morphology and behavior. *J. Morphol.* 181: 175-204.

Evans, H. E. 1986. Reptiles, introduction and anatomy. In *Zoo and Wild Animal Medicine,* ed. M. E. Fowler, pp. 107-132. Philadelphia: W. B. Saunders.

Faichney, G. J. 1975. The use of markers to partition digestion within the gastro-intestinal tract of ruminants. In *Digestion and Metabolism in the Ruminant,* ed. I. W. McDonald and A. C. I. Warner, pp. 277-291. Armidale, N.S.W.: University of New England Publishing Unit.

Faichney, G. J., and Black, J. L. 1974. Passage of markers through the gastrointestinal tract of the milk-fed lamb. *Proc. Aust. Physiol. Pharmacol. Soc.* 5: 67-68.

Faichney, G. J., and Griffiths, D. A. 1978. Behaviour of solute and particle markers in the stomach of sheep given a concentrate diet. *Br. J. Nutr.* 40: 71-82.

Falkmer, S., Fahrenkrug, J., Alumets, J., Hakanson, R., and Sundler, F. 1980. Vasoactive intestinal polypeptide (VIP) in epithelial cells of the gut mucosa of an elasmobranchian cartilaginous fish, the ray. *Endocrinol. Jpn.* 1: 31-35.

Falkmer, S., and Ostberg, Y. 1977. Comparative morphology of pancreatic islets in animals. In *The Diabetic Pancreas,* ed. B. W. Volk and K. F. Wellmann, pp. 15-59. New York: Plenum Publishing.

Fänge, R., Lundblad, G., and Lind, J. 1976. Lysozyme and chitinase in blood and lymphomyeloid tissues of marine fish. *Marine Biol.* 36: 277-282.

Fänge, R., Lundblad, G., Lind, J., and Slettengren, K. 1979. Chitinolytic enzymes in the digestive system of marine fishes. *Marine Biol.* 53: 317-321.

Fanning, J. C., Tyler, M. J., and Shearman, D. J. C. 1982. Converting a stomach to a uterus: the microscopic structure of the stomach of the gastric brooding frog *Rheobatrachus silus. Gastroenterology* 82: 62-70.

Farner, D. S. 1943. Biliary amylase in the domestic fowl. *Biol. Bull.* 84: 240-243.

Farner, D. S., King, J. R., and K. C. Parkes, eds. 1971. *Avian Biology,* pp. 1-19, 564-580. New York: Academic Press.

Feldman, M., and Fordtran, J. S. 1978. Rumination in adults: rumination in infants. In *Gastrointestinal Disease,* 2nd ed., ed. M. H. Sleisenger and J. S. Fordtran, pp. 207-208. Philadelphia: W. B. Saunders.

Fenna, L., and Boag, D. A. 1974. Filling and emptying of the galliform caecum. *Can. J. Zool.* 52: 537-540.

Ferreira, H. G., Harrison, F. A., and Keynes, R. D. 1964. Studies with isolated rumen eipthelium of the sheep. *J. Physiol. (Lond.)* 175: 28-29.

Ferreira, H. G., Harrison, F. A., Keynes, R. D., and Zurich, L. 1972. Ion transport across an isolated preparation of sheep rumen epithelium. *J. Physiol. (Lond.)* 222: 77-93.

Field, M. 1978. Some speculations on the coupling between sodium and chloride transport processes in mammalian and teleost intestine. In *Membrane Transport Processes,* Vol. 1, ed. J. F. Hoffman, pp. 277-292. New York: Raven Press.

Finegold, S. M., Sutter, V. L., and Mathisen, G. E. 1983. Normal indigenous intestinal flora. In *Human Intestinal Microflora in Health and Disease,* ed. D. J. Hentges, pp. 3-32. New York: Academic Press.

Fioramonti, J., and Bueno, L. 1977. Electrical activity of the large intestine in normal and megacolon pigs. *Ann. Rech. Vétér.* 8: 275-283.

Fioramonti, J., and Ruckebusch, Y. 1976. La motricité caecale chez le lapin. III. Dualité de l'excrétion fécale. *Ann. Rech. Vétér.* 7: 281-295.

Fioramonti, J., and Ruckebusch, Y. 1979. Diet and caecal motility in sheep. *Ann. Rech. Vétér.* 10: 593-599.

Fish, G. R. 1951. Digestion in *Tilapia esculenta. Nature* 167: 900-901.

Fishelson, L., Montgomery, W. L., and Myrberg, A. A., Jr. 1985. A unique symbiosis in the gut of tropical herbivorous surgeonfish (Acanthuridae: Teleostei) from the Red Sea. *Science* 229: 49-51.

Fitzgerald, R. J., Gustafsson, B. E., and McDaniel, E. G. 1964. Effects of coprophagy prevention on intestinal microflora in rats. *J. Nutr.* 84: 155-160.

Fleisher, D. R. 1979. Infant rumination syndrome. *Am. J. Dis. Child.* 133: 266-269.

Flemström, G. 1977. Active alkalinization by amphibian gastric fundic mucosa in vitro. *Am. J. Physiol.* 233: E1-E12.

Flemström, G., and Frenning, B. 1968. Migration of acetic acid and sodium acetate and their effects on the gastric transmucosal ion exchange. *Acta Physiol. Scand.* 74: 521-532.

Flemström, G., and Garner, A. 1980. Stimulation of HCO_3^- transport by gastric inhibitory peptide (GIP) in proximal duodenum of the bullfrog. *Acta Physiol. Scand.* 109: 231-232.

Flemström, G., Heylings, J. R., and Garner, A. 1982. Gastric and duodenal HCO_3^- transport in vitro: effects of hormones and local transmitters. *Am. J. Physiol.* 242: G100-G110.

Flower, W. H. 1872. Lectures on the comparative anatomy of the organs of digestion of the mammalia. *Medical Times and Gazette,* Feb. 24-Dec.14.

Floyd, J. C., Jr., Fajans, S. S., Pek, S., and Chance, R. E. 1977. A newly recognized pancreatic polypeptide: plasma levels in health and disease. *Recent Prog. Horm. Res.* 33: 519-570.

Foltmann, B. 1981. Gastric proteinases—structure, function, evolution and mechanism of action. In *Essays in Biochemistry,* Vol. 17, ed. P. N. Campbell and R. D. Marshall, pp. 52-84. New York: Academic Press.

Foltmann, B., Jensen, A. L., Lonblad, P., Smidt, E., and Axelsen, N. H. 1981. A developmental analysis of the production of chymosin and pepsin in pigs. *Comp. Biochem. Physiol.* 68B: 9-13.

Fontaine-Perus, J., Chanconie, M., Polak, J. M., and Le Douarin, N. M. 1981. Origin and development of VIP and substance P containing neurons in the embryonic avian gut. *Histochemistry* 71: 313-323.

Foot, J. Z., and Romberg, B. 1965. The utilization of roughage by sheep and the red kangaroo, *Macropus rufus* (Desmarest). *Aust. J. Agric. Res.* 16: 429-435.

Forbes, D. K., and Tribe, D. E. 1970. The utilization of roughages by sheep and kangaroos. *Aust. J. Zool.* 18: 247-256.

Fordtran, J. S. 1973. Diarrhea. In *Gastrointestinal Disease,* ed. M. H. Sleisenger and J. S. Fordtran, pp. 291-319. Philadelphia: W. B. Saunders.

Fordtran, J. S., Scroggie, W. B., and Polter, D. E. 1964. Colonic absorption of tryptophan metabolites in man. *J. Lab. Clin. Med.* 64: 125-132.

Forman, G. L. 1972. Comparative morphological and histochemical studies of the stomachs of selected American bats. *Univ. Kansas Sci. Bull.* 49: 594-729.

Forte, G. M., Limlomwongse, L., and Forte, J. G. 1969. The development of intracellular membranes concomitant with the appearance of HCl secretion in oxyntic cells of the metamorphosing bullfrog tadpole. *J. Cell. Sci.* 4: 709-727.

Forster, E., Hayslett, J., and Binder, H. 1984. Mechanism of active potassium absorption and secretion in the rat colon. *Am. J. Physiol.* 246: G611-G617.

Fouchereau-Peron, M., Laburthe, M., Besson, J., Rosselin, G., and Le Gal, Y. 1980. Characterization of the vasoactive intestinal polypeptide (VIP) in the gut of fishes. *Comp. Biochem. Physiol.* 65A: 489-492.

Francis-Smith, K., and Wood-Gush, D. G. 1977. Coprophagia as seen in thoroughbred foals. *Equine Vet. J.* 9: 155-157.

Frankignoul, M., and Jeuniaux, C. 1965. Distribution des chitinases chez les mammifères rongeurs. *Ann. Soc. R. Zool. Belg.* 95: 1-8.

Frechkop, J. 1955. Sous-ordre des suiformes. In *Traité de Zoologie,* Vol. 12, ed. P.-P. Grasse, pp. 507-567. Paris: Masson.

Frizzell, R., Halm, D., Musch, M., Stewart, C., and Field, M. 1984. Potassium transport by flounder intestinal mucosa. *Am. J. Physiol.* 246: F946-F951.

Frizzell, R., Koch, M., and Schultz, S. 1976. Ion transport by rabbit colon: I. Active and passive components. *J. Memb. Biol.* 27: 297-316.

Frizzell, R., Markscheid-Kaspi, L., and Schultz, S. 1974. Oxidative metabolism of rabbit ileal mucosa. *Am. J. Physiol.* 226: 1142-1148.

Frizzell, R., Nellans, H., Rose, R., Markscheid-Kaspi, L., and Schultz, S. 1973. Intracellular Cl concentrations and influxes across the brush border of rabbit ileum. *Am. J. Physiol.* 224: 328-337.

Frizzell, R. A., Smith, P. L., Vosburgh, E., and Field, M. 1979. Coupled sodium-chloride influx across brush border of flounder intestine. *J. Memb. Biol.* 46: 27-39.

Fruton, J. S., and Simmonds, S. 1958. *General Biochemistry,* 2nd ed. New York: John Wiley.

Fujita, T., Yui, R., Iwanaga, T., Nishiitsutsuji-Uwo, J., Endo, Y., and Yanaihara, N. 1981. Evolutionary aspects of "brain-gut peptides": an immunohistochemical study. *Peptides* 2(Suppl. 2): 123-131.

Gabella, G. 1981. Structure of muscles and nerves in the gastrointestinal tract. In *Physiology of the Gastrointestinal Tract,* Vol. 1, ed. L. R. Johnson, J. Christensen, M. I. Grossman, E. D. Jacobson, and S. G. Schultz, pp. 197-242. New York: Raven Press.

Galef, B. G., Jr. 1979. Investigation of the functions of coprophagy in juvenile rats. *J. Comp. Physiol. Psychol.* 93: 295-305.

Gamble, J. I. 1954. *Chemical Anatomy, Physiology and Pathology of Extracellular Fluid,* 6th ed. Cambridge, Mass.: Harvard University Press.

Gans, C., ed. 1969. *Biology of the Reptilia,* Vol. 1. New York: Academic Press.

Gans, C., and Gans, K. A., eds. 1978. *Biology of the Reptilia,* Vol. 18. New York: Academic Press.

Garcia, J., Campos, M., and Lopez, M. 1984. Bicarbonate secretion in the rabbit colon. Its relationship with sodium, potassium and chloride movements. *Mol. Physiol.* 5: 159-164.

Gardner, M. L. 1982. Absorption of intact peptides: studies on transport of protein digests and dipeptides across rat small intestine in vitro. *J. Exp. Physiol.* 67: 629-637.

Garner, A., and Flemström, G. 1978. Gastric HCO_3^- secretion in the guinea pig. *Am. J. Physiol.* 234: E535-E541.

Garner, R. J., Jones, H. G., and Ekman, L. 1960. Fission products and the dairy cow. 1. The fate of orally administered cerium-144. *J. Agric. Sci.* 55: 107-108.

Gas, N., and Noaillac-Depeyre, J. 1974. Renouvellement de l'épithélium intestinal de la carpe (*Cyprinus carpio* L.). Influence de la saison. *C. R. Séances Acad. Sci.* [D] 279: 1085-1088.

Gasaway, W. C. 1976. Seasonal variation in diet, volatile fatty acid production and size of the cecum of rock ptarmigan. *Comp. Biochem. Physiol.* 53A: 109-114.

Gasaway, W. C., Holleman, D. F., and White, R. G. 1975. Flow of digesta in the intestine and cecum of the rock ptarmigan. *Condor* 77: 467-474.

Gay, C. V., Schraer, H., and Shanabrook, V. M. 1981. Sites of carbonic anhydrase in avian gastric mucosa identified by electron microscope autoradiography. *Am. J. Physiol.* 241: G382-G388.

Geis, A. D. 1957. Coprophagy in the cottontail rabbit. *J. Mammal.* 38: 136.

Geistdoerfer, P. 1973. Sur quelques particularites histologiques de l'intestin de *Chalinura mediterranea* (Macrouridae, Gadiformes). *C. R. Séances Acad. Sci.* [D] 276: 331-333.

Giannella, R. A., Broitman, S. A., and Zamcheck, N. 1971. Vitamin B_{12} uptake by intestinal microorganisms: mechanism and relevance to syndromes of intestinal bacterial overgrowth. *J. Clin. Invest.* 50: 1100-1107.

Gibson, R. and Barker, P. L. 1979. The decapod hepatopancreas. *Oceanogr. Mar. Biol. Annu. Rev.* 17: 285-346.

Gibson, S. J., Polak, J. M., Anand, P., Blank, M. A., Morrison, J. F. B., Kelly, J. S., and Bloom, S. R. 1984. The distribution and origin of VIP in the spinal cord of six mammalian species. *Peptides* 5: 201-207.

Giddings, R. F., Argenzio, R. A., and Stevens, C. E. 1974. Sodium and chloride transport across the equine cecal mucosa. *Am. J. Vet. Res.* 35: 1511-1514.

Gill, J. 1959. Die durchgangszeiten der Nahrung durch den Verdauungskanal des Elches. *Trans. IV Cong. Int. Union Game Biologists* 4: 155-164.

Gill, J. 1960. Szybkość przechodzenia treści przez przewód pokarmowy slonia indyjskiego (*Elephas maximus* L.) w warunkach ogrodu zoologicznego. *Acta Physiol. Pol.* 11: 277-289.

Gill, J., and Bieguszewski, H. 1960. Die Durchgangzeiten der Nahrung durch den Verdauungskanal der Nutria, *Myocastor coypus* Molina 1782. *Acta Theriologica* 4: 11-26.

Gilmore, D. P. 1970. The rate of passage of food in the brushtailed possum, *Trichosurus vulpecula. Aust. J. Biol. Sci.* 23: 515-518.

Goodall, E. D., and Kay, R. N. B. 1965. Digestion and absorption in the large intestine of the sheep. *J. Physiol. (Lond.)* 176: 12-23.

Gordon, H. A., and Bruckner, G. 1984. Anomalous lower bowel function and related phenomena in germ-free animals. In *The Germ-Free Animal in Bio-*

medical Research, ed. M. E. Coates and B. E. Gustafsson, pp. 193-213. London: Laboratory Animals.

Gordon, J. G. 1968. Rumination and its significance. *World Rev. Nutr. Diet.* 9: 251-273.

Gossling, J., Loesche, W. J., and Nace, G. W. 1982. Large intestine bacterial flora of nonhibernating and hibernating leopard frogs (*Rana pipiens*). *Appl. Environ. Microbiol.* 44: 59-66.

Govoni, J. J., Boehlert, G. U., and Watanabe, Y. 1986. The physiology of digestion in fish larvae. *Environ. Biol. Fishes* 16: 59-77.

Grassé, P.-P. 1955. Ordre des édentés. In *Traité de Zoologie,* Vol. 17, ed. P.-P. Grassé, pp. 1182-1266. Paris: Masson.

Gregory, P. C. 1982. Forestomach motility in the chronically vagotomized sheep. *J. Physiol. (Lond.)* 328: 431-447.

Griffiths, M. 1965. Digestion, growth and nitrogen balance in an egg-laying mammal, *Tachyglossus aculeatus* (Shaw). *Comp. Biochem. Physiol.* 14: 357-375.

Griffiths, M., and Davies, D. 1963. The role of the soft pellets in the production of lactic acid in the rabbit stomach. *J. Nutr.* 80: 171-180.

Grovum, W. L., and Chapman, H. W. 1982. Pentagastrin in the circulation acts directly on the brain to depress motility of the stomach in sheep. *Regul. Pept.* 5: 35-42.

Grovum, W. L., and Williams, V. J. 1973. Rate of passage of digesta in sheep. 3. Differential rates of passage of water and dry matter from the reticulo-rumen, abomasum and caecum and proximal colon. *Br. J. Nutr.* 30: 231-240.

Guan, D., Yoshioka, M., Erickson, R. H., Heizer, W., and Kim, Y. S. In press. Digestion and absorption of protein by rat and human small intestine. *Gastroenterology.*

Guard, C. L. 1980. The reptilian digestive system: general characteristics. In *Comparative Physiology: Primitive Mammals,* ed. K. Schmidt-Nielsen, L. Bolis, and C. R. Taylor, pp. 43-51. Cambridge: Cambridge University Press.

Gunter-Smith, P., and White, J. 1979. Response of *Amphiuma* small intestine to theophylline: effect on bicarbonate transport. *Am. J. Physiol.* 236: E775-E783.

Gunther, R., and Wright, E. 1983. Na^+-Li^+ and Cl^- transport by brush border membranes from rabbit jejunum. *J. Memb. Biol.* 74: 85-94.

Gustafsson, B. 1948. Germ-free rearing of rats; general technique. *Acta Pathol. Microbiol. Scand. [Suppl.]* 73: 1-130.

Haga, R. 1960. Observations on the ecology of the Japanese pika. *J. Mammal.* 41: 200-212.

Hamilton, W. J. 1955. Coprophagy in the swamp rabbit. *J. Mammal.* 36: 303-304.

Hammond, P. B., Dziuk, H. E., Usenik, E. A., and Stevens, C. E. 1964. Experimental intestinal obstruction in calves. *J. Comp. Pathol. Ther.* 74: 210-222.

Hamosh, M. 1979. A review. Fat digestion in the newborn: role of lingual lipase and preduodenal digestion. *Pediatr. Res.* 13: 615-622.

Hansen, N. E. 1978. The influence of sulfuric acid preserved herring on the passage time through the gastro-intestinal tract in mink. *Z. Tierphysiol. Tiernahr. Futtermittelkd.* 40: 285-291.

Harder, W. 1950. Zur Morphologie und Physiologie des Blinddarms der Nagetiere. *Verh. Dtsch. Zool. Ges.* 2: 95-109.

Harder, W. 1975a. *Anatomy of Fishes,* Part 1. Stuttgart: E. Schweizerbart'sche Verlagsbuchhandlung.

Harder, W. 1975b. *Anatomy of Fishes,* Part 2. Stuttgart: E. Schweizerbart'sche Verlagsbuchhandlung.

Harding, R. S. O., and Strum, S. C. 1976. The predatory baboons of Kekopey. *Nat. Hist.* 85: 46-53.

Harrison, F. A. 1962. Bile secretion in the sheep. *J. Physiol. (Lond.)* 162: 212-224.

Harrison, F. A., Keynes, F. D., and Zurich, L. 1968. The active transport of chloride across the rumen epithelium of the sheep. *J. Physiol. (Lond.)* 194: 48-49.

Harrop, C. J. F., and Barker, S. 1972. Blood chemistry and gastro-intestinal changes in the developing red kangaroo (*Megaleia rufa,* Desmarest). *Aust. J. Exp. Biol. Med. Sci.* 50: 245-249.

Harrop, C. J. F., and Hume, I. D. 1980. Digestive tract and digestive function in monotremes and nonmacropod marsupials. In *Comparative Physiology: Primitive Mammals,* ed. K. Schmidt-Nielsen, L. Bolis, and C. R. Taylor, pp. 63-77. Cambridge: Cambridge University Press.

Hartley, B. S., and Shotton, D. M. 1971. Pancreatic elastase. In *The Enzymes,* 3rd ed., Vol. 3, ed. P. Boyer, pp. 323-373. New York: Academic Press.

Hartshorne, D. J. 1981. Biochemistry of the contractile process in smooth muscle. In *Physiology of the Gastrointestinal Tract,* Vol. 1, ed. L. R. Johnson, J. Christensen, M. I. Grossman, E. D. Jacobson, and S. G. Schultz, pp. 243-267. New York: Raven Press.

Haslewood, G. A. D. 1964. The biological significance of chemical differences in bile salts. *Biol. Rev.* 39: 537-574.

Haslewood, G. A. D. 1967. Bile salt evolution. *J. Lipid Res.* 8: 535-550.

Hawker, P. C., Mashiter, K. E., and Turnberg, L. A. 1978. Mechanisms of transport of Na, Cl, and K in the human colon. *Gastroenterology* 74: 1241-1247.

Heading, R. C. 1984. Role of motility in the upper digestive tract. *Scand. J. Gastroenterol. [Suppl.]* 19: 39-44.

Heisinger, J. F. 1965. Analysis of the reingestion rhythm of confined cottontails. *Ecology* 46: 197-201.

Heller, R., Gregory, P. C., and Engelhardt, W. v. 1984. Pattern of motility and flow of digesta in the forestomach of the llama (*Lama guanacoe* f. *glama*). *J. Comp. Physiol. Psychol.* 154B: 529-533.

Henning, S. J., and Hird, F. J. R. 1972. Diurnal variations in the concentrations of volatile fatty acids in the alimentary tracts of wild rabbits. *Br. J. Nutr.* 27: 57-64.

Herpol, C. 1964. Activité protéolytique de l'appareil gastrique d'oiseaux granivores et carniovres. *Ann. Biol. Anim. Biochim. Biophys.* 4: 239-244.

Herschel, D. A., Argenzio, R. A., Southworth, M., and Stevens, C. E. 1981. Absorption of volatile fatty acid, Na, and H_2O by the colon of the dog. *Am. J. Vet Res.* 42: 1118-1124.

Herwig, R. P., Staley, J. T., Nerini, M. K., and Braham, H. W. 1984. Baleen whales: preliminary evidence for forestomach microbial fermentation. *Appl. Environ. Microbiol.* 47: 421-423.

Hespell, R. B. 1981. Ruminal microorganisms—their significance and nutritional value. *Dev. Industr. Microbiol.* 22: 266.

Hewitt, E. A., and Schelkopf, R. L. 1955. pH values and enzymatic activity of the digestive tract of the chicken. *Am. J. Vet. Res.* 16: 576-579.

Hill, K. J. 1965. Abomasal secretory function in the sheep. In *Physiology of Digestion in the Ruminant,* ed. R. W. Dougherty, pp. 221-230. Washington, D.C.: Butterworths.

Hill, W. C. O. 1952. The external and visceral anatomy of the olive colobus monkey *Procolobus verus. Proc. Zool. Soc. Lond.* 122: 127-186.

Hill, W. C. O. 1966a. *Primates: Comparative Anatomy and Taxonomy,* Vol. 6: *Cercopithecoidea.* New York: John Wiley.

Hill, W. C. O. 1966b. *Primates: Comparative Anatomy and Taxonomy,* Vol. 7: *Cynopithecinae.* New York: John Wiley.

Hill, W. C. O., Porter, A., Bloom, R. T., Seago, J., and Southwide, M. D. 1957. Field and laboratory studies on the naked mole rat, *Heterocephalus glaber. Proc. Zool. Soc. Lond.* 128: 455-514.

Hinderer, S. 1978. Kinetik des Harnstoff-Stoffwechsels beim Llama bei proteinarmen Diäten. Ph.D. thesis, Universität Hohenheim, Stuttgart.

Hintz, H. F., Hogue, D. E., Walker, E. F., Lowe, J. E., and Schryver, H. F. 1971. Apparent digestion in various segments of the digestive tract of ponies fed diets with varying roughage-grain ratios. *J. Anim. Sci.* 32: 245-248.

Hintz, H. F., and Loy, R. G. 1966. Effects of pelleting on the nutritive value of horse rations. *J. Anim. Sci.* 25: 1059-1062.

Hirano, T., and Mayer-Gostan, N. 1976. Eel esophagus as an osmoregulatory organ. *Proc. Natl. Acad. Sci. USA* 73: 1348-1350.

Hirji, K. N. 1982. Fine structure of the oesophageal and gastric glands of the red-legged pan frog *Kassina maculata* Dumeril. *S. Afr. J. Zool.* 17: 28-31.

Hladik, C. M., Charles-Dominique, P., Valdebouze, P., Delort-Laval, J., and Flanzy, J. 1971. La caecotrophie chez un Primate phyllophage du genre *Lepilemur* et les correlations avec les particularités de son appareil digestif. *C. R. Séances Acad. Sci.*[D] 272: 3191-3194.

Hoar, W. S. 1983. Nutrition and digestion. In *General Comparative Physiology,* 3rd ed., pp. 407-453. Englewood Cliffs, N.J.: Prentice-Hall.

Hobson, P. N., and Wallace, R. J. 1982. Microbial ecology and activities in the rumen. *CRC Crit. Rev. Microbiol.* 9: 165-225.

Hodgkiss, J. P. 1984. Evidence that enteric cholinergic neurones project orally in the intestinal nerve of the chicken. *Q. J. Exp. Physiol.* 69: 797-807.

Hoelzel, F. 1930. The rate of passage of inert materials through the digestive tract. *Am. J. Physiol.* 92: 466-497.

Hofmann, R. 1968. Comparisons of the rumen and omasum structure in East African game ruminants in relation to their feeding habits. *Symp. Zool. Soc. Lond.* 21: 179-194.

Hofmann, R. 1973. *The Ruminant Stomach: Stomach Structure and Feeding Habits of East African Game Ruminants.* Nairobi, Kenya: East African Literature Bureau.

Hofmann, R. 1983. Adaptive changes of gastric and intestinal morphology in response to different fibre content in ruminant diets. *R. Soc. N.Z. Bull.* 20: 51-58.

Hogan, J. P., and Phillipson, A. T. 1960. The rate of flow of digesta and their removal along the digestive tract of the sheep. *Br. J. Nutr.* 14: 147-155.

Hogben, C. A. M., Kent, T. H., Woodward, P. A., and Sill, A. J. 1974. Quantitative histology of the gastric mucosa: man, dog, cat, guinea pig, and frog. *Gastroenterology* 67: 1143-1154.

Höller, H. 1970a. Untersuchungen über Sekret und Sekretion der Cardiadrüsenzone im Magen des Schweines. I. Sekretionsvolumina und -rhythmik, Eigenschaften der Sekrete. *Zentralbl. Veterinaermed. Reihe A* 17: 685-711.

Höller, H. 1970b. Untersuchungen über Sekret und Sekretion der Cardiadrüsenzone im Magen des Schweines. II. Versuche zur Beeinflussung der Spontansekretion der isolierten Cardiadrüsenzone, Flüssigkeits- und Elektrolytsekretion in den mit Verschiedenen Flüssigkeiten gefüllten isolierten kleinen Magen. *Zentralbl. Veterinaermed. Reihe A* 17: 857-873.

Holmes, J. H. G., Bayley, H. S., Leadbeater, R. A., and Horney, F. D. 1974. Digestion of protein in small and large intestine of the pig. *Br. J. Nutr.* 32: 479-489.

Holmgren, S., and Nilsson, S. 1983. Bombesin-, gastrin/CCK-, 5-hydroxytryptamine-, neurotensin-, somatostatin-, and VIP-like immunoreactivity and catecholamine fluorescence in the gut of the elasmobranch, *Squalus acanthius. Cell Tissue Res.* 234: 595-618.

Holmgren, S., Vaillant, C., and Dimaline, R. 1982. VIP-, substance P-, gastrin/CCK-, bombesin-, somatostatin- and glucagon-like immunoreactivities in the gut of the rainbow trout, *Salmo gairdneri. Cell Tissue Res.* 223: 141-153.

Holmquist, A. L., Dockray, C. J., Rosenquist, G. L., and Walsh, J. H. 1979. Immunochemical characterization of cholecystokinin-like peptides in lamprey gut and brain. *Gen. Comp. Endocrinol.* 37: 474-481.

Holstein, B., and Humphrey, C. S. 1980. Stimulation of gastric acid secretion and suppression of VIP-like immunoreactivity by bombesin in the Atlantic codfish, *Gadus morhua. Acta Physiol. Scand.* 109: 217-223.

Holtenius, K., and Björnhag, G. 1985. The colonic separation mechanism in the guinea-pig (*Cavia porcellus*) and the chinchilla (*Chinchilla laniger*). *Comp. Biochem. Physiol.* 82A: 537-542.

Honde, C., and Bueno, L. 1984. Evidence for central neuropeptidergic control of rumination in sheep. *Peptides* 5: 81-83.

Honigmann, H. 1936. Studies on nutrition in mammals, Part I. *Proc. Zool. Soc. Lond.* 106: 517-530.

Hoogkamp-Korstanje, J. A. A., Lindner, J. G. E. M., Marcelis, J.H., denDaas-Slagt, H., and deVos, N. M. 1979. Composition and ecology of the human intestinal flora. *Antonie Van Leeuwenhoek J. Microbiol. Serol.* 45: 33-40.

Hoover, W. H., and Heitmann, R. N. 1972. Effects of dietary fiber levels on weight gain, cecal volume and volatile fatty acid production in rabbits. *J. Nutr.* 102: 375-379.

Hoover, W. H., and Heitmann, R. N. 1975. Cecal nitrogen metabolism and amino acid absorption in the rabbit. *J. Nutr.* 105: 245-252.

Hopfer, U. 1987. Membrane transport mechanisms for hexoses and amino acids in the small intestine. In *Physiology of the Digestive Tract,* 2nd ed., Vol. 2, ed. L. R. Johnson, J. Christensen, M. Jackson, E. D. Jacobson, and J. Walsh, pp. 1499-1526. New York: Raven Press.

Hoppe, P. P., and Gwynne, M. D. 1978. Food retention times in the digestive tract of the suni antelope (*Nesotragus moschatus*). *Säugetierkd. Mitt.* 26: 236-237.

Hörnicke, H. 1981. Utilization of caecal digesta by caecotrophy (soft faeces ingestion) in the rabbit. *Livest. Prod. Sci.* 8: 361-366.

Hörnicke, H., and Björnhag, G. 1979. Coprophagy and related strategies for digesta utilization. In *Digestive Physiology and Metabolism in Ruminants*, ed. Y. Ruckebusch and P. Thivend, pp. 707-730. Westport, Conn.: AVI Publishing.

Hotton, N. 1955. A survey of adaptive relationships of dentition to diet in the North American Iguanidae. *Am. Midl. Naturalist* 53: 88-114.

Houpt, T. R., and Houpt, K. A. 1968. Transfer of urea nitrogen across the rumen wall. *Am. J. Physiol.* 214: 1296-1303.

Hourdry, J., Chabot, J.-G., Menard, D., and Hugon, J. S. 1979. Intestinal brush border enzyme activities in developing amphibian *Rana catesbeiana*. *Comp. Biochem. Physiol.* 63A: 121-125.

Huang, K., and Chen, T. 1971. Ion transport across intestinal mucosa of winter flounder, *Pseudopleuronectes americanus*. *Am. J. Physiol.* 220: 1734-1738.

Hubel, K. 1974. The mechanism of bicarbonate secretion in rabbit ileum exposed to choleragen. *J. Clin. Invest.* 53: 964-970.

Hukuhara, T., Naitoh, T., and Kameyama, H. 1975. Observations on the gastrointestinal movements of the tortoise (*Geoclemys reevsii*) by means of the abdominal window technique. *Jpn. J. Smooth Musc. Res.* 11: 39-46.

Hukuhara, T., and Neya, T. 1968. The movements of the colon of rats and guinea pigs. *Jpn. J. Physiol.* 18: 551-562.

Humbert, W., Kirsch, R., and Meister, M. F. 1984. Scanning electron microscopic study of the oesophageal mucous layer in the eel, *Anguilla anguilla* L. *J. Fish Biol.* 25: 117-122.

Hume, I. D. 1977. Production of volatile fatty acids in two species of wallaby and in sheep. *Comp. Biochem. Physiol.* 56A: 299-304.

Hume, I. D. 1982. *Digestive Physiology and Nutrition of Marsupials.* Cambridge: Cambridge University Press.

Hume, I. D. 1984. Microbial fermentation in herbivorous marsupials. *Bioscience* 34: 435-440.

Hume, I. D., and Dellow, D. W. 1980a. Form and function of the macropod marsupial digestive tract. In *Comparative Physiology: Primitive Mammals*, ed. K. Schmidt-Nielsen, L. Bolis, and C. R. Taylor, pp. 78-89. Cambridge: Cambridge University Press.

Hume, I. D., and Dellow, D. W. 1980b. Field and laboratory estimates of fermentation rates in the digestive tract of the eastern grey kangaroo. *Bull. Aust. Mammal Soc.* 6: 43.

Hume, I. D., Rübsamen, K., and Engelhardt, W. v. 1980. Nitrogen metabolism and urea kinetics in the rock hyrax (*Procavia habessinica*). *J. Comp. Physiol. Psychol.* 138B: 307-314.

Hume, I. D., and Warner, A. C. I. 1980. Evolution of microbial digestion in mammals. In *Digestive Physiology and Metabolism in Ruminants*, ed. Y. Ruckebusch and P. Thivend, pp. 665-684. Westport, Conn.: AVI Publishing.

Hungate, R. E. 1966. *The Rumen and Its Microbes.* New York: Academic Press.

Hungate, R. E. 1968. Ruminal fermentation. In *Handbook of Physiology,* Sec. 6: *Alimentary Canal,* Vol. 5: *Bile; Digestion; Ruminal Physiology,* ed. C. F. Code and W. Heidel, pp. 2725-2745. Washington, D.C.: American Physiological Society.

Hungate, R. E., Phillips, G. D., McGregor, A., Hungate, D. P., and Buechner, H. K. 1959. Microbial fermentation in certain mammals. *Science* 130: 1192-1194.

Hurwitz, S., and Bar, A. 1966. Rate of passage of calcium-45 and yttrium-91 along the intestine and calcium absorption in the laying fowl. *J. Nutr.* 89: 311-316.

Husar, S. L. 1975. *A Review of the Literature of the Dugong (Dugong dugong),* Wildlife Research Report No. 4. Washington, D.C.: Fish and Wildlife Service.

Hydén, S. 1961. The use of reference substances and the measurement of flow in the alimentary tract. In *Digestive Physiology and Nutrition of Ruminants,* ed. D. Lewis, pp. 35-47. London: Butterworth.

Hyodo-Taguchi, Y. 1970. Effect of x-irradiation on DNA synthesis and cell proliferation in the intestinal epithelial cells of goldfish at different temperatures with special reference to recovery process. *Radiat. Res.* 41: 568-578.

Imon, M., and White, J. 1984. Association between HCO_3^- absorption and K^+ uptake by *Amphiuma* jejunum: relations among HCO_3^- absorption, luminal K^+, and intracellular K^+ activity. *Am. J. Physiol.* 246: G732-G744.

Imoto, S., and Namioka, S. 1978. VFA metabolism in the pig. *J. Anim. Sci.* 47: 479-487.

Ingelfinger, F. J. 1958. Esophageal motility. *Physiol. Rev.* 38: 533-584.

Ingles, L. G. 1961. Reingestion in the mountain beaver. *J. Mammal.* 42: 411-412.

Ito, S., and Winchester, R. J. 1960. Electron microscopic observations of the bat gastric mucosa. *Anat. Rec.* 136: 338-339.

Ito, S., and Winchester, R. J. 1963. The fine structure of the gastric mucosa in the bat. *J. Cell Biol.* 16: 541-577.

Jackson, D. C. 1971. Mechanical basis for lung volume variability in the turtle *Pseudemys scripta elegans. Am. J. Physiol.* 220: 754-758.

Jackson, M. 1973. Transport of short chain fatty acids. *Biomembranes* 4B: 673-709.

Jackson, M., Tai, C.-Y., and Steane, J. 1981. Weak electrolyte permeation in alimentary epithelia. *Am. J. Physiol.* 240: G191-G198.

Jacobshagen, E. 1937. Mittel- und Enddarm (Rumpfdarm). In *Handbuch der vergleichenden Anatomie der Wirbeltiere,* Vol. 3, ed. L. Bolk, E. Göppert, E. Kallius, and W. Lubosch, pp. 563-724. Berlin: Schwarzenburg.

James, P. S., and Smith, M. W. 1976. Methionine transport by pig colonic mucosa measured during early post-natal development. *J. Physiol. (Lond.)* 262: 151-168.

Janes, R. G. 1934. Studies on the amphibian digestive system. I. Histological changes in the alimentary tract of anuran larvae during involution. *J. Exp. Zool.* 67: 73-91.

Janis, C. 1976. The evolutionary strategy of the equidae and the origins of rumen and cecal digestion. *Evolution* 30: 757-774.

Jarvis, L. G., Morgan, G., Smith, M. W., and Wooding, F. B. P. 1977. Cell replacement and changing transport function in the neonatal pig colon. *J. Physiol. (Lond.)* 273: 717-729.

Jeuniaux, C. 1961. Chitinase: an addition to the list of hydrolases in the digestive tract of vertebrates. *Nature* 192: 135-136.

Jeuniaux, C. 1962a. Digestion de la chitine chez les oiseaux et les mammifères. *Ann. Soc. R. Zool. Belg.* 92: 27-45.

Jeuniaux, C. 1962b. Recherche de polysaccharidases dans l'estomac d'un paresseux, *Choloepus hoffmanni. Arch. Int. Physiol. Biochim.* 70: 407-408.

Jeuniaux, C. 1963. *Chitine et Chitinolyse.* Paris: Masson.

Jeuniaux, C., and Cornelius, C. 1978. Distribution and activity of chitinolytic enzymes in the digestive tract of birds and mammals. In *Proceedings of the First International Conference on Chitin/Chitosan,* ed. R. A. A. Muzzarelli and E. R. Pariser, pp. 542-549. Cambridge, Mass.: MIT Press.

Jilge, B. 1979. Zur circadianen Caecotrophie des Kaninchens. *Z. Versuchstierkd.* 21: 302-312.

Jilge, V. B. 1980. The gastrointestinal transit time in the guinea pig. *Z. Versuchstierkd.* 22: 204-210.

Josefsson, L., and Lindberg, T. 1965. Intestinal dipeptidases. I. Spectrophotometric determination and characterization of dipeptidase activity in pig intestinal mucosa. *Biochim. Biophys. Acta* 105: 149-161.

Junqueira, L. C. U., and de Moraes, F. F. 1965. Comparative aspects of the vertebrate major salivary glands biology. In *Funktionelle und morphologische Organisation der Zelle.* Sekretion und Exkretion, ed. W. Bothermann, pp. 36-48. Berlin: Springer-Verlag.

Junqueira, L. C. U., Malnic, G., and Monge, C. 1966. Reabsorptive function of the ophidian cloaca and large intestine. *Physiol. Zool.* 39: 151-159.

Kandatsu, M., Yoshihara, I., and Yoshida, T. 1959. Studies on cecal digestion. II. Excretion of hard and soft feces and fecal composition in rabbits. *Jpn. J. Zootechol. Sci.* 29: 366-371.

Kanou, T. 1984. Morphological studies of the mucous membrane of the small intestine of vertebrates with an emphasis on comparative anatomy. *Kawaski Med. J.* 10: 49-61.

Kapoor, B. G., Smit, H., and Verighina, I. A. 1975. The alimentary canal and digestion in teleosts. *Adv. Mar. Biol.* 13: 109-239.

Karasov, W. H., Petrossian, E., Rosenberg, L., and Diamond, J. M. 1986. How do food passage rate and assimilation differ between herbivorous lizards and nonruminant mammals? *J. Comp. Physiol. Psychol.* 156: 599-609.

Karasov, W. H., Solberg, D. H., and Diamond, J. M. 1985. What transport adaptations enable mammals to absorb sugars and amino acids faster than reptiles? *Am. J. Physiol.* 249: G271-G283.

Karpov, L. V. 1919. O perevarivanii nekrotorykh rastitelnykh i zhivotnykh belkou gusiiym zheludochwom sokum. *Fiziol. Zh. SSSR* 2: 185-196.

Karr, M. R., Little, C. O., and Mitchell, G. E. 1966. Starch disappearance from different segments of the digestive tract of steers. *J. Anim. Sci.* 25: 652-654.

Kasper, H. 1962. Methode fur Prüfung der Resorption in Caecum und Colon. *Z. Versuchstierkd.* 1: 104-106.

Kauffman, G. L., Reeve, J. J., and Grossman, M. I. 1980. Gastric bicarbonate secretion: effect of topical and intravenous 16,16-dimethyl prostaglandin E_2. *Am. J. Physiol.* 239: G44-G48.

Kawai, S., and Ikeda, S. 1971. Studies on digestive enzymes of fishes. I. Carbohydrases in digestive organs of several fishes. *Bull. Jpn. Soc. Sci. Fish.* 37: 333-
337.

Kay, R. N. B. 1960. I. The rate of flow and composition of various salivary secretions in sheep and calves. II. The development of parotid salivary secretion in young goats. *J. Physiol. (Lond.)* 150: 515-545.

Kay, R. N. B., and Pfeffer, E. 1970. Movements of water and electrolytes into and
from the intestine of sheep. In *Physiology of Digestion and Metabolism in
the Ruminant,* ed. A. T. Phillipson, pp. 390-402. Newcastle upon Tyne: Oriel
Press.

Kear, J. 1972. Feeding habits of birds. In *Biology of Nutrition,* Vol. 18, ed. R. N.
T-W-Fiennes, pp. 471-503. Elmsford, N.Y.: Pergamon Press.

Keast, D., and Walsh, L. G. 1979. The use of ruthenium-103 for the determination
of the rate of passage of food through the gut of captive wild birds. *Int. J.
Appl. Radiat. Isot.* 30: 463-468.

Keast, J. R., Furness, J. B., and Costa, M. 1985. Distribution of certain peptide-
containing nerve fibres and endocrine cells in the gastrointestinal mucosa in
five mammalian species. *J. Comp. Neurol.* 236: 403-422.

Kelly, K. A. 1981. Motility of the stomach and gastroduodenal junction. In *Physiol-
ogy of the Gastrointestinal Tract,* Vol. 1, ed. L. R. Johnson, J. Christensen,
M. I. Grossman, E. D. Jacobson, and S. G. Schultz, pp. 393-410. New York:
Raven Press.

Kempton, T. J., Murray, R. M., and Leng, R. A. 1976. Methane production and
digestibility measurements in the grey kangaroo and sheep. *Aust. J. Biol. Sci.*
29: 209-214.

Kenchington, R. A. 1972. Observations on the digestive system of the dugong,
Dugong dugon Erxleben. *J. Mammal.* 53: 884-887.

Kennedy, P. M., and Hume, I. D. 1978. Recycling of urea nitrogen to the gut of
the tammar wallaby (*Macropus eugenii*). *Comp. Biochem. Physiol.* 61A: 117-
121.

Kennedy, P. M., Young, B. A., and Christopherson, R. J. 1977. Studies on the
relationship between thyroid function, cold acclimation and retention time
of digesta in sheep. *J. Anim. Sci.* 45: 1084-1090.

Kerry, K. R. 1969. Intestinal disaccharidase activity in a monotreme and eight
species of marsupials (with an added note on the disaccharidases of five
species of sea birds). *Comp. Biochem. Physiol.* 29: 1015-1022.

Kerry, K. R., and Messer, M. 1968. Intestinal glycosidases of three species of seals.
Comp. Biochem. Physiol. 25: 437-446.

Keynes, R. D. 1969. From frog skin to sheep rumen: a survey of transport of salts
and water across multicellular structures. *Q. Rev. Biophys.* 2: 177-281.

Keys, J. E., Jr., and DeBarthe, J. V. 1974. Site and extent of carbohydrate, dry matter, energy and protein digestion and the rate of passage of grain diets in
swine. *J. Anim. Sci.* 39: 57-62.

Kidder, D. E., and Manners, M. J. 1978. *Digestion in the Pig.* Bristol, U.K.:
Scientechnica.

Kim, Y. S., and Erickson, R. H. 1985. Role of peptidases of the human small intestine in protein digestion. *Gastroenterology* 88: 1071-1073.

Kimmel, J. R., Hayden, L. J., and Pollock, H. G. 1975. Isolation and characterization of a new pancreatic polypeptide hormone. *J. Biol. Chem.* 24: 9369-9376.

Kimmel, J. R., Pollock, H. G., and Hayden, L. J. 1978. Biological activity of the avian PP in the chicken. In *Gut Hormones,* ed. S. R. Bloom, pp. 234-241. New York: Churchill Livingstone.

King, K. W., and Moore, W. E. C. 1957. Density and size as factors affecting passage rate of ingesta in the bovine and human digestive tracts. *J. Dairy Sci.* 40: 528-536.

Kinnear, J. E., and Main, A. R. 1975. The recycling of urea nitrogen by the wild tammar wallaby (*Macropus eugenii*)—a "ruminant-like" marsupial. *Comp. Biochem. Physiol.* 51A: 593-610.

Kirsch, R. 1978. Role of the esophagus in osmoregulation in teleost fishes. In *Osmotic and Volume Regulation,* ed. C. B. Jorgensen and E. Skadhauge, pp. 138-154. New York: Academic Press.

Kitamikado, M., and Tachino, S. 1960. Studies on the digestive enzymes of rainbow trout. I. Carbohydrases. *Bull. Jpn. Soc. Sci. Fish.* 26: 679-684.

Knickelbein, R., Aronson, P., Schron, C., Seifter, J., and Dobbins, J. 1985. Sodium and chloride transport across rabbit ileal brush border. II. Evidence for Cl-HCO₃ exchange and mechanism of coupling. *Am. J. Physiol.* 249: G236-G245.

Kochva, E. 1978. Oral glands of the reptilia. In *Biology of the Reptilia,* Vol. 8, ed. C. Gans and K. A. Gans, pp. 43-162. New York: Academic Press.

Koefed-Johnson, V., and Ussing, H. H. 1958. On the nature of the frog skin potential. *Acta Physiol. Scand.* 42: 298-308.

Koike, T. I., and McFarland, L. Z. 1966. Urography in the unanesthetized hydropenic chicken. Am. J. Vet. Res. 27: 1130-1133.

Koldovský, O. 1970. Digestion and absorption during development. In *Physiology of the Perinatal Period,* ed. U. Stave, pp. 379-415. New York: Appleton-Century-Crofts.

Komori, S., and Ohashi, H. 1982. Some characteristics of transmission from non-adrenergic, non-cholinergic excitatory nerves to the smooth muscle of the chicken. *J. Auton. Nerv. Syst.* 6: 199-210.

Kostelecka-Myrcha, A., and Myrcha, A. 1964a. The rate of passage of foodstuffs through the alimentary tracts of certain *Microtidae* under laboratory conditions. *Acta Theriol.* 9: 37-53.

Kostelecka-Myrcha, A., and Myrcha, A. 1964b. Choice of indicator in the investigation of the passage of foodstuffs through the alimentary tract of rodents. *Acta Theriol.* 9: 55-65.

Kostelecka-Myrcha, A., and Myrcha, A. 1964c. Rate of passage of foodstuffs through the alimentary tract of *Neomys fodiens* (Pennant 1771) under laboratory conditions. *Acta Theriol.* 9: 371-373.

Kostelecka-Myrcha, A., and Myrcha, A. 1965. Effect of the kind of indicator on the results of investigations of the rate of passage of foodstuffs through the alimentary tract. *Acta Theriol.* 10: 229-242.

Kostuch, T. E., and Duke, G. E. 1975. Gastric motility in great horned owls (*Bubo virginianus*). *Comp. Biochem. Physiol.* 51A: 201-205.

Krause, W. J. 1970. Brunner's glands of the echidna. *Anat. Rec.* 167: 473-487.

Krawielitzki, K., Schadereit, R., Wünsche, J., Völker, T., and Bock, H.-D. 1983. Untersuchungen über Resorption und Verwertung von ins Zäkum wachsender Schweine infundierten Aminosäuren. *Arch. Tierernähr. (Berl.)* 33: 731-742.

Krawielitzki, K., Schadereit, R., Zebrowska, T., Wünsche, J., and Bock, H.-D. 1984. Untersuchungen über Resorption und Verwertung von ins Zäkum wachsender Schweine infundierten Aminosäuren. *Arch. Tierernähr. (Berl.)* 34: 1-18.

Krieger, D. T. 1983. Brain peptides: what, where, and why? *Science* 222: 975-985.

Kronfeld, D. S. 1973. Diet and the performance of racing sled dogs. *J. Am. Vet. Med. Assoc.* 162: 470-473.

Kronfeld, D. S., and Van Soest, P. J. 1976. Carbohydrate nutrition. In *Comparative Animal Nutrition,* ed. M. Rechcigl, Jr., pp. 23-73. Basel: S. Karger.

Kuhn, H.-J. 1964. Zur Kenntnis von Bau and Funktion des Magens der Schlankaffen (*Colobinae*). *Folia Primatol.* 2: 193-221.

Kumar, B. A., and Raghavan, G. V. 1974. Effect of level of intake on the rate of passage of food, and its effect on the digestibility of nutrients in Murrah buffaloes and Hariana cattle. *Indian J. Anim. Sci.* 44: 953-958.

Landry, S. O. 1970. The rodentia as omnivores. *Q. Rev. Biol.* 45: 351-372.

Langer, M., Van Noorden, S., Polak, J., and Pearse, A. G. E. 1979. Peptide hormone-like immunorectivity in the gastrointestinal tract and endocrine pancreas of eleven teleost species. *Cell Tissue Res.* 199: 493-508.

Langer, P. 1974. Stomach evolution in the artiodactyla. *Mammalia* 38: 295-314.

Langer, P. 1980. Anatomy of the stomach in three species of Potorinae (Marsupialia:Macropodidae). *Aust. J. Zool.* 28: 19-31.

Langer, P. 1987. Formenmannigfaltigkeit mehrkammeriger Mägen bei Säugetieren. *Natur und Museum* 117: 47-60.

Langer, P., Dellow, D. W., and Hume, I. D. 1980. Stomach structure and function in three species of macropodine marsupials. *Aust. J. Zool.* 28: 1-18.

Langslow, D. R., Kimmel, J. R., and Pollock, H. G. 1973. Studies of the distribution of a new avian pancreatic polypeptide and insulin among birds, reptiles, amphibians and mammals. *Endocrinology* 93: 558-565.

Laplace, J. P., and Lebas, F. 1975. Le transit digestif chez le lapin. III. Influence de l'heure et du mode d'administration sur l'excrétion fécale du cérium-141 chez le lapin alimenté ad libitum. *Ann. Zootech.* 24: 255-265.

Larsen, L. O. 1984. Feeding in adult toads: physiology, behaviour, ecology. *Vidensk. Meddr. Dansk. Naturh. Foren.* 145: 97-116.

Larsson, L. I., and Rehfeld, J. F. 1977. Evidence for a common evolutionary origin of gastrin and cholecystokinin. *Nature* 269: 335-338.

Larsson, L. I., and Rehfeld, J. F. 1978. Evolution of CCK-like hormones. In *Gut Hormones,* ed. S. R. Bloom, pp. 68-73. New York: Churchill Livingstone.

Larsson, L.-I., Sundler, F., Hakanson, R., Rehfeld, J. F., and Stadil, F. 1974. Distribution and properties of gastrin cells in the gastrointestinal tract of the chicken. *Cell Tissue Res.* 154: 409-422.

Lawrence, A. L. 1963. Specificity of sugar transport by the small intestine of the bullfrog, *Rana catesbeiana. Comp. Biochem. Physiol.* 9: 69-73.

Laws, B. M., and Moore, J. H. 1963. The lipase and esterase activities of the pancreas and small intestine of the chick. *Biochem. J.* 87: 632-638.

Lee, K. Y., Park, H. J., Chang, T.-M., and Chey, W. Y. 1983. Cholinergic role on release and action of motilin. *Peptides* 4: 375-380.

Lee, R. F., Hirota, J., Nevenzel, J. C., Sauerheber, R., Lewis, A., and Benson, A. A. 1972. Lipids in the marine environment. *Calif. Mar. Res. Comm., CalCOFI Rep.* 16: 95-102.

Leeson, C. R. 1967. Structure of salivary glands. In *Handbook of Physiology,* Sec. 6: *Alimentary Canal,* Vol. 2: *Secretion,* ed. C. F. Code and W. Heidel, pp. 463-495. Washington, D.C.: American Physiological Society.

Léger, C. 1979. La lipase pancréatique. In *Nutrition des Poissons,* ed. M. Fontaine, pp. 69-77. Paris: Actes du Colloque C.N.E.R.N.A.

Lehninger, A. L. 1975. *Biochemistry.* New York: Worth Publishers.

Leon, M. A. 1974. Maternal pheromone. *Physiol. Behav.* 13: 441-453.

Leopold, A. S. 1953. Intestinal morphology of gallinaceous birds in relation to food habits. *J. Wildlife Mgmt.* 17: 197-203.

Lester, R., and Grim, E. 1975. Substrate utilization and oxygen consumption by canine jejunal mucosa in vitro. *Am. J. Physiol.* 229: 139-143.

Levenson, S. M., Crowley, L. V., Horowitz, R. E., and Malm, O. J. 1959. The metabolism of carbon-labeled urea in the germfree rat. *J. Biol. Chem.* 234: 2061-2062.

Levitt, M. D., and Bond, J. H. 1970. Volume, composition, and source of intestinal gas. *Gastroenterology* 59: 921-929.

Liedtke, C., and Hopfer, U. 1977. Anion transport in brush border membranes isolated from rat small intestine. *Biochem. Biophys. Res. Commun.* 76: 579-585.

Liedtke, C., and Hopfer, U. 1982. Mechanism of Cl^- translocation across small intestinal brush border membrane. I. Absence of Na^+-Cl^- cotransport. *Am. J. Physiol.* 242: G263-G271.

Lintern-Moore, S. 1973. Utilization of dietary urea by the Kangaroo Island wallaby—*Protemnodon eugenii* (Desmarest). *Comp. Biochem. Physiol.* 46A: 345-351.

Livingston, H. G., Payne, W. J. A., and Friend, M. T. 1962. Urea excretion in ruminants. *Nature* 194: 1057-1058.

Lönnberg, E. 1902. On some points of relation between the morphological structure of the intestine and the diet of reptiles. *Bih. Suensk Vet. Ak. Handl.* 28: 1-51.

Lubinsky, G. 1957. Studies on the evolution of the Ophryoscolecidae (Ciliata: Oligotricha). III. Phylogeny of the Ophryoscolecidae based on their comparative morphology. *Can. J. Zool.* 35: 141-159.

Lucas, M. 1976. The association between acidification and electrogenic events in the rat proximal jejunum. *J. Physiol. (Lond.)* 257: 645-662.

Luckey, T. D., Hartman, R., Knox, T., Palmer, S., Kay, M., and Terry, B. 1979. Lanthanide marker transit times, and rate of flow and passage for three meals in humans. *Nutr. Rep. Int.* 19: 561-571.

Luckey, T. D., Pleasants, J. R., Wagner, M., Gordon, H. A., and Reyniers, J. A. 1955. Some observations on vitamin metabolism in germ-free rats. *J. Nutr.* 57: 169-182.

Lundgren, O. 1984. Microcirculation of the gastrointestinal tract and pancreas. In *Handbook of Physiology,* Sec. 2: *The Cardiovascular System,* Vol. 4: *Microcirculation,* Part 2, ed. E. M. Renkin, C. C. Michel, and S. R. Geiger pp. 799-863. Bethesda: American Physiological Society.

Luppa, H. 1977. Histology of the digestive tract. In *Biology of the Reptilia,* Vol. 6., ed. C. E. Gans and T. S. Parsons, pp. 225-314. New York: Academic Press.

Madsen, H. 1939. Does the rabbit chew its cud? *Nature* 143: 981.

Magee, D. F. 1961. An investigation into the external secretion of the pancreas in sheep. *J. Physiol. (Lond.)* 158: 132-143.

Mäkelä, A. 1956. Studies on the question of bulk in the nutrition of farm animals with special reference to cattle. *Suomen Maataloust. Seuran Julk.* 85: 1-139.

Marsh, H., Heinsohn, G. E., and Spain, A. V. 1977. The stomach and duodenal diverticula of the dugong (*Dugong dugong*). In *Functional Anatomy of Marine Mammals,* Vol. 3, ed. R. T. Harrison. New York: Academic Press.

Marshall, J. A., and Dixon, K. E. 1978. Cell proliferation in the intestinal epithelium of *Xenopus laevis* tadpoles. *J. Exp. Zool.* 203: 31-40.

Martin, R. 1971. Étude autoradiographique du renouvellement de l'épithélium intestinal de l'axolotl (amphibien urodele). *C. R. Séances Acad. Sci.[D]* 272: 2816-2819.

Masson, M. J., and Phillipson, A. T. 1951. The absorption of acetate, propionate and butyrate from the rumen of sheep. *J. Physiol. (Lond.)* 113: 189-206.

Materazzi, G., and Menghi, G. 1975. Il canale alimentare in embrioni ed avannotti di *Salmo irideus*. *Ann. Ist. Mus. Zool. Napoli* 21: 21-37.

Mathias, J. R., and Sninsky, C. A. 1985. Motility of the small intestine: a look ahead. *Am. J. Physiol.* 248: G495-G500.

Mattson, F. H., and Volpenheim, R. A. 1968. Hydrolysis of primary and secondary esters of glycerol by pancreatic juice. *J. Lipid Res.* 9: 79-84.

Mautz, M. W., and Petrides, G. A. 1971. Food passage rate in the white-tailed deer. *J. Wildlife Mgmt.* 35: 723-731.

McAvoy, J. W., and Dixon, K. E. 1977. Cell proliferation and renewal in the small intestinal epithelium of metamorphosing and adult *Xenopus laevis. J. Exp.Zool.* 202: 128-138.

McBee, R. H. 1977. Fermentation in the hindgut. In *Microbial Ecology of the Gut,* ed. R. T. J. Clarke and T. Bauchop, pp. 185-222. New York: Academic Press.

McBee, R. H., and West, G. C. 1969. Cecal fermentation in the willow ptarmigan. *Condor* 71: 54-58.

McCuistion, W. R. 1966. Coprophagy, a quest for digestive enzymes. *Vet. Med. Small Anim. Clin.* 61: 445-447.

McDonald, T. J., Jörnvall, H., Nilsson, G., Vagne, M., Ghatei, M., Bloom, S. R., and Mutt, V. 1979. Characterization of a gastrin releasing peptide from porcine non-antral gastric tissue. *Biochem. Biophys. Res. Commun.* 90: 227-233.

McGeachin, R. L., and Bryan, J. A. 1964. Amylase in the cottonmouth water moccasin, *Agkistrodon piscivorus. Comp. Biochem. Physiol.* 13: 473-475.

McIntosh, D. L. 1966. The digestibility of two roughages and the rates of passage of their residues by the red kangaroo *Megaleia rufa* (Desmarest) and the merino sheep. *CSIRO Wildlife Res.* 11: 125-135.

McLeay, L. M., and Titchen, D. A. 1970. Effects of pentagastrin in gastric secretion and motility in the sheep. *Proc. Aust. Physiol. Pharmacol. Soc.* 1: 33-34.

McLeay, L. M., and Titchen, D. A. 1975. Gastric, antral and fundic pouch secretion in sheep. *J. Physiol. (Lond.)* 248: 595-612.

McMillan, G. L., and Churchill, E. P. 1947. The gross and histological structure of the digestive system of the little brown bat. *Proc. So. Dakota Acad. Sci.* 26: 103-109.

Mead, G. C. 1974. Anaerobic utilization of uric acid by some group D streptococci. *J. Gen. Microbiol.* 82: 421-423.

Mead, G. C., and Adams, B. W. 1975. Some observations on the caecal microflora of the chick during the first two weeks of life. *Br. Poult. Sci.* 16: 169-176.

Meldrum, L. A., and Burnstock, G. 1985. Investigations into the identity of the non-adrenergic, non-cholinergic excitatory transmitter in the smooth muscle of chicken rectum. *Comp. Biochem. Physiol.* 81C: 307-309.

Menking, M., Wagnitz, J. G., Burton, J. J., Coddington, R. D., and Sotos, J. F. 1969. Rumination—a near fatal psychiatric disease of infancy. *N. Engl. J. Med.* 280: 802-804.

Merrett, T. G., Bar-Eli, E., and Van Vunakis, H. 1969. Pepsinogens A, C, and D from the smooth dogfish. *Biochemistry* 8: 3696-3702.

Merritt, A. M., and Brooks, F. P. 1970. Basal and histamine-induced gastric acid and pepsin secretion in the conscious guinea pig. *Gastroenterology* 58: 801-814.

Micha, J. C., Dandrifosse, G., and Jeuniaux, C. 1973a. Activités des chitinases gastriques de reptiles en fonction du pH. *Arch. Int. Physiol. Biochim.* 81: 629-637.

Micha, J. C., Dandrifosse, G., and Jeuniaux, C. 1973b. Distribution et localisation tissulaire de la synthèse des chitinases chez les vertébrés inférieurs. *Arch. Int. Physiol. Biochim.* 81: 439-451.

Mikel'Saar, M. E., Tjuri, M. E., Väljaots, M. E., and Lencner, A. A. 1984. Anaerobe Inhalts- und Wandmikroflora des Magen-Darm-Kanals. *Nahrung* 28: 727-733.

Milchunas, D. G., Dyer, M. I., Wallmo, O. C., and Johnson, D. E. 1978. In vivo/in vitro relationships of Colorado mule deer forages. *Colo. Div. Wildlife Special Rep.* 43: 1-43.

Milenov, K., and Rakovska, A. 1983. The role of prostaglandins in the spontaneous and cholinergic nerve-mediated motility of guinea-pig gastric muscle. *Methods Find. Exp. Clin. Pharmacol.* 5: 121-126.

Miller, J. K., Perry, S. C., Chandler, P. T., and Cragle, R. G. 1967. Evaluation of radiocerium as a nonabsorbed reference material for determining gastrointestinal sites of nutrient absorption and excretion in cattle. *J. Dairy Sci.* 50: 355-361.

Miller, J. K., Swanson, E. W., Lyke, W. A., Moss, B. R., and Byrne, W. F. 1974. Effect of thyroid status on digestive tract fill and flow rate of undigested residues in cattle. *J. Dairy Sci.* 57: 193-197.

Milne, J. A., MacRae, J. C., Spence, A. M., and Wilson, S. 1978. A comparison of the voluntary intake and digestion of a range of forages at different times of the year by the sheep and the red deer. *Br. J. Nutr.* 40: 347-357.

Minchin, K. 1973. Notes on the weaning of a young koala (*Phascolarctos cinereus*). *Records South Aust. Mus.* 6: 1-3.

Minnich, J. E. 1970. Water and electrolyte balance of the desert iguana, *Dipsosaurus dorsalis,* in its natural habitat. *Comp. Biochem. Physiol.* 35: 921-933.

Mircheff, A. K., van Os, C. H., and Wright, E. M. 1980. Pathways for alanine transport in intestinal basal lateral membrane vesicles. *J. Memb. Biol.* 52: 83-92.

Mishra, S. K., and Raviprakash, V. 1981. Non-adrenergic inhibitory and non-cholinergic excitatory neural involvement in DMPP and nicotine action in fowl rectum. *Arch. Int. Pharmacodyn.* 253: 210-219.

Mitchell, H. K., and Isbel, E. R. 1942. Intestinal bacterial synthesis as a source of B vitamin for the rat. *Studies of the Vitamin Content of Tissues.* Univ. of Tex. Biochem. Inst. Publ. No. 4237, Part 2, pp. 125-134.

Mitchell, P. C. 1901. On the intestinal tract of birds; with remarks on the valuation and nomenclature of zoological characters. *Trans. Linn. Soc. Lond.* [*Zool.*] 8: 173-275.

Mitchell, P. C. 1905. On the intestinal tract of mammals. *Trans. Zool. Soc. Lond.* 17: 437-536.

Moir, R. J. 1965. The comparative physiology of ruminant-like animals. In *Physiology of Digestion in the Ruminant,* ed. R. W. Dougherty, pp. 1-14. Washington, D.C.: Butterworths.

Moir, R. J. 1968. Ruminant digestion and evolution. In *Handbook of Physiology,* Sec. 6: *Alimentary Canal,* Vol. 5: *Bile; Digestion; Ruminal Physiology,* ed. C. F. Code and W. Heidel, pp. 2673-2694. Washington, D.C.: American Physiological Society.

Moir, R. J., Somers, M., and Waring, H. 1956. Studies on marsupial nutrition. I. Ruminant-like digestion in a herbivorous marsupial *Setonix brachyurus* (Quoy and Gairmard). *Aust. J. Biol. Sci.* 9: 293-304.

Montgomery, G. G., and Sunquist, M. E. 1978. Habitat selection and use by two-toed and three-toed sloths. In *Ecology of Arboreal Folivores,* ed. G. G. Montgomery, pp. 329-359. Washington, D.C.: Smithsonian Institution Press.

Morii, H. 1972. Bacteria in the stomach of marine little toothed whales. *Bull. Jpn. Soc. Sci. Fish.* 38: 1177-1183.

Morii, H. 1979. The viable counts of microorganisms, pH values, amino acid contents, ammonia contents and volatile fatty acid contents in the stomach fluid of marine little toothed whales. *Bull. Fac. Fish. Nagasaki Univ.* 47: 55-60.

Morii, H., and Kanazu, R. 1972. The free volatile fatty acids in the blood and stomach fluid from porpoise, *Neomeris phocaenoides. Bull. Jpn. Soc. Sci. Fish.* 38: 1035-1039.

Morot, M. Ch. 1882. Des pelotes stomachales des léporidés. *Mém. Soc. Centr. Méd. Vét.* 12: 139-239. Paris.

Moss, R., and Parkinson, J. A. 1972. The digestion of heather (*Calluna vulgaris*) by red grouse (*Lagopus lagopus scoticus*). *Br. J. Nutr.* 27: 285-298.

Mousa, H. M., Ali, K. E., and Hume, I. D. 1983. Effects of water deprivation on urea metabolism in camels, desert sheep and desert goats fed dry desert grass. *Comp. Biochem. Physiol.* 74A: 715-720.

Murer, H., Hopfer, U., and Kinne, R. 1976. Sodium/proton antiport in brush-border-membrane vesicles isolated from rat small intestine and kidney. *Biochem. J.* 154: 597-604.

Murray, R. M., Bryant, A. M., and Leng, R. A. 1976. Rates of production of methane in the rumen and large intestine of sheep. *Br. J. Nutr.* 36: 1-14.

Murray, R. M., Marsh, H., Heinsohn, G. E., and Spain, A. V. 1977. The role of the midgut caecum and large intestine in the digestion of sea grassses by the dugong (Mammalia: Sirenia). *Comp. Biochem. Physiol.* 56A: 7-10.

Nakaya, M., Takahashi, I., Suzuki, T., Takeuchi, S., Arai, H., Wakabayashi, K., and Ito, Z. 1983. Regulation of interdigestive contractions in the denervated stomach. *Gastroenterol. Jpn.* 18: 417-427.

Nellans, H., Frizzell,R., and Schultz, S. 1973. Coupled sodium-chloride influx across the brush border of rabbit ileum. *Am. J. Physiol.* 225: 467-475.

Nelson, J. S. 1984. *Fishes of the World,* 2nd ed. New York: John Wiley.

Nicholson, T. 1982. The mode of action of intravenous pentagastrin injections on forestomach motility of sheep. *Q. J. Exp. Physiol.* 67: 537-542.

Niiyama, M., Dequchi, E., Kagota, K., and Namioka, S. 1979. Appearance of ^{15}N-labeled intestinal microbial amino acids in the venous blood of the pig colon. *Am. J. Vet. Res.* 40: 716-718.

Nilsson, A. 1970. Gastrointestinal hormones in the holocephalian fish *Chimaera monstrosa* (L.). *Comp. Biochem. Physiol.* 32: 387-390.

Nilsson, A. 1973. Secretin-like and cholecystokinin-like activity in *Myxine glutinosa* L. *Acta Regiae Soc. Sci. Litt. Gothub.* 8: 30-32.

Nilsson, A., Carlquist, M., Jörnvall, H., and Mutt, V. 1980. Isolation and characterization of chicken secretin. *Eur. J. Biochem.* 112: 383-388.

Nilsson, A., and Fänge, R. 1969. Digestive proteases in the holocephalian fish *Chimaera monstrosa* (L.). *Comp. Biochem. Physiol.* 31: 147-165.

Nilsson, A., and Fänge, R. 1970. Digestive proteases in the cyclostome *Myxine glutinosa* (L.). *Comp. Biochem. Physiol.* 32: 237-250.

Noaillac-Depeyre, J., and Hollande, E. 1981. Evidence for somatostatin, gastrin and pancreatic polypeptide-like substances in the mucosa cells of the gut in fishes with and without stomach. *Cell Tissue Res.* 216: 192-203.

Norin, K. E., Gustafsson, B. E., Lindblad, B. S., and Midtvedt, T. 1985. The establishment of some microflora associated biochemical characteristics in feces of children during the first years of life. *Acta Paediatr. Scand.* 74: 207-212.

Nowak, R. M., and Paradiso, J. L. 1983. *Walker's Mammals of the World,* Vols. 1 and 2, 4th ed. Baltimore: John's Hopkins University Press.

Ochi, Y., Mitsuoka, T., and Sega, T. 1964. Untersuchungen über die Darmflora des Huhnes. III. Die entwicklung der Darmflora von Küken bis zum Huhn. *Zentralbl. Bakteriol. Parasitenkd. Infektionskr. Hyg. Abt. 1 Orig.* 193: 80-95.

Oftedal, O. T. 1980. Milk and mammalian evolution. In *Comparative Physiology: Primitive Mammals,* ed. K. Schmidt-Nielsen, L. Bolis, and C. R. Taylor, pp. 31-42. Cambridge: Cambridge University Press.

Ogimoto, K., and Imai, S. 1981. *Atlas of Rumen Microbiology.* Tokyo: Japan Scientific Societies Press.

Ohmart, R. D., McFarland, L. Z., and Morgan, J. P. 1970. Urographic evidence that urine enters the rectum and ceca of the roadrunner (*Geococcyx californianus*) Aves. *Comp. Biochem. Physiol.* 35: 487-489.

Ohwaki, K., Hungate, R. E., Lotter, L., Hofmann, R. R., and Maloiy, G. 1974. Stomach fermentation in East African colobus monkeys in their natural state. *Appl. Microbiol.* 27: 713-723.

Okutani, K. 1966. Studies of chitinolytic systems in the digestive tracts of *Lateolabrax japonicus. Bull. Misaki Mar. Biol. Inst. Kyoto Univ.* 10: 1-47.

Olsen, H. M., and Madsen, H. 1943. Investigations on pseudo-rumination rabbits. *Vidensk. Meddr. Dans. Naturhist. Foren.* 107: 37-58.

Oppel, A. 1897. *Lehrbuch der vergleichenden mikroskopischen Anatomie der Wirbeltiere.* Zweiter Teil. Schlund und Darm. Jena: Gustav Fischer.

Orskov, E. R., Fraser, C., and McDonald, I. 1971. Digestion of concentrates in sheep. Effects of rumen fermentation on barley and maize diets on protein digestion. *Br. J. Nutr.* 26: 477-486.

Osborne, T. B., and Mendel, L. B. 1911. *Feeding Experiments with Isolated Food Substances,* Part II. Carnegie Institute of Washington Publ. No. 156.

Östberg, Y., Van Noorden, S., Pearse, A. G. E., and Thomas, N. W. 1976. Cytochemical, immunofluorescence, and ultrastructural investigations on polypeptide hormone containing cells in the intestinal mucosa of a cyclostome, *Myxine glutinosa. Gen. Comp. Endocrinol.* 28: 213-227.

Ostrom, J. H. 1963. Further comments on herbivorous lizards. *Evolution* 17: 368-369.

Ottaviani, G., and Tazzi, A. 1977. The lymphatic system. In *Biology of the Reptilia,* Vol. 6, ed. C. G. Gans and T. S. Parsons, pp. 315-462. New York: Academic Press.

Owen, R. 1835. On the sacculated form of stomach as it exists in the genus *Semnopithecus,* F. Cuv. *Trans. Zool. Soc. Lond.* 1: 65-70.

Pairet, M., Bouyssou, T., and Ruckebusch, Y. 1986. Colonic formation of soft feces in rabbits: a role for endogenous prostaglandins. *Am. J. Physiol.* 250: G302-G308.

Palfrey, H. C., and Rao, M. C. 1983. Na/K/Cl co-transport and its regulation. *J. Exp. Biol.* 106: 43-54.

Parmelee, J. T., and Renfro, J. L. 1983. Esophageal desalination of seawater in flounder: role of active sodium transport. *Am. J. Physiol.* 245: R888-R893.

Parra, R. 1978. Comparison of foregut and hindgut fermentation in herbivores. In *The Ecology of Arboreal Folivores,* ed. G.G. Montgomery, pp. 205-229. Washington, D.C.: Smithsonian Institution Press.

Parsons, T. S., and Cameron, J. E. 1977. Internal relief of the digestive tract. In *Biology of the Reptilia,* Vol. 6, ed. C. Gans and T. S. Parsons, pp. 159-224. New York: Academic Press.

Patton, J. S. 1975. High levels of pancreatic nonspecific lipase in rattlesnake and leopard shark. *Lipids* 10: 562-564.

Patton, J. S., and Benson, A. A. 1975. A comparative study of wax ester digestion in fish. *Comp. Biochem. Physiol.* 52B: 111-116.

Patton, J. S., Nevenzel, J. C., and Benson, A. A. 1975. Specificity of digestive lipases in hydrolysis of wax esters and triglycerides studied in anchovies and other selected fish. *Lipids* 10: 575-583.

Paul, R. J. 1981. Smooth muscle: mechanochemical energy conversion relations between metabolism and contractility. In *Physiology of the Gastrointestinal Tract,* Vol. 1, ed. L. R. Johnson, J. Christensen, M. I. Grossman, E. D. Jacobson, and S. G. Schultz, pp. 269-288. New York: Raven Press.

Pearse, A. G. E. 1969. The cytochemistry and ultrastructure of polypeptide hormone producing cells of the APUD series and the embryological, physiological, and pathological implications of the concept. *J. Histochem. Cytochem.* 17: 303-313.

Pehrson, A. 1983. Caecotrophy in caged mountain hares (*Lepus timidus*). *J. Zool. (Lond.)* 199: 563-574.

Pennington, R. J. 1952. The metabolism of short-chain fatty acids in the sheep. 1. Fatty acid utilization and ketone body production by rumen epithelium and other tissues. *Biochem. J.* 51: 251-258.

Pennington, R. J., and Sutherland, T. M. 1956. The metabolism of short-chain fatty acids in the sheep. 4. The pathway of propionate metabolism in rumen epithelial tissue. *Biochem. J.* 63: 618-628.

Pernkopf, E. 1937. B. Die Vergleichung der verscheidenen Formtypen des Vorderdarmes der Kranioten. In *Handbuch der Vergleichenden den Anatomie der Wirbeltiere,* Vol. 3, ed. L. Bolk, E. Göppert, E. Kallius, and W. Lubosch, pp. 477-562. Berlin: Urban and Schwarzenberg.

Pernkopf, E., and Lehner, J. 1937. A Vergleichende Beschreibung des Vorderdarmes bei den einzelnen Klassen der Kranioten. In *Handbuch der Vergleichenden den Anatomie der Wirbeltiere,* Vol. 3, ed. L. Bolk, E. Göppert, E. Kallius, and W. Lubosch, pp. 349-476. Berlin: Urban and Schwarzenberg.

Phillips, G. D. 1961. Physiological comparisons of European and Zebu steers. II. Effects of restricted water intake. *Res. Vet. Sci.* 2: 209-216.

Phillips, S. F., and Code, C. E. 1966. Sorption of potassium in the small and the large intestine. *Am. J. Physiol.* 211: 607-613.

Phillipson, A. T. 1947. The production of fatty acids in the alimentary tract of the dog. *J. Exp. Biol.* 23: 346-349.

Phillipson, A. T. 1970. Ruminant digestion. In *Dukes' Physiology of Domestic Animals,* 8th ed., ed. M. J. Swenson, pp. 424-483. Ithaca: Cornell University Press.

Phillipson, A. T. 1977. Ruminant digestion. In *Dukes' Physiology of Domestic Animals,* 9th ed., ed. M. J. Swenson, pp. 250-286. Ithaca: Cornell University Press.

Pickard, D. W., and Stevens, C. E. 1972. Digesta flow through the rabbit large intestine. *Am. J. Physiol.* 222: 1161-1166.

Piekarz, R. 1963. The influence of coprophagy on the time of passing of food down the alimentary tract in the domestic rabbit. *Acta Physiol. Pol.* 14: 337-348.

Pitts, R. F. 1968. *Physiology of the Kidney and Body Fluids,* 2nd ed. Chicago: Year Book Medical Publishers.

Ponnappa, C. G., Uddin, M. N., and Raghavan, G. V. 1971. Rate of passage of food and its relation to digestibility of nutrients in Murrah buffaloes and Hariana cattle. *Indian J. Anim. Sci.* 41: 1026-1031.

Pough, F. H. 1973. Lizard energetics and diet. *Ecology* 54: 837-844.

Powell, D. W. 1978. Transport in large intestine. In *Membrane Transport in Biology,* Vol. 4A and B: *Transport Organs,* ed. G. Giebisch, D. C. Tosteson, and H. H. Ussing, pp. 781-809. New York: Springer-Verlag.

Powell, D. W. 1987. Intestinal water and electrolyte transport. In *Physiology of the Gastrointestinal Tract,* 2nd ed., Vol. 2, ed. L. R. Johnson, J. Christensen, M. J. Jackson, E. D. Jacobson, and J. H. Walsh, pp. 1267-1306. New York: Raven Press.

Powell, D. W., and Fan, C.-C. 1984. Coupled NaCl transport: cotransport or parallel ion exchange? In *Mechanisms of Intestinal Electrolyte Transport and Regulation by Calcium,* ed. M. Donowitz and E. W. G. Sharp, pp. 13-26. New York: Alan R. Liss.

Powell, D. W., Johnson, P., Bryson, J., Orlando, R., and Fan, C.-C. 1982. Effect of phenolphthalein on monkey intestinal water and electrolyte transport. *Am. J. Physiol.* 243: G268-G275.

Powell, D. W., and Tapper, E. J. 1979. Autonomic control of intestinal electrolyte transport. In *Frontiers of Knowledge in the Diarrheal Diseases,* ed. H. J. Janowitz and D. B. Sachar, pp. 37-52. Upper Montclair, N.J.: Projects in Health.

Prahl, J. W., and Neurath, H. 1966. Pancreatic enzymes of the spiny pacific dogfish. I. Cationic chymotrypsinogen and chymotrypsin. *Biochemistry* 5: 2131-2146.

Prior, R. L., Hintz, H. F., Lowe, J. E., and Visek, W. J. 1974. Urea recycling and metabolism of ponies. *J. Anim. Sci.* 38: 565-571.

Rao, M., Dubinsky, W., Vosburgh, E., Field, M., and Frizzell, R. 1981. Sodium proton antiport in intestinal brush border vesicles of the flounder, *Pseudopleuronectes americanus. Bull. Mt. Desert Island Biol. Lab.* 21: 99-103.

Rapley, S., Lewis, W. H. P., and Harris, H. 1971. Tissue distribution, substrate specificities, and molecular sizes of human peptidases determined by separate gene loci. *Ann. Hum. Genet.* 34: 307-320.

Rask-Madsen, J., and Hjelt, K. 1977. Effect of amiloride on electrical activity and electrolyte transport in human colon. *Scand. J. Gastroenterol.* 12: 1-6.

Raven, H. C. 1950. *The Anatomy of the Gorilla.* New York: Columbia University Press.

Rawdon, B. B. 1984. Gastrointestinal hormones in birds: morphological, chemical, and developmental aspects. *J. Exp. Zool.* 232: 659-670.

Rawdon, B. B., and Andrew, A. 1981. An immunocytochemical survey of endocrine cells in the gastrointestinal tract of chicks at hatching. *Cell Tissue Res.* 220: 279-292.

Reeck, G. R., Winter, W. P., and Neurath, H. 1970. Pancreatic enzymes of the African lungfish *Protopterus aethiopicus. Biochemistry* 9: 1398-1403.

Reeder, W. G. 1964. The digestive system. In *Physiology of the Amphibia,* Vol. 1, ed. J. A. Moore, pp. 99-149. New York: Academic Press.

Reenstra, W. W., Bettencourt, J. D., and Forte, J. G. 1987. Mechanisms of active Cl⁻ secretion by frog gastric mucosa. *Am. J. Physiol.* 252: 543-547.

Regoeczi, E., Irons, L., Koj, A., and McFarlane, A. S. 1965. Isotopic studies of urea metabolism in rabbits. *Biochem. J.* 95: 521-535.

Rehfeld, J. F. 1980. Cholecystokinin. *Clin. Gastroenterol.* 9: 593-607.

Reinecke, M., Carraway, R. E., Falkmer, S., Feurle, G. E., and Forssmann, W. G. 1980. Occurrence of neurotensin-immunoreactive cells in the digestive tract of lower vertebrates and deuterostomian invertebrates. *Cell Tissue Res.* 212: 173-183.

Reinecke, M., Schlüter, P., Yanaihara, N., and Forssmann, W. G. 1981. VIP immunoreactivity in enteric nerves and endocrine cells of the vertebrate gut. *Peptides* 2 (Suppl. 2): 149-156.

Rérat, A. 1978. Digestion and absorption of carbohydrates and nitrogenous matters in the hindgut of the omnivorous nonruminant animal. *J. Anim. Sci.* 46: 1808-1837.

Richard, P.-B. 1959. La caecotrophie chez le castor du Rhône (castor-fiber). *C. R. Séances Acad. Sci.* 248: 1424-1426.

Richardson, K. C., and Creed, K. 1981. Aspects of haustral motility of the stomach of macropods. *Proc. Aust. Physiol. Pharmacol. Soc.* 12: 24P.

Richardson, K., and Wyburn, R. S. 1983. The electromyographic events in the stomach and small intestine of a small kangaroo, the tammar wallaby (*Macropus eugenii*). *J. Physiol.* (*Lond.*) 342: 453-463.

Richardson, K. C., and Wyburn, R. S. In press. Electromyography of the stomach and small intestine of the tammar wallaby (*Macropus eugenii*) and the quokka (*Setonix brachyurus*). *Aust. J. Zool.*

Ridgway, S. H. 1972. *Mammals of the Sea.* Springfield, Ill.: Charles C. Thomas.

Roberts, M. C. 1975. The development and distribution of mucosal enzymes in the small intestine of the foetus and the young foal. *J. Reprod. Fertil.* [*Suppl.*] 23: 717-723.

Robertson, D. R. 1982. Fish feces as fish food on a Pacific coral reef. *Mar. Ecol. Prog. Ser.* 7: 253-265.

Robinson, G. B. 1960. The hydrolysis of dipeptides by rat intestinal extracts. *Biochim. Biophys. Acta* 44: 386-387.

Robinson, I. M., Allison, M. J., and Bucklin, J. A. 1981. Characterization of the cecal bacteria of normal pigs. *Appl. Environ. Microbiol.* 41: 950-955.

Robinson, J. W. L. 1976. Inhibition of transport processes in the dog colon. In *Intestinal Ion Transport,* ed. J. W. L. Robinson, pp. 287-299. Lancaster: MTP Press.

Roche, M., and Ruckebusch, Y. 1978. A basic relationship between gastric and duodenal motilities in chickens. *Am. J. Physiol.* 235: E670-E677.

Rogerson, A. 1958. Diet and digestion in sections of the alimentary tract of the sheep. *Br. J. Nutr.* 12: 164-176.

Roman, C., and Gonella, J. 1981. Extrinsic control of digestive tract motility. In *Physiology of the Gastrointestinal Tract,* Vol. 1, ed. L. R. Johnson, J. Christensen, M. I. Grossman, E. D. Jacobson, and S. G. Schultz, pp. 289-334. New York: Raven Press.

Rombout, J. H. W. M., and Reinecke, M. 1984. Immunohistochemical localization of (neuro)peptide hormones in endocrine cells and nerves of the gut of a stomachless teleost fish, *Barbus conchonius* (Cyprinidae). *Cell Tissue Res.* 237: 57-65.

Romer, A. S. 1966. *Vertebrate Paleontology,* 3rd ed. Chicago: University of Chicago Press.

Rose, R., Nahrwold, D., and Koch, M. 1977. Electrical potential profile in rabbit ileum: role of rheogenic Na transport. *Am. J. Physiol.* 232: E5-E12.

Roth, J., LeRoith, D., Collier, E. S., Weaver, N. R., Watkinson, A., Cleland, C. F., and Glick, S. M. 1985. Evolutionary origins of neuropeptides, hormones, and receptors: Possible applications to immunology. *J. Immunol.* 135: 816s-819s.

Rothney, W. B. 1969. Rumination and spasmus nutans. *Hosp. Pract.* 4: 102-106.

Rouk, C. S., and Glass, B. P. 1970. Comparative gastric histology of five North and Central American bats. *J. Mammal.* 51: 455-490.

Rübsamen, K., and Engelhardt, W. v. 1978. Bicarbonate secretion and solute absorption in forestomach of the llama. *Am. J. Physiol.* 234: E1-E6.

Rübsamen, K., and Engelhardt, W. v. 1981. Absorption of Na, H ions and short-chain fatty acids from the sheep colon. *Pflügers Arch. Eur. J. Physiol.* 391: 141-146.

Ruckebusch, Y. 1971. The effects of pentagastrin on the motility of the ruminant stomach. *Experientia* 27: 1185-1186.

Ruckebusch, Y. 1975. Motility of the ruminant stomach associated with states of sleep. In *Digestion and Metabolism in the Ruminant,* ed. I. W. McDonald and A. C. I. Warner, pp. 77-90. Armidale, N.S.W.: University of New England Publishing Unit.

Ruckebusch, Y. 1981. Motor functions of the intestine. In *Advances in Veterinary Science and Comparative Medicine,* Vol.25, ed. C. E. Cornelius and C. F. Simpson, pp. 345-369. New York: Academic Press.

Ruckebusch, Y. 1986. Development of digestive motor patterns during perinatal life: mechanism and significances. *J. Pediatr. Gastroenterol.* 5: 523-536.

Ruckebusch, Y., Fargeas, J., and Dumas, J.-P. 1970. Recherches sur le comportement alimentaire des ruminants. IX. La mastication mérycique. *Rev. Méd. Vét.* 121: 345-357.

Ruckebusch, Y., and Fioramonti, J. 1976. The fusus coli of the rabbit as a pacemaker area. *Experientia* 32: 1023-1024.

Ruckebusch, Y., and Hörnicke, H. 1977. Motility of the rabbit's colon and cecotrophy. *Physiol. Behav.* 18: 871-888.

Ruckebusch, Y., and Tomov, T. 1973. The sequential contractions of the rumen associated with eructation in sheep. *J. Physiol. (Lond.)* 235: 447-458.

Ruckebusch, Y., and Vigroux, P. 1974. Etude électromyographique de la motricité du caecum chez le Cheval (*Equus caballus*). *C. R. Soc. Biol. (Paris)* 168: 887-892.

Ruppin, H., Bur-Meir, S., Soergel, K. H., Wood, C. M., and Schmitt, M. G. 1980. Absorption of short-chain fatty acids by the colon. *Gastroenterology* 78: 1500-1507.

Ryan, C. A. 1965. Chicken chymotrypsin and turkey trypsin, Part 1. Purification. *Arch. Biochem. Biophys.* 110: 169-174.

Sack, W. O., and Ballantyne, J. H. 1965. Anatomical observations on a musk-ox calf (*Ovibos moschatus*) with particular reference to thoracic and abdominal topography. *Can. J. Zool.* 43: 1033-1047.

Saffrey, M. J., Marcus, N., Jessen, K. R., and Burnstock, G. 1983. Distribution of neurons with high-affinity uptake sites for GABA in the myenteric plexus of the guinea-pig, rat and chicken. *Cell Tissue Res.* 234: 231-235.

Saffrey, M. J., Polak, J. M., and Burnstock, G. 1982. Distribution of vasoactive intestinal polypeptide-, substance P-, enkephalin-, and neurotensin-like immunoreactive nerves in the chicken gut during development. *Neuroscience* 7: 279-293.

Salanitro, J. P., Blake, I. G., Muirhead, P. A., Maglio, M., and Goodman, J.R. 1978. Bacteria isolated from the duodenum, ileum, and cecum of young chicks. *Appl. Environ. Microbiol.* 35: 782-790.

Salse, A., Crampes, F., and Raynaud, P. 1977. Measurement of N urea dietary value by intracaecal perfusion in rabbit. *Ann. Biol. Anim. Biochim. Biophys.* 17: 559-566.

Sanson, G. D. 1978. The evolution and significance of mastication in the Macropodidae. *Aust. Mamm.* 2: 23-28.

Sarna, S. K., Bardakjian, B. L., Waterfall, W. E., and Lind, J. F. 1980. Human colonic electrical control activity (ECA). *Gastroenterology* 78: 1526-1536.

Sauer, W. C., Stothers, S. C., and Parker, R. J. 1977. Apparent and true availabilities of amino acids in wheat and milling by-products for growing pigs. *Can. J. Anim. Sci.* 57: 775-784.

Savage, D. C. 1977. Interactions between the host and its microbes. In *Microbial Ecology of the Gut,* ed. R. T. J. Clarke and T. Bauchop, pp. 277-310. New York: Academic Press.

Savage, D. C. 1986. Gastrointestinal microflora in mammalian nutrition. *Annu. Rev. Nutr.* 6: 155-178.

Savory, J. C., and Hodgkiss, J. P. 1984. Influence of vagotomy in domestic fowls on feeding activity, food passage, digestibility and satiety effects of two peptides. *Physiol. Behav.* 33: 937-944.

Scapin, S., and Lambert-Gardini, S. 1979. Digestive enzymes in the exocrine pancreas of the frog *Rana esculenta. Comp. Biochem. Physiol.* 62A: 691-697.

Schaefer, A. L., Young, B. A., and Chimwano, A. M. 1978. Ration digestion and retention times in domestic cattle (*Bos taurus*), American bison (*Bison bison*) and Tibetan yak (*Bos grunniens*). *Can. J. Zool.* 56: 2355-2358.

Schalk, A. F., and Amadon, R. S. 1921. Gastric motility studies in the stomach of the goat and horse. *J. Am. Vet. Med. Assoc.* 59: 151-172.

Schalk, A. F., and Amadon, R. S. 1928. *Physiology of the Ruminant Stomach.* Bull. No. 216, North Dakota Agric. Exp. Station, Fargo.

Schmidt-Nielsen, B., Schmidt-Nielsen, K., Houpt, T. R., and Jarnum, S. A. 1957. Urea excretion in the camel. *Am. J. Physiol.* 188: 477-483.

Schmidt-Nielsen, K. 1983. Food and energy. In *Animal Physiology: Adaptation and Environment,* 3rd ed., pp. 137-176. Cambridge: Cambridge University Press.

Schneyer, L. H., and Schneyer, C. A. 1967. Inorganic composition of saliva. In *Handbook of Physiology,* Sec. 6: *Alimentary Canal,* Vol. 2: *Secretion,* ed. C. F. Code and W. Heidel, pp. 497-530. Washington, D.C.: American Physiological Society.

Schultz, S. G., Yu-tu, L., Alvarez, O. O., and Curran, P. F. 1970. Dicarboxylic amino acid influx across brush border of rabbit ileum. *J. Gen. Physiol.* 56: 621-639.

Schurg, W. A., Frei, D. L., Cheeke, P. R., and Holtan, D. W. 1977. Utilization of whole corn plant pellets by horses and rabbits. *J. Anim. Sci.* 45: 1317-1321.

Seino, Y., Porte, D., and Smith, P. H. 1979a. Immunohistochemical localization of somatostatin-containing cells in the intestinal tract: a comparative study. *Gen. Comp. Endocrinol.* 38: 229-233.

Seino, Y., Porte, D., Yanaihara, N., and Smith, P. H. 1979b. Immunocytochemical localization of motilin-containing cells in the intestines of several vertebrate species and a comparison of antisera against natural and synthetic motilin. *Gen. Comp. Endocrinol.* 38: 234-237.

Sellers, A. F., Lowe, J. E., and Brondum, J. 1979. Motor events in equine large colon. *Am. J. Physiol.* 237: E457-E464.

Sellers, A. F., Lowe, J. E., Drost, C. J., Rendano, V. T., Georgi, J. R., and Roberts, M. C. 1982. Retropulsion-propulsion in equine large colon. *Am. J. Vet. Res.* 43: 390-396.

Sellers, A. F., and Stevens, C. E. 1966. Motor functions of the ruminant forestomach. *Physiol. Rev.* 46: 634-661.

Sellers, A. F., and Titchen, D. A. 1959. Responses to localized distension of the oesophagus in decerebrate sheep. *Nature* 184: 645.

Sellin, J., and De Soignie, R. 1984. Rabbit proximal colon: a distinct transport epithelium. *Am. J. Physiol.* 246: G603-G610.

Shcherbina, M. A., and Kazlauskene, O. P. 1971. The reaction of the medium and the rate of absorption of nutrients in the intestines of the carp [*Cyprinus carpio* (L.)]. *J. Ichthyol.* 11: 81-85.

Shcherbina, M. A., Mochul'skaya, V. F., and Erman, Y. Z. 1970. Study of the digestibility of the nutrients in artificial foods by pond fishes. Communication I. Digestibility of the nutrients of peanut oil meal, peas, barley and feed mixture by two-year-old carp. *J. Ichthyol.* 10: 662-667.

Shearman, D. J. C., Taylor, P., Tyler, M. J., O'Brien, P., Laidler, P., and Seamark, R. F. 1984. An update on the role of prostaglandins in the stomach and intestine of the gastric brooding frog *Rheobatrachus silus.* In *Mechanisms of Mucosal Protection in the Upper Gastrointestinal Tract,* Ed. A. Allen et al., pp. 323-327. New York: Raven Press.

Shellenberger, P. R., and Kesler, E. M. 1961. Rate of passage of feeds through the digestive tract of Holstein cows. *J. Anim. Sci.* 20: 416-419.

Shkolnik, A., Maltz, E., and Choshniak, I. 1980. The role of the ruminants digestive tract as a water reservoir. In *Digestive Physiology and Metabolism in Ruminants,* ed. Y. Ruckebusch and P. Thivend, pp. 731-742. Lancaster, U.K.: MTP Press.

Sibbald, I. R., Sinclair, D. G., Evans, E. V., and Smith, D. L. T. 1962. The rate of passage of feed through the digestive tract of the mink. *Can. J. Biochem. Physiol.* 40: 1391-1394.

Silk, D. B. A., Grimble, G. K., and Rees, R. G. 1985. Protein digestion and amino acid and peptide absorption. *Proc. Nutr. Soc.* 44: 63-72.

Simnet, J., and Spray, G. 1961. The influence of diet on the vitamin B_{12} activity in the cecum, urine and faeces of rabbits. *Br. J. Nutr.* 15: 555-556.

Simpson, G. G. 1945. The principals of classification and a classification of mammals. *Bull. Am. Mus. Nat. Hist.* 85: 1-XVI, 1-350.

Simpson, G. G. 1950. *The Meaning of Evolution.* New York: Oxford University Press.

Sisson, S. 1975. Porcine digestive system. In *Sisson and Grossman's The Anatomy of the Domestic Animals,* 5th ed., ed. R. Getty, pp. 1268-1282. Philadelphia: W. B. Saunders.

Skadhauge, E. 1968. The cloacal storage of urine in the rooster. *Comp. Biochem. Physiol.* 24: 7-18.

Skadhauge, E. 1973. Renal and cloacal salt and water transport in the fowl (*Gallus domesticus*). *Dan. Med. Bull.* 20: 1-82.

Skadhauge, E. 1980. Water transport in the vertebrate intestine. In *Animals and Environmental Fitness,* ed. R. Giles, pp. 79-90. Elmsford, N.Y.: Pergamon Press.

Skadhauge, E., Warüi, C. N., Kamau, J. M. Z., and Maloiy, G. M. O. 1984. Function of the lower intestine and osmoregulation in the ostrich: Preliminary anatomical and physiological observations. *Quart J. Exp. Physiol.* 69: 809-818.

Sklan, D., Dubrov, D., Eisner, U., and Hurwitz, S. 1975. ^{51}CrEDTA, ^{91}Y, ^{141}Ce as nonabsorbed reference substances in the gastrointestinal tract of the chicken. *J. Nutr.* 105: 1549-1552.

Skoczylas, R. 1978. Physiology of the digestive tract. In *Biology of the Reptilia,* Vol. 8, ed. C. G. Gans and K. A. Gans, pp. 589-717. New York: Academic Press.

Slade, L. M., Bishop, R., Morris, J. G., and Robinson, D. G. 1971. Digestion and absorption of ^{15}N-labeled microbial protein in the large intestine of the horse. *Br. Vet. J.* 127: XI-XII.

Slijper, E. J. 1962. *Whales,* pp. 253-293. New York: Basic Books.

Smith, C. R., and Richmond, M. E. 1972. Factors influencing pellet egestion and gastric pH in the barn owl. *Wilson Bull.* 84: 179-183.

Smith, H. W. 1943. The evolution of the kidney. In *Lectures on the Kidney,* Porter Lectures Series No. 9. Lawrence: University of Kansas Press.

Smith, H. W. 1965. The development of the flora of the alimentary tract in young animals. *J. Pathol. Bacteriol.* 90: 495-513.

Smith, H., Farinacci, N., and Breitweiser, A. 1930. The absorption and excretion of water and salts by marine teleosts. *Am. J. Physiol.* 93: 480-505.

Smith, P. L., Cascairo, M., and Sullivan, S. 1985. Sodium dependence of luminal alkalinization by rabbit ileal mucosa. *Am. J. Physiol.* 249: G358-G368.

Smith, P. L., and McCabe, R. D. 1984. Mechanism and regulation of transcellular potassium transport by the colon. *Am. J. Physiol.* 247: G445-G456.

Smith, R. H., and McAllan, A. B. 1971. Nucleic acid metabolism in the ruminant. *Br. J. Nutr.* 25: 181-190.

Snipes, R. L. 1984. Anatomy of the cecum of the West Indian manatee, *Trichechus manatus* Mammalia, Sirenia. *Zoomorphology* 104: 67-78.

Soergel, K. H., and Hofmann, A. F. 1972. Absorption. In *Pathophysiology: Altered Regulatory Mechanisms in Disease,* ed. E.D. Frohlich, pp. 423-453. Philadelphia: J. B. Lippincott.

Sokol, O. 1965. Herbivory in lizards. *Evolution* 211: 192-194.

Solcia, E., Capella, C., Buffa, R., Usellini, L., Fiocca, R., and Sessa, F. 1981. Endocrine cells of the digestive system. In *Physiology of the Gastrointestinal Tract,* Vol. 1, ed. L. R. Johnson, J. Christensen, M. I. Grossman, E. D. Jacobson, and S. G. Schultz, pp. 39-58. New York: Raven Press.

Sorrell, M. F., Frank, O., Thomson, A. D., Aquino, H., and Baker, H. 1971. Absorption of vitamins from the large intestine in vivo. *Nutr. Rep. Int.* 3: 143-148.

Southern, N. H. 1940. Coprophagy in the wild rabbit. *Nature* 145: 262.

Southern, N. H. 1942. Periodicity of refection in the wild rabbit. *Nature* 149: 553.

Sperber, I. 1968. Physiological mechanisms in herbivores for retention and utilization of nitrogenous compounds. In *Isotope Studies on the Nitrogen Chain,* Symposium on the use of isotopes in the studies of nitrogen metabolism in the soil-plant-animal system, pp. 209-219. Vienna: International Atomic Energy Agency.

Sperber, I., Björnhag, G., and Ridderstrale, Y. 1983. Function of proximal colon in lemming and rat. *Swed. J. Agric. Res.* 13: 243-256.

Sperber, I., and Hydén, S. 1952. Transport of chloride through the ruminal mucosa. *Nature* 169: 587.

Steenbock, H., Seel, M. T., and Nelson, E. M. 1923. Vitamin B. *J. Biol. Chem.* 60: 399.

Steven, G. A. 1930. Bottom fauna and the food of fishes. *J. Mar. Biol. Assoc. U. K.* 16: 677-706.

Stevens, C. E. 1964. Transport of sodium and chloride by the isolated rumen epithelium. *Am. J. Physiol.* 206: 1099-1105.

Stevens, C. E. 1973. Transport across rumen epithelium. In *Transport Mechanisms in Epithelia,* ed. H. H. Ussing and N. A. Thorn, pp. 404-426. New York: Academic Press.

Stevens, C. E. 1977. Comparative physiology of the digestive system. In *Duke's Physiology of Domestic Animals,* ed. M. J. Swenson, 9th ed., pp. 216-232. Ithaca: Cornell University Press.

Stevens, C. E. 1980. The gastrointestinal tract of mammals: major variations. In *Comparative Physiology: Primitive Mammals,* ed. K. Schmidt-Nielsen, L. Bolis, and C. R. Taylor, pp. 52-62. Cambridge: Cambridge University Press.

Stevens, C. E. 1983. Comparative anatomy and physiology of the herbivore digestive tract. In *Proceedings of the Second Annual Dr. Scholl Conference on the Nutrition of Captive Wild Animals,* ed. T. P. Meehan, B. A. Thomas, and K. Bell, pp. 8-16. Chicago: Lincoln Park Zoological Society.

Stevens, C. E., Argenzio, R. A., and Clemens, E. T. 1980. Microbial digestion: rumen versus large intestine. In *Digestive Physiology and Metabolism in Ruminants,* ed. Y. Ruckebusch and P. Thivend, pp. 685-706. Lancaster U.K.: MTP Press.

Stevens, C. E., Argenzio, R. A., and Roberts, M. C. 1986. Comparative physiology of the mammalian colon and suggestions for animal models of human disorders. *Clin. Gastroenterol.* 15: 763-786.

Stevens, C. E., Dobson, A., and Mammano, J. H. 1969. A transepithelial pump for weak electrolytes. *Am. J. Physiol.* 216: 983-987.

Stevens, C. E., and Sellers, A. F. 1960. Pressure events in bovine esophagus and reticulorumen associated with eructation, deglutition and regurgitation. *Am. J. Physiol.* 199: 598-602.

Stevens, C. E., and Sellers, A. F. 1968. Rumination. In *Handbook of Physiology,* Sec. 6: *Alimentary Canal,* Vol. 5: *Bile; Digestion; Ruminal Physiology,* ed. C. F. Code and W. Heidel, pp. 2699-2704. Washington, D.C.: American Physiological Society.

Stevens, C. E., Sellers, A. F., and Spurrell, F. A. 1960. Function of the bovine omasum in ingesta transfer. *Am. J. Physiol.* 198: 449-455.

Stevens, C. E., and Stettler, B. K. 1966a. Factors affecting the transport of volatile fatty acids across rumen epithelium. *Am. J. Physiol.* 210: 365-372.

Stevens, C. E., and Stettler, B. K. 1966b. Transport of fatty acid mixtures across rumen epithelium. *Am. J. Physiol.* 211: 264-271.

Stevens, C. E., and Stetter, B. K. 1967. Evidence for active transport of acetate across bovine rumen epithelium. *Am. J. Physiol.* 213: 1335-1339.

Stickney, R. R., and Shumway, S. E. 1974. Occurrence of cellulose activity in the stomachs of fishes. *J. Fish Biol.* 6: 779-790.

Stoebel, D. P., and Goldner, A. M. 1975. Ion transport across the rat colon. *Physiologist* 18: 410.

Storer, R. W. 1971a. Classification of birds. In *Avian Biology,* Vol. 1, ed. D. S. Farner, J. R. King, and K. C. Parkes, pp. 1-18. New York: Academic Press.

Storer, R. W. 1971b. Adaptive radiation of birds. In *Avian Biology,* Vol.1, ed. D. S. Farner, J. R. King, and K. C. Parkes, pp. 150-188. New York: Academic Press.

Strauss, E. W., and Ito, S. 1969. A fine structural study of lipid-uptake in the terminal intestine of the scup, *Stenotomus chrysops. Biol. Bull.* 137: 414-415.

Stroband, H. W. J., and Debets, F. M. H. 1978. The ultrastructure and renewal of the intestinal epithelium of the juvenile grass carp, *Ctenopharyngodon idella* (Val.). *Cell Tissue Res.* 187: 181-200.

Sturkie, P. D. 1965. *Avian Physiology,* 2nd ed. Ithaca: Cornell University Press.

Sturkie, P. D. 1970. Avian digestion. In *Dukes' Physiology of Domestic Animals,* 8th ed., ed. M. J. Swenson, pp. 526-537. Ithaca: Cornell University Press.

Sundler, F., Alumets, J., Håkanson, R., Carraway, R. E., and Leeman, S. E. 1977a. Ultrastructure of the gut neurotensin cell. *Histochemistry* 53: 25-34.

Sundler, F., Håkanson, R., Hammer, R. A., Alumets, J., Carraway, R. E., Leeman, S. E., and Zimmerman, E. A. 1977b. Immunohistochemical localization of neurotensin in endocrine cells of the gut. *Cell Tissue Res.* 178: 313-322.

Svendsen, P. 1969. Etiology and pathogenesis of abomasal displacement in cattle. *Nord. Vet. Med.* 21 (Suppl. 1).

Symons, L. E. A., and Jones, W. O. 1966. The distribution of dipeptidase activity in the small intestine of the sheep (*Ovis aries*). *Comp. Biochem. Physiol.* 18: 71-82.

Szarski, H. 1962. Some remarks on herbivorous lizards. *Evolution* 16: 529.

Takahashi, S., and Seifter, S. 1974. An enzyme with collagenolytic activity from dog pancreatic juice. *Israel J. Chem.* 12: 557-571.

Tanaka, H. 1942. Chemie der Meerschildkröten. II. Über die fermente der meerschildkrötenorgane. *J. Biochem.* (*Tokyo*) 36: 301-312.

Tansy, M. F., Kendall, F. M., and Murphy, J. J. 1972. The reflex nature of the gastrocolic propulsive response in the dog. *Surg. Gynecol. Obstet.* 135: 404-410.

Taverner, M. R., Hume, I. D., and Farrell, D. J. 1981a. Availability to pigs of amino acids in cereal grains. 1. Endogenous levels of amino acids in ileal digesta and faeces of pigs given cereal diets. *Br. J. Nutr.* 46: 149-158.

Taverner, M. R., Hume, I. D., and Farrell, D. J. 1981b. Availability to pigs of amino acids in cereal grains. 2. Apparent and true ileal availability. *Br. J. Nutr.* 46: 159-171.

Taylor, R. B. 1962. Pancreatic secretion in the sheep. *Res. Vet. Sci.* 3: 63-77.

Teather, R. M., Erfle, J. D., Boila, R. J., and Sauer, F. D. 1980. Effect of dietary nitrogen on the rumen microbial population in lactating dairy cattle. *J. Appl. Bacteriol.* 49: 231-238.

Thacker, E. J., and Brandt, C. S. 1955. Coprophagy in the rabbit. *J. Nutr.* 55: 375.

Thaysen, J. H., Thorn, N. A., and Schwartz, I. L. 1954. Excretion of sodium, potassium, chloride and carbon dioxide in human parotid saliva. *Am. J. Physiol.* 178: 155-159.

Thewis, A., Francois, E., Debouche, C., and Thielemans, M.-F. 1976. Utilisation des radiolanthides dans la détermination du transit gastro-intestinal chez les petits ruminants. Comparaison des techniques directe (abbatage) et indirecte. *Ann. Zootechnol.* 25: 373-385.

Thewis, A., Francois, E., and Thill, N. 1975. Le transit gastrointestinal chez le ruminant mesuré à l'aide des radiolanthides. Signification des paramètres de la phase exponentielle des courbes de concentration fécale en traceur. *Bull. Recherches Agron. Gembloux* 10: 307-320.

Thompson, R. C., and Hollis, O. L. 1958. Irradiation of the gastrointestinal tract of the rat by ingested ruthenium-106. *Am. J. Physiol.* 194: 308-312.

Thurston, J. P., Noirot-Timothée, C., and Arman, P. 1968. Fermentative digestion in the stomach of *Hippopotamus amphibius* (Artiodactyla: Suiformes), and associated ciliate protozoa. *Nature* 218: 882-883.

Timson, C. M., Polak, J. M., Wharton, J., Ghatei, M. A., Bloom, S. R., Usellini, L., Capella, C., Solicia, E., Brown, M. R., and Pearse, A. G. E. 1979. Bombesin-like immunoreactivity in the avian gut and its localisation to a distinct cell type. *Histochemistry* 61: 213-222.

Tindall, A. R. 1979. The innervation of the hind gut of the domestic fowl. *Br. Poult. Sci.* 20: 473-480.

Titchen, D. A. 1968. Nervous control of motility of the forestomach of ruminants. In *Handbook of Physiology, Sec. 6: Alimentary Canal,* Vol. 5: *Bile; Digestion; Ruminal Physiology,* ed. C. F. Code and W. Heidel, pp. 2705-2724. Washington, D.C.: American Physiological Society.

Titchen, D. A. 1986. Gastrointestinal peptide hormone distribution, release and action in ruminants. In *Control of Digestion and Metabolism in Ruminants,* ed. L. P. Milligan, W. L. Grovum, and A. Dobson, pp. 227-248. Engelwood Cliffs, N.J.: Prentice-Hall.

Tobey, N., Heizer, W., Yek, R., Huang, T., and Hoffner, C. 1985. Human intestinal brush border peptidases. *Gastroenterology* 88: 913-926.

Torrey, T. W. 1971. *Morphogenesis of the Vertebrates,* 3rd ed., pp. 44-45. New York: John Wiley.

Troyer, K. 1984. Behavioral acquisition of the hindgut fermentation system by hatchling *Iguana iguana. Behav. Ecol. Sociobiol.* 14: 189-193.

Turcek, F. J. 1963. Beitrag zur Ökologie des Ziesels (*Citellus citellus* L.) II. *Biológia (Bratisl.)* 18: 419-432.

Turnberg, L. A. 1971. Potassium transport in the human small bowel. *Gut* 12: 811-818.

Turnberg, L. A., Bieberdorf, F. A., Morawski, S. G., and Fordtran, J. S. 1970. Interrelationships of chloride, bicarbonate, sodium and hydrogen transport in the human ileum. *J. Clin. Invest.* 49: 557-567.

Tuttle, M. D. 1987. Bats. In *Wild Animals of North America,* ed. R. W. Nowak. Washington, D.C.: National Geographic Society.

Uden, P. 1978. Comparative studies on rate of passage, particle size and rate of digestion in ruminants, equines, rabbits and man. Ph.D. thesis, Cornell University, Ithaca, N.Y.

Umesaki, Y., Yajima, T., Yokokura, T., and Mutai, M. 1979. Effect of organic acid absorption on bicarbonate transport in rat colon. *Pflügers Arch. Eur. J. Physiol.* 379: 43-47.

Usellini, L., Buchan, A. M. J., Polak, J. M., Capella, C., Cornaggia, M., and Solcia, E. 1984. Ultrastructural localization of motilin in endocrine cells of human and dog intestine by the immunogold technique. *Histochemistry* 81: 363-368.

Ushiyama, H., Fujimori, T., Shibata, T., and Yoshimura, K. 1965. Studies on carbohydrases in the pyloric caeca of the salmon, *Oncorhynchus keta. Bull. Fac. Fish Hokkaido Univ.* 16: 183-188.

Ussing, H. H. 1975. Epithelial transport phenomena. In *Intestinal Absorption and Malabsorption,* ed. T. Z. Csaky, pp. 1-7. New York: Raven Press.

Utley, P. R., Bradley, N. W., and Boling, J. A. 1970. Effect of water restriction on nitrogen metabolism in bovine fed two levels of nitrogen. *J. Nutr.* 100: 551-556.

Vadivel, G., and Leibach, F. H. 1985. Is intestinal peptide transport energized by a proton gradient? *Am. J. Physiol.* 249: G153-G160.

Vaillant, C., Dimaline, R., and Dockray, G. J. 1980. The distribution and cellular origin of vasoactive intestinal polypeptide in the avian gastrointestinal tract and pancreas. *Cell Tissue Res.* 211: 511-523.

Vaillant, C., Dockray, G. J., and Walsh, J. H. 1979. The avian proventriculus is an abundant source of endocrine cells with bombesin-like immunoreactivity. *Histochemistry* 64: 307-314.

Vallenas, A. P., Cummings, J. F., and Munnell, J. F. 1971. A gross study of the compartmentalized stomach of two New-World camelids, the llama and guanaco. *J. Morphol.* 134: 399-423.

Vallenas, A. P., and Stevens, C. E. 1971a. Motility of the llama and guanaco stomach. *Am. J. Physiol.* 220: 275-282.

Vallenas, A. P., and Stevens, C. E. 1971b. Volatile fatty acid concentrations and pH of llama and guanaco forestomach digesta. *Cornell Vet.* 61: 239-252.

Vanderhaeghen, J. J., Signeau, J. C., and Gept, W. 1975. New peptide in the vertebrate CNS reacting with antigastrin antibodies. *Nature* 257: 604-605.

Vander Noot, G. W., Symons, L. D., Lydman, R. K., and Fonnesbeck, P. V. 1967. Rate of passage of various feedstuffs through the digestive tract of horses. *J. Anim. Sci.* 26: 1309-1311.

van Lawick-Goodall, J. 1968. The behaviour of free-living chimpanzees in the Gombe stream reserve. *Anim. Behav. Monogr.* 1: 161-311.

van Lennep, E. W., and Lanzing, W. J. R. 1967. The ultrastructure of glandular cells in the external dendritic organ of some marine catfish. *J. Ultrastruct. Res.* 18: 333-344.

van Lennep, E. W., and Young, J. A. 1979. Part II: Salt glands. In *Membrane Transport in Biology,* Vol. 4B: *Transport Organs,* ed. G. Giebisch, D. C. Tosteson, H. H. Ussing, and M. T. Tosteson, pp. 675-692. New York: Springer-Verlag.

Van Noorden, S., and Polak, J. M. 1979. Hormones of the alimentary tract. In *Hormones and Evolution,* Vol. 2, ed. E. J. W. Barrington, pp. 791-828. New York: Academic Press.

Van Soest, P. J. 1982. *Nutritional Ecology of the Ruminant.* Corvallis, Ore.: O & B Books.

van Weel, P. B. 1974. "Hepatopancreas"? *Comp. Biochem. Physiol.* 47A: 1-9.

Varga, F. 1976. Transit time changes with age in the gastrointestinal tract of the rat. *Digestion* 14: 319-324.

Vaughan, T. A. 1986. *Mammology.* Philadelphia: W. B. Saunders.

Vercellotti, J. R., Salyers, A. A., and Wilkins, T. D. 1978. Complex carbohydrate breakdown in the human colon. *Am. J. Clin. Nutr.* 31: S86-S89.

Vigna, S. R. 1979. Distinction between cholecystokinin-like and gastrin-like biological activities extracted from gastrointestinal tissues of some lower vertebrates. *Gen. Comp. Endocrinol.* 39: 512-520.

Vigna, S. R. 1984. Radioreceptor and biological characterization of cholecystokinin and gastrin in the chicken. *Am. J. Physiol.* 246: G296-G304.

Vigna, S. R., Fischer, B. L., Morgan, J. L. M., and Rosenquist, G. L. 1985. Distribution and molecular heterogeneity of cholecystokinin-like immunoreactive peptides in the brain and gut of the rainbow trout, *Salmo gairdneri. Comp. Biochem. Physiol.* 82C: 143-146.

Vigna, S. R., and Gorbman, A. 1977. Effects of cholecystokinin, gastrin, and related peptides on coho salmon gallbladder contraction in vitro. *Am. J. Physiol.* 232: E485-E491.

Vigna, S. R., and Gorbman, A. 1979. Stimulation of intestinal lipase secretion by porcine cholecystokinin in the hagfish, *Eptatretus stouti. Gen. Comp. Endocrinol.* 38: 356-359.

von Euler, U. S., and Gaddum, J. H. 1931. An unidentified depresser substance in certain tissue extracts. *J. Physiol. (Lond.)* 72: 74-87.

Vonk, H. J. 1927. Die Verdauuing bei den Fischen. *Z. Vergl. Physiol.* 5:445-546.

Vonk, H. J., and Western, J. R. H. 1984. *Comparative Biochemistry and Physiology of Enzymatic Digestion.* New York: Academic Press.

Walsh, J. H. 1981. Gastrointestinal hormones and peptides. In *Physiology of the Gastrointestinal Tract,* Vol. 1, ed. L. R. Johnson, J. Christensen, M. I. Grossman, E. D. Jacobson, and S. G. Schultz, pp. 59-144. New York: Raven Press.

Walter, W. G. 1939. Bedingte Magensaftsekretion bei der ente. *Acta Brev. Neerl. Physiol.* 9: 56-57.

Waring, H., Moir, R. J., and Tyndale-Biscoe, C. H. 1966. Comparative physiology of marsupials. *Adv. Comp. Physiol. Biochem.* 2: 237-376.

Warner, A. C. I. 1981. Rate of passage of digesta through the gut of mammals and birds. *Nutr. Abstr. Rev. [B]* 51: 789-820.

Washburn, L. E., and Brody, S. 1937. *Growth and Development, with Special Reference to Domestic Animals. XLII. Methane, Hydrogen, and Carbon Dioxide Production in the Digestive Tract of Ruminants in Relation to Respiratory Exchange.* Mo. Agric. Exp. Stat. Bull. No. 263.

Watari, N. 1968. Fine structure of nervous elements in the pancreas of some vertebrates. *Z. Zellforsch. Mikrosk. Anat.* 85: 291-314.

Watson, J. S., and Taylor, R. H. 1955. Reingestion in the hare, *Lepus europaeus* Pal. *Science* 121: 314.

Webb, K. E. 1986. Amino acid and peptide absorption from the gastrointestinal tract. *Fed. Proc.* 45: 2268-2271.

Weisbrodt, N. W. 1981. Motility of the small intestine. In *Physiology of the Gastrointestinal Tract,* Vol. 1, ed. L. R. Johnson, J. Christensen, M. I. Grossman, E. D. Jacobson, and S. G. Schultz, pp. 411-443. New York: Raven Press.

Wellard, G. A., and Hume, I. D. 1981. Digestion and digesta passage in the brushtail possum, *Trichosurus vulpecula* (Kerr). *Aust. J. Zool.* 29: 157-166.

Welsh, M., Smith, P., Fromm, M., and Frizzell, R. 1982. Crypts are the site of intestinal fluid and electrolyte secretion. *Science* 218: 1219-1221.

Wester, J. 1926. *Die Physiologie und Pathologie der Vormägen beim Rinde.* Berlin: Richard Schoetz.

Westra, R., and Christopherson, R. J. 1976. Effects of cold on digestibility, retention time of digesta, reticulum motility and thyroid hormones in sheep. *Can. J. Anim. Sci.* 56: 699-708.

Wharton, C. H. 1950. Notes on the life history of the flying lemur. *J. Mammal.* 31: 269-273.

White, J. 1982. Intestinal electrogenic HCO_3^- absorption localized to villus epithelium. *Biochim. Biophys. Acta* 687: 343-345.

White, J. 1985. Omeprazole inhibits H+ secretion by *Amphiuma* jejunum. *Am. J. Physiol.* 248: G256-G259.

White, J., and Imon, M. 1981. Bicarbonate absorption by in vitro amphibian small intestine. *Am. J. Physiol.* 241: G389-G396.

White, J., and Imon, M. 1983. A role for basolateral anion exchange in active jejunal absorption of HCO_3^-. *Am. J. Physiol.* 244: G397-G405.

Whiteside, C. H., and Prescott, J. M. 1962. Activities of chicken pancreatic proteinases toward synthetic substrates. *Proc. Soc. Exp. Biol. Med.* 110: 741-744.

Wiesbrodt, N. W. 1981. Motility of the small intestine. In *Physiology of the Gastrointestinal Tract,* 2nd ed., Vol. 2, ed. L. R. Johnson, J. Christensen, M. J. Jackson, E. D. Jacobson, and J. H. Walsh, pp. 631-664. New York: Raven Press.

Wigglesworth, V. B. 1972. *The Principles of Insect Physiology,* 7th ed. London: Chapman and Hall.

Wigglesworth, V. B. 1984. *Insect Physiology,* 8th ed. New York: Chapman and Hall.

Williams, V. J. 1963. Rumen function in the camel. *Nature* 197: 1221.

Wolin, M. J. 1979. The rumen fermentation: a model for microbial interactions in anaerobic ecosystems. *Adv. Microb. Ecol.* 3: 49-77.

Wolin, M. J. 1981. Fermentation in the rumen and human large intestine. *Science* 213: 1463-1468.

Wolter, R., Durix, A., and Letourneau, J.-C. 1974. Influence du mode de présentation du fourrage sur la vitesse du transit digestif chez le poney. *Ann. Zootech.* 23: 293-300.

Wolter, R., Durix, A., and Letourneau, J.-C. 1976. Influence du mode de présentation d'un aliment complet sur la vitesse du transit et la digestibilité chez le poney. *Ann. Zootech.* 25: 181-188.

Wood, J. D. 1981. Physiology of the enteric nervous system. In *Physiology of the Gastrointestinal Tract,* Vol. 1, ed. L. R. Johnson, J. Christensen, M. I. Grossman, E. D. Jacobson, and S. G. Schultz, pp. 1-38. New York: Raven Press.

Wootton, J. F., and Argenzio, R. A. 1975. Nitrogen utilization within equine large intestine. *Am. J. Physiol.* 229: 1062-1067.

Wostmann, B. S., and Knight, P. L. 1961. Synthesis of thiamine in the digestive tract of the rat. *J. Nutr.* 74: 103-110.

Wostmann, B. S., Knight, P. L., and Reyniers, J. A. 1958. The influence of orally administered penicillin upon growth and liver thiamine of growing germ-free and normal stock rats fed a thiamine deficient diet. *J. Nutr.* 66: 577-586.

Wright, R. D., Florey, H. W., and Sanders, A. G. 1957. Observations on the gastric mucosa of Reptilia. *Q. J. Exp. Physiol.* 42: 1-14.

Wrong, O. M., Edmonds, L. J., and Chadwick, V. S. 1981. *The Large Intestine.* Lancaster, U.K.: MTP Press.

Wrong, O. M., and Vince, A. J. 1984. Urea and ammonia metabolism in the human large intestine. *Proc. Nutr. Soc.* 43: 77-86.

Wrong, O. M., Vince, A. J., and Waterlow, J. C. 1985. The contribution of endogenous urea to faecal ammonia in man, determined by ^{15}N labelling of plasma urea. *Clin. Sci.* 68: 193-199.

Wurth, M. A., and Musacchia, X. J. 1964. Renewal of intestinal epithelium in the freshwater turtle, *Chrysemys picta. Anat. Rec.* 148: 427-439.

Yamamoto, M., and Hirano, T. 1978. Morphological changes in the esophageal epithelium of the eel, *Anguilla japonica,* during adaptation to seawater. *Cell Tissue Res.* 192: 25-38.

Yoshida, T., Pleasants, J. R., Reddy, B. S., and Wostmann, B. S. 1968. Efficiency of digestion in germ-free and conventional rabbits. *Br. J. Nutr.* 22: 723-737.

Youson, J. H., and Horbert, W. R. 1982. Transformation of the intestinal epithelium of the larval anadromous sea lamprey *Petromyzon marinus* L. during metamorphosis. *J. Morphol.* 171: 89-117.

Youson, J. H., and Langille, R. M. 1981. Proliferation and renewal of the epithelium in the intestine of young adult anadromous sea lampreys, *Petromyzon marinus* L. *Can. J. Zool.* 59: 2341-2349.

Zendzian, E. N., and Barnard, E. A. 1967a. Distributions of pancreatic ribonuclease, chymotrypsin, and trypsin in vertebrates. *Arch. Biochem. Biophys.* 122: 699-713.

Zendzian, E. N., and Barnard, E. A. 1967b. Reactivity evidence for homologies in pancreatic enzymes. *Arch. Biochem. Biophys.* 122: 714-720.

Ziswiler, V., and Farner, D. S. 1972. Digestion and the digestive system. In *Avian Biology,* Vol. 2, ed. D. S. Farner, J. R. King, and K. C. Parkes, pp. 343-430. New York: Academic Press.

Zoppi, G., and Schmerling, D. H. 1969. Intestinal disaccharidase activities in some birds, reptiles and mammals. *Comp. Biochem. Physiol.* 29: 289-294.

Index

aardvark, 8t, 46
aardwolf, 52
absorption, 1, 2, 3, 4, 5, 6, 13, 14, 16, 17,
    20, 21, 28, 48
    amino acids, 28, 152, 155, 158, 180-1,
        205, 206
    ammonia, 177f, 210
    bicarbonate, 191, 208
    bile salts, 144
    chloride, 175, 176f, 192-3, 201, 202, 203,
        206, 208, 215, 219, 225, 239
    galactose, 140, 155, 205
    glucosamine, 140
    glucose, 28, 140, 155, 158, 205
    glycerol, 143-4
    hydrogen, 214
    immunoglobulins, 153, 154t
    mannose, 140
    peptides, 152-3, 155, 206
    phospholipids, 143-4
    potassium, 208, 209f, 210
    protein, 153
    sodium, 153, 172, 174, 175, 175t, 176f,
        192-3, 201, 202, 204, 205, 206, 208,
        215, 219, 225, 239
    triglycerides, 142-3
    volatile fatty acids, 162, 171-6, 176f, 180,
        208, 210, 215, 219
    water, 172, 175t, 178, 179f, 193, 202,
        205, 215, 219, 238
    xylose, 140, 155
albacore, 145
algae, 5, 24, 35, 81, 128, 130, 131t, 141,
    159, 170
alligators, 8t, 31, 33, 229, 233
alpacas, 75
amphioxus, 6
Amphiuma, 148, 150t, 157t, 208
amylopectin, *see* starch
amylose, *see* starch
anchovys, 143
annelids, 3, 5, 6, 131f
anteaters, 8t, 9, 42, 57, 196, 239
    aardvarks, 46
    scaly, 44, 46f

South American giant, 57
spiny (*see also* echidna), 44, 145
Tamandua, 57
two-toed, 57
antelope (*see also* dik-diks; suni), 76
Aotus, 66, 68f, 138t
apes, 8t, 65, 67, 70, 85
archer fish, 23
Arctocephalus, 139t, 142
Arenicola marina, 3
armadillos, 8t, 57, 58f, 59f, 156t
arthropods (*see also* crustaceans; insects), 2,
    3, 4-5, 6, 131f
Athene, 145
aye-aye, 65

babirusa (*see also Babyrousa babyrussa;*
    pig deer), 73, 164
baboons (*see also Papio*), 64f, 66, 85, 166,
    168
Babyrousa babyrussa, 71f
bacteria, 5, 30, 122, 123, 128, 148, 154,
    155, 159, 161, 161t, 162, 164, 165,
    166, 167, 170, 176, 184t
bandicoots, 55, 56f, 133, 136t, 137
barbs, 23
barracuda, 148, 151t, 157t
basal electrical rhythm (BER), *see* myoelec-
    tric activity
basking sharks, 24
bass, 28, 148, 149
bats, 8t
    anatomy of, 10, 11, 12, 42, 49f, 50, 51f,
        88, 89, 196
    digestion (endogenous enzymes) in, 134t,
        137, 138t
beak, 33, 34
bears (*see also Ursus maritimus*), 8t, 52
beavers, 62f, 63, 182, 183, 196
Bettongia, 187t, 188
bilbies, 55
bile, 141-2, 204, 215, 218, 226, 230, 231,
    232f
bill, *see* beak
bison, 113t, 156t

blackbirds, 132, 135t
boa constrictors, 33
body fluid compartments, 191-2, 193f, 216-7
Bos (*see also* cattle), 71f, 112-13t
Bos grunniens, 113t
Bos indicus, 113t
Bos taurus, 71f, 112t
bovine (*see also Bos taurus;* cattle; ox), 77, 94, 97, 149, 165, 203f
bowfin, 8t, 23, 145, 148
Brunner's glands, 16, 44, 51f
brush border, 3, 14, 15, 18, 133, 137, 140, 149, 152, 153, 155
budgerigar (*see also* parakeets), 38, 39
buffalo (*see also* bison), 76, 112t
Bufo americanus, 32f, 157t
Bufo marinus, 135t, 157t
bush babies
  anatomy of, 64f, 65
  digesta transit in, 108, 109, 109f, 114
  microbial fermentation and metabolism in, 166, 167f, 168
Buteo, 145
Buteo buteo, 36
Buteo jamaicensis, 37f
buzzards, 36

Caiman crocodilus, 32f, 117t, 156t, 170
caimans, 8t, 31, 32f, 116, 156t, 170
calves, 103, 103f, 137, 153
camels (*see also* dromedary), 12, 70, 74, 75, 79, 164, 177
capuchin monkey (*see also Cebus*), 66, 67
capybara, 62f, 63, 182, 188
carp (*see also Cyprinus carpio*), 23, 30, 148, 149, 169, 170, 229
catfish, 28, 170, 192, 229
cats (*see also* panther; *Panthera Leo*), 8t
  absorption in, 153, 154t
  anatomy of, 10, 11, 52, 53f, 88, 89, 90
  coprophagy in, 182
  digesta transit in, 111t
  digestion (endogenous enzymes) in, 133, 134t, 138t, 145, 154t
  electrolyte secretion and absorption in, 196, 198, 200, 201, 202, 214, 215
  motility in, 102f, 104, 105, 106f
  neurohumoral control in, 225, 240
cattle (*see also Bos;* ox), 8t
  absorption in, 153
  anatomy of, 11, 12, 76, 77, 78
  digesta transit in, 94-7, 112t, 115, 116, 118
  digestion (endogenous enzymes) in, 132, 133, 142, 145, 149
  electrolyte secretion and absorption in,

198, 202, 203
microbial fermentation and metabolism in, 155, 160, 161, 163, 164, 165, 168, 169, 175, 181
motility in, 90, 92, 94, 97, 99, 103f
neurohumoral control in, 99, 101, 229, 233, 236
Cavia procellus, 61f, 138t
Cebus, 66,134t, 138t
cecotrophy, 182, 183, 184t, 185, 189
cellulose, 128, 130f, 131, 160, 162, 164, 168, 186
Cervus, 112t, 156t
chameleons, 34f
chelonians (*see also* terrapins; tortoises; turtles), 8t, 9, 10, 31, 33
Chelydra, 148, 150t, 156t
chevrotains, 71, 71f, 74, 74f
chickens
  anatomy of, 36, 37f, 88
  digesta transit in, 113t, 116
  digestion (endogenous enzymes) in, 132, 135t, 137, 142, 145, 148, 149, 151t, 156t
  microbial fermentation and metabolism in, 166, 170, 178
  motility in, 102, 107
  neurohumoral control in, 225, 228, 229, 230, 232, 233, 235t
Chimaera, 23, 148, 149
chimaeras (*see also Chimaera*), 8t, 132, 137, 145, 149, 152
chimpanzees, 67, 69f, 167, 225
chinchilla, 61f, 63, 123, 182
chitin, 126t, 128, 130f, 131, 131t, 166
cholesterol, 126t, 144, 184t
Chrysemys, 148
Chrysemys picta, 117t, 150t, 157t, 170
Chrysemys scripta, 16
chubs, 23, 25f
chuckwalla, 158
civits, 52
cod, 28, 145, 149, 223
coelacanths, 192
coelenterates, 3, 131f, 237
colobus monkeys (*see also Procolobus*), 66, 68f, 164, 165, 189
colugos, *see* flying lemurs
congo eels (*see also Amphiuma*), 8t, 30, 150t, 157t, 208
conies (*see also* hyrax; rabbits), 8t
Cottus, 152
cows (*see also Bos taurus;* cattle; ox), 151t, 156t, 163f, 198, 202
cranes, 137
Cricetus cricetus, 61f, 63f, 134t, 138t
crocodiles, 8t, 31, 33, 229, 230

crop, 12, 36, 238
Crossopterygii (*see also* coelacanths), 8t,
  20, 23, 28
crows, 135t
crustaceans, 4-5, 141, 195
crypts of Lieberkühn, 15, 15f, 16, 205
cuscus, 55
cyclostomes (*see also* hagfish; lamprey), 6
  anatomy of, 9, 16, 19, 23, 24
  digestion (endogenous enzymes) in, 132,
    145, 149, 155
  neurohumoral control in, 225, 226, 228,
    230, 231, 233, 237, 238
Cyprinus carpio, 137, 148, 229

Dasypus, 59f
Dasypus novemcinctus, 57, 156t
Dasypus sabanicola, 58f
Dasypus septemcinctus, 57
deer (*see also Cervus*), 17, 76, 112t, 132,
  168
defecation, 182, 220
deglutition, 9, 87-8, 91f, 95f, 98f, 220
dik-diks, 78, 79f
dingoes, 182
dinosaurs, 33, 35, 185, 188, 189
dogfish (*see also Squalus*), 88, 143, 145,
  148, 149, 151t, 157t, 229, 231
dogs, 8t
  absorption in, 153, 154t, 172, 173t, 174,
    174t, 206
  anatomy of, 10, 11, 12, 13f, 18f, 52, 53f,
    88, 89, 90
  coprophagy in, 182
  digesta transit in, 108, 109, 109f, 111t,
    114
  digestion (endogenous enzymes) in, 132,
    134t, 138t, 141, 145, 149, 150t, 157t
  electrolytes secretion and absorption in,
    196, 198, 199f, 200, 201, 202, 205,
    206, 210, 211, 211f, 214, 215
  microbial fermentation and metabolism
    in, 161, 166, 167f, 168, 169, 178
  motility in, 87, 92, 93, 98, 101, 103, 104,
    106
  neurohumoral control in, 221, 225, 230,
    231, 233, 240
dolphins, 8t, 45f, 46, 47, 157t
donkeys, 121
dormouse, 63
dromedary (*see also* camels), 75, 76
ducks, 38, 225, 229
dugongs, 8t, 81, 82, 84, 84f, 165, 168

echidna, 8t, 44, 45f, 132, 133, 136t, 137
echinoderms, 3, 5
eels, 24, 27f, 192

elephants (*see also Elephas maximus*), 8t,
  11, 81, 83f, 111t, 137, 157t, 166, 167,
  168, 188
elephant shrews, 8t, 48, 49f
Elephas maximus, 8t, 111t, 139t
elk (*see also Cervus*), 76, 156t
emesis, 1, 90, 92
Enophrys, 152
enzymes, digestive, 1, 2, 3, 4, 9, 16, 30, 31,
  40, 125, 126-7, 128, 143, 155, 158,
  160, 238, 239
  amylase, 36, 126t, 132-3, 133f, 182
  carboxypeptidases, 4, 127t, 148, 149, 155
  cellobiase, 133, 136t, 137, 138-9t
  cellulase, 170
  chitinase, 126t, 131, 132, 134-5t, 140, 155
  chitobiase, 126t, 133, 137, 140
  cholesterol esterase, 126t, 143, 155
  chymosin, 144, 145
  chymotrypsin, 127t, 144, 148, 150-1t
  colaginase, 144, 148, 149
  colipase, 126t, 142, 143
  deaminase, 127t, 154
  elastase, 127t, 144, 148-9, 155
  enterokinase, 148
  esterases, 126t, 141, 143, 155
  lactase, 126t, 133, 136t, 137, 138-9t, 155,
    240
  lipases, 126t, 141-3
  maltase, isomaltase, 126t, 133, 133f, 136t,
    137, 138-9t, 155
  nucleotidase, 127t, 154
  pepsin, 10, 11-12, 21, 36, 127t, 144, 145,
    155, 199, 201, 225, 229, 231, 233, 238
  peptidases, 4, 127t, 152, 155
  phosphatase, 126t
  phospholipase, 126t, 143, 155
  phosphorylase, 154
  rennin, 144, 145
  ribonuclease, 127t, 148, 150-1t, 154-5,
    156-7t
  sucrase, 126t, 133, 136t, 137, 138-9t, 155
  trehalase, 126t, 133, 136t, 137, 138-9t,
    155
  trypsin, 127t, 144, 145, 148, 150-1t, 153
Equus, 82f
Equus burchelli, 79f
Equus caballus, 79f, 112t, 167f
eructation, 91f, 92, 95f, 97, 98f, 99, 162
esophagus, 10, 30, 36, 39, 59, 73f, 74, 74f,
  75f, 84f, 85, 238
  absorption by, 192-3, 239
  anatomy of, 10-11, 24, 26f, 27f, 30, 31,
    33, 47, 86, 88, 196-7
  motility in, 87-8, 89-92, 95, 97, 123
  neurohumoral control in, 221, 229
  secretion by, 192

Eumetopias jubatus, 139t
euro, *see Macropus robustus erubescens*

Falco, 36, 145
ferrets, 11, 52, 88, 89, 90, 104
flamingos, 35
flounder, 192-3, 209f
flying lemurs, 8t, 80
foals, 103, 182
foxes, 52, 88, 134t
frogs (*see also Rana* spp.), 8t, 9, 10
  absorption in, 140, 153, 194
  anatomy of, 10, 14, 17, 30, 31, 88
  body fluids of, 192
  digestion (endogenous enzymes) in, 135t,
    137, 143, 145, 150t, 151t, 157t
  electrolyte secretion in, 14, 200, 240
  microbial fermentation in, 170
  neurohumoral control in, 229, 230, 231,
    233, 239
fungi, 131f, 159, 164, 236f

galagos, 65
gallbladder, 17, 27f, 204, 218, 220, 221,
  230, 232f, 236, 240
gar, 8t, 23
gases
  carbon dioxide, 160, 162, 163, 163f, 165,
    169, 175, 176f
  hydrogen, 160, 162, 163f, 165, 169
  methane, 160, 162, 163f, 165, 169
  nitrogen, 162, 163, 163f, 169
  oxygen, 159, 162, 163f, 169
gastric epithelium
  cardiac glandular, 10, 12, 13f, 44, 46f, 51f,
    59f, 60, 63f, 71f, 73, 74, 75, 77, 78, 80,
    81, 82, 82f, 85, 201, 208
  proper gastric glandular, 10, 12, 13f, 44,
    46f, 51f, 55, 57, 59f, 60, 63f, 71f, 73, 74,
    75, 80, 81, 82, 82f, 85, 199, 200
  pyloric glandular, 12, 13f, 46f, 47, 51f, 57,
    59f, 63f, 71f, 73, 74, 75, 80, 81, 82, 82f
  stratified squamous, 12, 13f, 44, 46f, 51f,
    55, 57, 59, 59f, 63, 63f, 66, 67, 71f, 73,
    74, 75, 78, 81, 82f, 84, 85, 160, 202,
    239
gastroesophageal junction, 88-90
gavials, 8t, 31
geese, 36, 37f, 38, 113t, 123, 132, 170, 225
genets, 52
gerbils, 225
gibbons, 65, 67
giraffes, 76
gizzard (*see also* ventriculus), 12, 14, 35,
  36, 47, 57, 145, 229
gliders, 55, 58f, 168
glycogen, 126t, 130, 131, 132

glycolipids, 140
goats
  absorption in, 153, 154t, 172, 174, 174t,
    202
  anatomy of, 71f, 76, 77
  digesta transit in, 113t, 115
  digestion (endogenous enzymes) in, 132,
    139t, 151t, 156t
  microbial fermentation and metabolism
    in, 178
  motility in, 96, 103f
goldfish, 148
gorillas, 67, 167
grebs, 38
groupers, 148, 157t
grouse, 35, 38, 171, 189
guanaco, 75, 75f, 100f
guinea fowl, 123
guinea pigs (*see also Cavia porcellus*)
  absorption in, 154t, 158, 215
  anatomy of, 60, 61f, 63, 88, 89
  coprophagy in, 183
  digesta transit in, 120, 123
  digestion (endogenous enzymes) in, 132,
    156t
  electrolyte secretion by, 200
  microbial fermentation and metabolism
    in, 168, 169
  motility in, 102, 104, 105
  neurohumoral control in, 93, 225, 236
gulls, 36

hagfish (*see also* cyclostomes), 8t, 9, 20, 23,
  25, 132, 137, 143, 148, 152, 192, 232,
  236
hake, 145
hamsters (*see also Cricetus cricetus*) 61f, 63,
  63f, 88, 89, 134t, 156t, 158, 168, 182
Hapala penicillata, 66
hares, 8t, 80, 182
hawks, 10, 36, 37f, 39, 90
hedgehogs, 48, 88, 104, 132, 134t, 153,
  154t
hemicellulose, 128, 130f, 131, 160, 162,
  168
hepatopancreas, 4-5, 148, 149
herons, 36, 38
hippopotamuses, 70, 71f, 72f, 73, 79, 156t,
  164
hoatzin, 36
hormones, 6, 15, 21, 99, 220, 226, 228,
  234, 236f, 237, 238
  aldosterone, 199
  bombesin/gastrin-releasing polypeptide
    (GRP), 233, 234f, 235t
  caerulin, 230
  cholecystokinin (CCK), 93, 102, 230,

232t, 235t, 236
enkephalin, 234, 235t
enterogastrone, 231, 232f
gastric inhibitory polypeptide (GIP), 93,
   231, 232f
gastrin, 93, 201, 214, 225, 228, 229, 230,
   232f, 233, 234, 234f, 235t, 236, 237
motilin, 93, 228, 231-2, 235t
neurotensin, 234f, 235t
pancreatic polypeptide (PP), 232-3, 234f,
   235t, 237
pentagastrin, 99, 229
polypeptide YY(PYY), 232-3, 235t
prostaglandins, 93, 120, 233
secretin, 93, 230-1, 232t, 234f, 235t, 236-
   7
serotonin, 15, 235
somatostatin, 93, 228, 234f, 235t, 237
horses (*see also* donkeys; *Equus* entries;
      ponies), 8t
   absorption in, 153, 154t
   anatomy of, 11, 13f, 17, 19f, 81, 82f, 85,
      88, 89
   coprophagy in, 182
   digesta transit in, 112t, 115, 116
   digestion (endogenous enzymes) in, 132,
      139t, 142, 145, 150t, 156t
   electrolyte secretion and absorption in,
      202, 204, 215
   microbial fermentation and metabolism
      in, 164, 167, 169, 180
   motility, 90, 103f, 106
howling monkeys, 66
Hucho, 29f
humans, *see* man
hummingbirds, 38
Hydra, 3, 237
hyenas, 52
hyrax, 83f, 84, 108, 109, 111t, 114, 115,
   115f, 166, 178

iguanas (*see also Iguana iguana*), 34f, 107,
   158, 182
Iguana iguana, 34f, 117t, 156t, 170, 182
immunoglobulins (*see also* absorption), 145
Indri brevicaudatus, 65
insects, 2, 4, 5, 6, 128, 195, 196, 237, 240

jackels, 52
javelines, *see* peccaries
jerboa, 63

kangaroos (*see also Macropus* spp. pade-
      melons; quokkas; rat kangaroo), 8t
   anatomy of, 14, 55, 58f, 59f
   digesta transit in, 110t, 118, 199f
   digestion (endogenous enzymes) in, 133,

136t, 145, 156t, 164
   microbial fermentation and metabolism
      in, 164, 165, 189
kestrels, 36
kingfishers, 36
knifefish, 28
koalas, 55, 56f, 108, 110t, 120, 133, 136t,
   166, 168, 182, 183, 189
kob, 156t

Lacerta, 135t, 152
Lacerta muralis, 117t
Lacerta sicula, 117t
Lacerta viridis, 117t
lactic acid, 166, 168, 211, 211f, 212f, 213f,
   214
lactose, 126t, 128
lampreys (*see also* cyclostomes), 8t, 9, 16,
   23, 25f, 229
langur monkeys (*see also Presbytis entellus*),
   164, 165, 189
lemmings, 63, 65, 121f, 122f, 123
lemurs, 8t, 65, 182, 183
lizards (*see also Iguana iguana; Lacerta* en-
      tries), 8t
   anatomy of, 31, 33, 35
   digesta transit in, 116, 117t
   digestion (endogenous enzymes) in, 135t,
      156t
   microbial fermentation and metabolism
      in, 170
   neurohumoral control in, 225, 232
llamas, 12, 13f, 71f, 72f, 75, 75f, 76f, 78, 99,
   100f, 101, 177, 201
loons, 38
lorises (*see also Nycticebus; Perodicticus* en-
      tries), 65
lungfish, 8t, 17, 20, 23, 28, 145, 148, 149,
   157t

Macaca (see also rhesus monkeys), 64f, 65,
   110t, 138t, 157t
macaques, 64f, 66, 85
mackerel, 148
macropod marsupials, *see* kangaroos; *Mac-
      ropus spp;* quokkas; rat kangaroo; wallaby
Macropus eugenii, 59f, 101, 119f, 151t,
   156t, 165
Macropus giganteus, 58f, 59f, 119f, 136t,
   156t, 164, 187f
Macropus robustus robustus, 164
Macropus robustus erubescens, 177
Macropus rufogriseus, 164
Macropus rufus, 156t
maltose and isomaltose, 126t, 130, 133f
man, 8t

man (*continued*)
absorption in, 153, 154t, 172, 174t, 205, 206
anatomy of, 12, 13f, 19, 65, 69f, 70, 85
body fluid compartments in, 192f, 193f, 216-17
digesta transit in, 110t
digestion (endogenous enzymes) in, 132, 134t, 138t, 142, 143, 145, 149, 152, 153, 154t, 157t
electrolyte secretion and absorption in, 196, 198, 201, 202, 203, 204, 207f, 208, 215, 216, 217, 218
microbial fermentation and metabolism in, 161, 166, 167, 168, 169, 177, 178, 181, 214
motility in, 87, 90, 92, 93, 97, 98, 101, 103, 104, 105, 106
neurohumoral control in, 225, 231, 233, 240
manatees, 8t, 81, 84, 165, 196
marsupial mole, 55
marsupial mouse, 136t
martens, 134t
mastication, 9, 41, 124, 220, 224, 239
Megaleia rufa (*see also* *Macropus rufus*), 110t
menhaden, 24
mice (*see also* dormouse), 9t
absorption by, 153, 154t
anatomy of, 11, 88, 89
coprophagy in, 183
digestion (endogenous enzymes) in, 132, 134t, 156t
microbial fermentation and metabolism in, 168
neurohumoral control in, 225
microvilli, *see* brush border
Midas, 66
migrating myoelectric complexes (MMC), *see* myoelectric activity
mink (*see also* *Mustela*), 52, 54f, 108, 111t
minnows, 23, 30
Mirounga leonina, 139t
Misqurnus, 152
mole rat, 182
moles, 8t, 11, 48, 49f, 88, 89, 132, 134t
mollusks, 3, 4, 5, 6, 131f, 236f, 237
mongoose, 52
monkeys (*see also* names of individual varieties), 8t, 65, 66, 67, 70, 88, 89, 134, 154t, 157t, 206
moose, 112t
mountain beaver, 20, 182, 183
mouse deer, *see* chevrotains
mud puppies (*see also* *Necturus*), 151t, 157t
mullet, 24, 148, 149

musk ox, 78
muskrats, 63, 63f
Mustela, 54f, 111t, 138t
myoelectric activity, 93, 101-2, 102f, 103, 103f, 105-7, 228, 231

Natrix, 152
Natrix natrix, 117t, 135t
Natrix taxi, 157t
Necturus, 31, 148, 151t, 157t
Neotoma, 63, 158
nerves (*see also* neurotransmitters and neuromodulators), 6, 21, 92, 97, 98, 101, 102, 103, 220-8, 236f, 237, 238
adrenergic, 93, 221, 222, 223f, 224f, 225
aminergic, 226
cholinergic, 93, 221, 222, 223f, 224f, 225
nonadrenergic inhibitory, 93, 222, 223, 223f, 224f, 225, 226
noncholinergic excitatory, 225
peptidergic, 226
reflexes, 87, 90, 106, 182, 221, 225
neurotransmitters and neuromodulators, 220, 222
acetylcholine, 221, 222, 228
adenosine triphosphate (ATP), 226, 227t, 228
bombesin/gastrin-releasing polypeptide (GRP), 226, 227t
cholecystokinin (CCK), 226, 227t
dopamine, 226, 227t
enkephalins, 226, 227t, 228
5-hydroxytryptamine (5-HT), 226, 227t
γ-aminobutyric acid (GABA), 226, 227t, 228
neuropeptide Y/pancreatic polypeptide (PP), 226, 227t
neurotensin, 226, 227t, 228
norepinephrine, 221, 228
somatostatin, 93, 227t, 228, 235t, 237
substance P, 226, 227t, 228, 235t, 237
vasoactive intestinal polypeptide (VIP), 93, 226, 227t, 228, 231, 235t, 236
newts, 8t, 30
nightingales, 135t
night monkeys, 66, 67, 68f
nitrogen recycling, *see* urea
numbats, 55
nutria, 111t
Nycticebus, 65, 138t

Ochotona rubescens, 80
Odobenus r. divergens, 139t
Oedipomidas, 138t
Oncorhynchus keta, 143
Oncorhynchus kisutch, 229
Oncorhynchus nerka, 137

opossums, 8t, 55, 56f, 107, 156t
orangutans, 67, 69f
Oryctolagus, 79f, 110t, 134t, 138t
ostrich, 38, 39, 107
otters, 52
owls (*see also Athene*), 10, 36, 90, 135t, 225
ox (*see also Bos;* cattle), 13, 19, 71f, 88, 154t, 168, 169, 173t, 207t

pacemakers, *see* myoelectric activity
pack rat, 63
paddlefish, 8t, 23
pademelons (*see also Thylogale thetis*), 119f, 164
pagolins (*see also* anteaters, scaly), 8t, 44
Pagrus, 137
pancreas, 3, 4, 6, 21, 132
  anatomy of, 10, 16-17, 238
  innervation of, 220, 221, 225
  neurohumoral control of, 220, 221, 225, 228, 230, 231, 233, 237
  secretion by, 127f, 131, 133f, 134-5t, 137, 141, 142, 143, 148, 149, 150-1t, 153, 154, 155, 156-7t, 202-4, 215, 217, 218, 228, 230, 231, 232f, 233
panther, 133
Panthera leo, 138t
Papio, 64f, 138t
parakeets (*see also* budgerigar), 38f
parrot fish, 24
parrots, 36, 38, 135t
partridge, 35
peccaries (*see also Tayassu*), 70, 71f, 73, 164
pectin, 128, 130f, 131, 162, 168
pelicans, 38
Pelomyxa, 5
penguins, 38, 39
Perca, 152, 229
perch (*see also Perca*), 24, 145, 229
Periodicticus, 138t
Periodicticus calabarensis, 65
Periodicticus potto, 65, 134t
petrels, 36
pH of secretions and digesta, 1, 44, 145, 153, 159, 160, 162, 163, 164, 165, 166, 171, 174, 184t, 191, 198, 210-15, 218, 219, 221
phagocytosis, 2-4, 30, 155
phascogales, 54f
pheasants, 35
Phoca vitulina, 52, 139t
phospholipids, 126t, 140, 141, 142, 143
pig deer, 71, 71f, 73
pigeons, 36, 38, 132, 135t, 157t
pigs (*see also* river hogs; swine; warthogs), 8t
  absorption in, 154t, 172, 173f, 174, 174f
  anatomy of, 11, 12, 13f, 15, 18f, 71, 71f, 72f, 73, 85, 88, 89
  digesta transit in, 103f, 106, 109, 112t, 114f, 115
  digestion (enzymes) in, 132, 134t, 137, 139t, 143, 149, 152, 153, 156
  electrolyte secretion and absorption in, 201, 204, 206, 207t, 208, 210, 211, 212f, 214, 215
  microbial fermentation and metabolism in, 166, 167f, 168, 169, 177, 180, 181
  motility in, 108
  neurohumoral control in, 231, 232, 233, 240
pig-tailed monkeys, 65
pika, 8t, 80, 182, 183
pike, 24, 25f, 26f, 145
pinnipeds (*see also* seals), 11, 52, 133, 137, 139t
pinocytosis, 153, 155
planigales, 55
platypus, 8t, 44, 45f
Plecoglossus, 137
Podocnemis, 148, 150t, 156t
ponies (*see also* donkeys; *Equus* entries; horses)
  absorption in, 172, 173t, 174t
  anatomy of, 79f
  body fluid compartments in, 217
  digesta transit in, 108, 109, 114, 114f, 115, 121, 178, 180
  electrolyte secretion and absorption in, 206, 207t, 210, 213f, 214, 216f, 217, 218
  microbial fermentation and metabolism in, 166, 167f, 168, 169, 178, 179f, 180, 211
porcupines, 62f, 63, 65, 145, 182
porpoises, 8t, 46
possums, 55, 110t, 120, 132, 136t, 137, 182
Presbytis entellus, 66, 67f
primates (*see also* apes, lemurs; lorises; man; monkeys; tarsiers), 11, 85, 110t, 132, 138t, 188, 240
prochordates, 6, 237
Procolobus, 66
protozoa, 2, 5, 234, 236f
  amoeba, 5
  parameceum, 2, 5
  gastrointestinal, 159, 161, 162, 164, 165, 167
proventriculus, 12, 36, 88, 238
Pseudemys, 148, 150t, 156t
ptarmigans, 113t, 116, 171
pyloric appendices, *see* pyloric ceca
pyloric ceca, 14, 26, 26f, 27f, 29f, 124, 132, 143, 148, 149, 158, 238

quail, 35, 107, 137, 232
quokkas, 101, 102, 110t, 154t, 164, 165

rabbits (*see also Oryctolagus*), 9t
  absorption in, 154t
  anatomy of, 11, 79f, 80, 88, 89
  coprophagy in, 182, 183, 184t
  digesta transit in, 108, 109, 110t, 114,
    115, 115f, 118, 119, 120
  digestion (endogenous enzymes) in, 132,
    134t, 151t, 157t
  electrolyte secretion and absorption in,
    202, 204, 205, 206, 207t, 208, 215
  microbial fermentation and metabolism
    in, 166, 168, 178, 180
  motility in, 89, 103f, 104, 105, 106, 107
  neurohumoral control in, 225, 233, 240
raccoons, 52, 53f, 108, 109f, 111t, 114, 166,
  167f, 168
rails, 38
Rana berlandieri, 229
Rana cancrivora, 192
Rana catesbeiana, 137, 140, 148, 150t,
  151t, 157t, 231, 234f
Rana esculenta, 143
Rana pipiens, 148, 150t, 153, 157t, 170
Rana temporaria, 135t
rat kangaroo (*see also Bettongia*), 55, 188
rats (*see also* pack rat, wood rat), 9t
  absorption in, 153, 154t, 172, 174, 197
  anatomy of, 13f, 17, 43f, 60, 63, 63f, 88, 89
  coprophagy in, 182, 183
  digesta transit in, 111t, 123
  digestion (endogenous enzymes) in, 132,
    134t, 138t, 141, 142, 143, 145, 149,
    152, 153, 156t
  electrolyte secretion and absorption in,
    196, 198, 201, 202, 204, 205, 206,
    207t, 208, 210
  microbial fermentation and metabolism
    in, 166, 168, 169, 178, 180, 181, 185
  motility in, 90, 103, 104, 105
  neurohumoral control in, 221, 225
rays, 8t, 23, 132, 143, 148, 149, 151t, 157t,
  192
regurgitation, 10, 36, 90-2, 97, 98f, 99, 124
rheas, 37f, 189
rhesus monkeys (*see also Macaca*), 65,
  88, 89, 110t
rhinoceros, 8t, 80, 81, 82f, 83f, 167, 168,
  188
ringtail cats, 52
river hogs, 71, 73
roach, 23, 196
roadrunners, 107
robins, 134t
ruffed lemur, 65

rumination, 66, 70, 74, 91-2, 97, 98f, 99,
  187, 229

Saimiri, 68f, 138t
salamanders (*see also Necturus*), 8t, 14, 16,
  30, 31, 32f, 135t, 229, 231, 233
salivary galnds, 10, 33, 36, 57, 59, 238
  anatomy of, 4, 9
  innervation of, 220-2, 225
  secretion by, 3, 132, 141, 142, 162, 196-
    9, 215, 217, 218, 220
Salmo facia, 30
Salmo fario, 27f
Salmo gairdneri, 28, 137, 229
salmon (*see also Onorhynchus spp.*), 143,
  145, 148, 229, 236
Sciurus, 138t
scup, 30
sea lions, 11
sea otters, 52
seals (*see also Arctocephalus; Eumtopias
    jubatus; Mirounga leonina; Odobenus
    r. divergens; Phoca vitulina; Zalophus
    californianus*), 11, 17, 52, 54f, 141,
    142
secretion, 1, 2, 3, 4, 6, 9, 10, 12, 13, 14, 15,
  16, 20, 21, 36, 217-18
  bicarbonate, 10, 162, 174, 175, 191, 200,
    201, 202, 203, 204, 205, 206, 208,
    210, 214, 215, 225, 233
  chloride, 176, 191, 202, 205, 206, 225
  enzymes, *see* enzymes, digestive
  HCl, 10, 12, 145, 153, 199-201, 214, 215,
    218, 225, 226, 229, 230, 231, 233
  hydrogen, 153, 175, 176f, 191, 202, 204,
    206, 208
  phosphate, 10, 162, 191, 210
  potassium, 191, 202, 205, 210, 215
  sodium, 191, 202, 205
  water, 178, 205, 206, 215, 217, 218, 219
shad, 24
sharks (*see also* dogfish), 8t, 23, 24, 132,
  140, 141, 143, 145, 148, 151t, 157t,
  192t
sheatfish, 17
sheep, 8t
  absorption in, 171, 174, 174t
  anatomy of, 11, 71f, 72f, 76, 77, 78, 88
  body fluid compartments in, 192f, 217
  digesta transit in, 113t, 115, 116, 118,
    119f
  digestion (endogenous enzymes) in, 132,
    133, 134t, 139t, 152, 153, 154t, 156t
  electrolyte secretion and absorption in,
    196, 198, 199, 202, 203f, 207t, 217,
    218
  microbial fermentation and metabolism

in, 160, 161, 164, 165, 168, 169, 177,
    178, 180, 181
motility in, 90, 92, 94, 97, 98, 99, 101,
    103, 106
neurohumoral control in, 229, 233
shrew opossums, 55
shrews, 8t, 48, 49f, 50, 108, 110t, 182
shrikes, 36
sirenians (*see also* dugongs; manatees), 188
skates, 8t, 23, 143
skunks, 52
sloths, 8t, 57, 58f, 59, 59f, 60, 108, 118,
    120, 134t, 164, 165, 188, 189
snakebirds, 36
snakes (*see also Lacerta* entries; *Natrix* en-
    tries), 8t, 10, 31, 33, 34f, 58f, 107,
    117t, 132, 135t, 143, 157t, 225
sparrows, 134t
spider monkeys, 66-7
spiral valves, 14, 25, 25f, 26f, 28f, 124, 238
sponges, 2, 6, 236f
Squalus, 132
Squalus acanthius, 145
Squalus suckleyi, 143, 151t, 157t
squirrel monkeys (*see also Saimiri*), 68f
squirrels (*see also Sciurus*), 9t, 63, 132,
    182, 183
starch, 128, 131, 132, 133f, 162, 163, 164,
    168, 169
    amylopectin, 126t, 128, 130, 132
    amylose, 126t, 128, 130, 132
starlings, 132, 135t
stomach, *see* enzymes, pepsin; gastric
    epithelium; secretion, HCl
storks, 38
sturgeons, 8t, 23, 24, 26f, 29f, 88
sucrose, 126t, 128
suni, 113t
surfperch, 30
surgeon fish, 169
swifts, 9, 36, 38
swine (*see also* pigs; river hogs; warthogs),
    70, 145, 201
Sykes monkey, 166, 168

Tamarinus, 138t
tanagers, 36
tapir, 8t, 80, 81, 82f, 167, 186
tarsiers, 65
Tasmanian devil, 55
Tasmanian tiger, 55
Tayassu, 73, 139t
teeth, 9, 23-4, 30, 31, 33, 35, 39, 40, 41-2,
    46, 57, 239
tenrecs, 48
terrapins (*see also* chelonians, *Chelydra;*
    Chrysemys entries; *Podocnemis; Pseud-*

emys), 8t, 9, 16, 20, 31, 32f, 107, 116,
    135t, 149, 150t, 156-7t, 158, 170
Testudo, 152
Testudo graeca, 225
Testudo hermanni, 135t, 137
Thylogale thetis, 59f, 119f, 164, 187f
tiger cat, 54f, 136t
Tilapia esqulenta, 170
toads, 8t, 9, 10, 11, 30, 32f, 135t, 145, 157t
tongue, 9, 23-4, 31, 33, 39, 42, 46, 59, 239
tortoises (*see also* chelonians; *Testudo*), 8t,
    9, 31, 33, 34f, 88, 116, 117t, 135t, 137,
    170, 180, 225
tree shrews, 8t, 50
trehalose, 126t, 128
triglycerides, 126t, 140, 141, 142, 143
trout (*see also Salmo spp.*), 24, 27f, 28, 30,
    143, 229
tuatara, 8t, 31
tuna, 145, 148, 149, 151t, 157t
Tupaia, 138t
Turbellaria lapidaria, 3
turkeys, 107, 113t, 116, 123, 132, 148,
    151t, 156t, 229
turtles (*see also* chelonians), 8t, 9, 10, 31,
    33, 132, 149, 170

urea, 155, 160, 175, 176, 177, 178, 179f,
    180
Ursus maritimus, 138t

ventriculus (*see also* gizzard), 36, 238
vervet monkeys, 64f, 66, 108, 109, 114f,
    166, 167f, 168
vicuna, 75
vitamins, 1, 142, 159, 160, 162, 181, 183,
    185, 188, 189
volatile fatty acids (*see also* absorption),
    162-75, 178-80, 183, 184t, 189, 191,
    205, 211-15, 216, 218
    acetate, 162, 164, 170, 171, 173t, 176f,
        184, 214
    butyrate, 162, 171, 173t, 183, 184t, 205
    propionate, 162, 164, 171, 173t, 184t
voles, 62f, 63, 65, 111t, 123, 132, 188
vomiting, *see* emesis

wallabies (*see also Macropus eugenii*); 55,
    101, 102, 119f, 151t, 156t, 164, 165,
    177
wallaroos, *see Macropus robustus robustus*
walruses, 11, 52
warthogs, 71, 73
waxes, 126t, 140-1, 143, 240
weasels, 8t, 52
whales, 8t, 12, 14, 42, 45f, 46-8, 140, 145,
    157t, 165-6, 188, 189

whistlers, 113t
wolverines, 52
wolves, 52
wombats, 55, 56f
woodchucks, 61f, 63
woodpeckers, 9, 36, 38
wood rat (*see also* pack rat), 158
woolly lemurs, 65

woolly monkeys, 67, 68f
worm lizards, 8t, 31

yaks, 76, 113t

Zalophus calfornianus, 139t
zebras, 79f, 145
zebu, 113t